Abigail E. Weeks
Memorial Library
Union College

presented by

Artificial Intelligence
Applications in Chemistry

ACS SYMPOSIUM SERIES **306**

Artificial Intelligence Applications in Chemistry

Thomas H. Pierce, EDITOR
Rohm and Haas Company

Bruce A. Hohne, EDITOR
Rohm and Haas Company

Developed from a symposium sponsored by
the Division of Computers in Chemistry
at the 190th Meeting
of the American Chemical Society,
Chicago, Illinois, September 8–13, 1985

American Chemical Society, Washington, DC 1986

Library of Congress Cataloging-in-Publication Data

Artificial intelligence applications in chemistry.
   (ACS symposium series, ISSN 0097–6156; 306)

   "Developed from a symposium sponsored by the
Division of Computers in Chemistry at the 190th
meeting of the American Chemical Society, Chicago,
Illinois, September 8–13, 1985."

   Includes bibliographies and indexes.

   1. Chemistry—Data processing—Congresses.
2. Artificial intelligence—Congresses.

   I. Pierce, Thomas H., 1952–   . II. Hohne, Bruce
A., 1954–   . III. American Chemical Society.
Division of Computers in Chemistry. IV. American
Chemical Society. Meeting (190th: 1985: Chicago, Ill.)
V. Series.

QD39.3.E46A78   1986   542'.8   86–3315
ISBN 0–8412–0966–9

ACS Symposium Series

M. Joan Comstock, *Series Editor*

Advisory Board

FOREWORD

The ACS SYMPOSIUM SERIES was founded in 1974 to provide a medium for publishing symposia quickly in book form. The format of the Series parallels that of the continuing ADVANCES IN CHEMISTRY SERIES except that, in order to save time, the papers are not typeset but are reproduced as they are submitted by the authors in camera-ready form. Papers are reviewed under the supervision of the Editors with the assistance of the Series Advisory Board and are selected to maintain the integrity of the symposia; however, verbatim reproductions of previously published papers are not accepted. Both reviews and reports of research are acceptable, because symposia may embrace both types of presentation.

CONTENTS

vii

viii

PREFACE

ARTIFICIAL INTELLIGENCE (AI) is not a new field, as AI dates back to the beginnings of computer science. It is not even new to the field of chemistry, as the DENDRAL project dates back to the early 1960s. AI is, however, just beginning to emerge from the ivory towers of academia. To many people it is still just a buzz word associated with no real applications. Because AI work involves people from multiple disciplines, the work is difficult to locate and the application is sometimes difficult to understand.

We decided that now would be a good time for an AI book for several reasons: (1) enough applications can now be presented to expose newcomers to many of the possibilities that AI has to offer, (2) showing what everyone else is doing with AI should generate new interest in the field, and (3) we felt an overview was needed to collect the different areas of AI applications to help people who are starting to apply AI techniques to their disciplines. The final and possibly most important reason is our personal interest in the field.

Chemistry is an ideal field for applications in AI. Chemists have been using computers for years in their day-to-day work and are quite willing to accept the aid of a computer. In addition, the DENDRAL project, throughout its long history, has graduated many chemists already trained in AI. It is not surprising that chemistry is one of the leading areas for AI applications. Scientists have been developing the theories of chemistry for centuries, but the standard approach taken by a chemist to solve a problem is heuristic; past experience and rules of thumb are used. AI offers a method to combine theory with these rules. These systems will not replace chemists, as is commonly thought; but rather, these programs will assist chemists in performing their daily work.

Computer applications developed from theoretical chemistry tend to be algorithmic and numerical by nature. AI applications tend to be heuristic and symbolic by nature. Multilevel expert systems combine these techniques to use the heuristic power of expert systems to direct numerical calculations. They can also use the results of numerical calculations in their symbolic processing. The problems faced by chemists today are so complex that most require the added power of the multilevel approach to solve them.

Defining exactly which applications constitute AI is difficult in any field. The problem in chemistry is even worse because chemical applications that use AI methods often use numerical calculations. Some applications that are strictly numerical accomplish tasks similar to AI programs. The key feature used to limit the scope of this book was symbolic processing. The work presented includes expert systems, natural language applications, and manipulation of chemical structures.

ix

The book is divided into five sections. The first chapter is outside this structure and is an overview of the technology of expert systems.

The book's first section, on expert systems, is a collection of expert-system applications. Expert systems can simplistically be thought of as computerized clones of an expert in a particular specialty. Various schemes are used to capture the expert's knowledge of the specialty in a manner that the computer can use to solve problems in that field. Expert-systems technology is the most heavily commercialized area in AI as shown by the wide variety of applications that use this technology. These applications help show the breadth of problems to which AI has been applied. Much of the work from other sections of the book also uses expert-system techniques in some manner.

The second section, on computer algebra, details chemical applications whose emphasis is on the mathematical nature of chemistry. As chemical theories become increasingly complex, the mathematical equations have become more difficult to apply. Symbolic processing simplifies the construction of mathematical descriptions of chemical phenomena and helps chemists apply numerical techniques to simulate chemical systems. Not only does computer algebra help with complex equations, but the techniques can also help students learn how to manipulate mathematical structures.

The third section, on handling molecular structures, presents the interface between algebra and chemical reactions. The storage of molecular representations in a computer gives the chemist the ability to manipulate abstract molecular structures, functional groups, and substructures. The rules that govern the changes in the molecular representations vary with each approach. Molecules can be described as connected graphs, and the theorems of graph theory can be used to define their similarity. Another approach uses heuristic rules for chemical substructures to define and display molecules.

The fourth section, on organic synthesis, discusses methods to construct complex organic syntheses using simple one-step reactions. Many groups have used the computer to search for synthetic pathways for chemical synthesis in the past. Each approach must deal with the problem of multiple possible pathways for each step in the reaction. The chapters in this section apply AI techniques to select "good" paths in the synthesis.

The final section, on analytical chemistry, is a combination of structure-elucidation techniques and instrumental optimizations. Instrumental analysis can be broken into several steps: method development, instrumental optimization, data collection, and data analysis. The trend today in analytical instrumentation is computerization. Data collection and analysis are the main reasons for this. The chapters in this section cover all aspects of the process except data collection. Organic structure elucidation is really an extension of data analysis. These packages use spectroscopic data to determine what structural fragments are present and then try to determine how those fragments are connected. Different people have used both individual spectroscopic techniques and combinations of techniques to solve this very difficult problem. This area holds great promise for future work in AI.

We gratefully acknowledge the efforts of all the authors who contributed their time and ideas to the symposium from which this book was developed. We also thank the staff of the ACS Books Department for their helpful advice. Finally, we acknowledge the encouragement and support we received from our management at Rohm and Haas Company.

THOMAS H. PIERCE
Rohm and Haas Company
Spring House, PA 19477

BRUCE A. HOHNE
Rohm and Haas Company
Spring House, PA 19477

January 13, 1986

Artificial Intelligence: The Technology of Expert Systems

Dennis H. Smith

Biotechnology Research and Development, IntelliGenetics, Inc., Mountain View, CA 94040

Expert systems represent a branch of artificial intelligence that has received enormous publicity in the last two to three years. Many companies have been formed to produce computer software for what is predicted to be a substantial market. This paper describes what is meant by the term expert system and the kinds of problems that currently appear amenable to solution by such systems. The physical sciences and engineering disciplines are areas for application that are receiving considerable attention. The reasons for this and several examples of recent applications are discussed. The synergism of scientists and engineers with machines supporting expert systems has important implications for the conduct of chemical research in the future; some of these implications are described.

Expert systems represent a sub-discipline of artificial intelligence (AI). Before beginning a detailed discussion of such systems, I want to outline my paper so that the focus and objectives are clear. The structure of the paper is simple. I will:

- Describe the technology of expert systems

- Discuss some areas of application related to chemistry

- Illustrate these areas with some examples

Although the structure of the paper is simple, my goal is more complex. It is simply stated, but harder to realize: I want to demystify the technology of applied artificial intelligence and expert systems.

The word mystify means "to involve in mystery, to make difficult to understand, to puzzle, to bewilder." Therefore, I will try to remove some of the mystery, to make things easier to understand, to clarify what the technology is and what it can (and cannot) do.

I am going to discuss a special kind of computer software, but software nonetheless.

Everything I will describe could be built from the ground up using assembly language, BASIC or any other computer language. In the future, some expert systems will certainly be built using languages such as Fortran, C or PASCAL as opposed to LISP and PROLOG which are currently in vogue. So there is no mystery here. What is different, but is still not mysterious, is the approach taken by AI techniques toward solving symbolic, as opposed to numeric, problems. I discuss this difference in more detail, below. Most readers of this collection of papers will be scientists and engineers, engaged in research, business or both. They expect new technologies to have some substantial practical value to them in their work, or they will not buy and use them. So I will stress the practicality of the technology.

Where is the technology currently? Several descriptions of the marketplace have appeared over the last year. Annual growth rates for companies involved in marketing products based on AI exceed 300%, far outstripping other new computer-based applications, such as control and management of information networks, private telephone networks, automation of the home and factory. Of course, those are growth rates, not market sizes or dollar volumes. The technology will ultimately be successful only to the extent that it does useful work, by some measure. In this paper I illustrate some areas where useful work can be, and is being, done. There are many expert systems under development at major corporations, in the areas of chemistry, chemical engineering, molecular biology and so forth. Because many of these systems are still proprietary, the examples I will discuss are drawn from work that is in the public domain. However, the casual reader will easily be able to generalize from my examples to his or her own potential applications.

The Technology of Expert Systems

I am going to begin my discussion of the technology of expert systems with two provocative statements. The first is:

Knowledge engineering is the technology base of the "Second Computer Age"

It is possible to use knowledge, for example, objects, facts, data, rules, to manipulate knowledge, and to cast it in a form in which it can be used easily in computer programs, thereby creating systems that solve important problems.

The second statement is:

What's on the horizon is not just the Second Computer Age, it's the important one!

We are facing a second computer revolution while still in the midst of the first one! And it's probably the important revolution.

Characteristics and Values of Expert Systems. What leads me to make such bold and risky statements? The answer can be summarized as follows. First, knowledge is power. You can't solve problems using any technology unless you have some detailed knowledge about the problem and how to solve it. This fact seems so obvious that it is unnecessary to state it. Many systems will fail, however, because the builders will attempt to build such systems to solve ill-defined problems.

Second, processing of this knowledge will become a major, perhaps dominant part of the computer industry. Why? Simply because most of the world's problem solving activities involve symbolic reasoning, not calculation and data processing. We have constructed enormously powerful computers for performing calculations, our *number crunchers*. We devote huge machines with dozens of disk drives to database management systems. Our need for such methods of computing will not disappear in the future. However, when we have to fix our car, or determine why a processing plant has shut down, or plan an organic synthesis, we don't normally solve sets of differential equations or pose queries to a large database. We might use such numerical solutions or the results of such queries to help solve the problem, but we are mainly reasoning, not calculating.

How do we construct programs that aid us in reasoning as opposed to calculating? AI is the underlying science. It has several sub-disciplines, including, for example, robotics, machine vision, natural language understanding and expert systems, each of which will make a contribution to the second computer age. My focus is on expert systems.

Knowledge engineering is the technology behind construction of expert systems, or knowledge systems, or expert support systems. Such systems are designed to advise, inform and solve problems. They can perform at the level of experts, and in some cases exceed expert performance. They do so not because they are "smarter" but because they represent the collective expertise of the builders of the systems. They are more systematic and thorough. And they can be replicated and used throughout a laboratory, company or industry at low cost.

There are three major components to an expert system:

- the knowledge base of facts and heuristics

- the problem-solving and inference engine

- an appropriate human-machine interface

The contents of a knowledge base, the facts and rules, or *heuristics*, about a problem will be discussed shortly. The problem-solving and inference *engine* is the component of the system that allows rules and logic to be applied to facts in the knowledge base. For example, in rule-based expert systems, "IF-THEN" rules (production rules) in a knowledge base may be analyzed in two ways:

- in the forward, or data-driven direction, to solve problems by asserting new facts, or conditions, and examining the consequences, or conclusions

- in the backward, or goal-driven direction, to solve problems by hypothesizing conclusions and examining the conditions to determine if they are true.

For the purposes of this paper, I will not describe the inference procedures further. I will also say very little about the human-machine interface. However, since expert systems are designed to be built by experts and used by experts and novices alike, the interface is of crucial importance. The examples discussed later illustrate how powerful interfaces are implemented through use of high resolution bit-mapped graphics, menu and "button" driven operations, a "mouse" as a pointing device, familiar icons to represent objects such as schematics, valves, tanks, and so forth.

The Knowledge Base.   The knowledge base holds symbolic knowledge.   To be sure, the knowledge base can also contain tables of numbers, ranges of numerical values, and some numerical procedures where appropriate.   But the major content consists of facts and heuristics.

The facts in a knowledge base include descriptions of objects, their attributes and corresponding data values, in the area to which the expert system is to be applied.   In a process control application, for example, the factual knowledge might include a description of a physical plant or a portion thereof, characteristics of individual components, values from sensor data, composition of feedstocks and so forth.

The heuristics, or rules, consist of the judgemental knowledge used to reason about the facts in order to solve a particular problem.   Such knowledge is often based on experience, is used effectively by experts in solving problems and is often privately held. Knowledge engineering has been characterized as the process by which this knowledge is "mined and refined" by builders of expert systems.   Again, using the motif of process control, such knowledge might include rules on how to decide when to schedule a plant or subsystem for routine maintenance, rules on how to adjust feedstocks based on current pricing, or rules on how to diagnose process failures and provide advice on corrective action.

Expert systems create value for groups of people, ranging from laboratory units to entire companies, in several ways, by:

- capturing, refining, packaging, distributing expertise; an "an expert at your fingertips";

- solving problems whose complexity exceeds human capabilities;

- solving problems where the required scope of knowledge exceeds any individual's;

- solving problems that require the knowledge and expertise of several fields (fusion);

- preserving the group's most perishable asset, the organizational memory;

- creating a competitive edge with a new technology.

The packaging of complex knowledge bases leads to powerful performance.   This performance is possible due to the thoroughness of the machine and the synthesis of expertise from several experts.   Similarly, if the knowledge base cuts across several disciplines, the *fusion* of such knowledge creates additional value.   An obvious value of expert systems is what is referred to above as preserving the organizational memory.   Many organizations will have to confront the loss of some of their most valuable experts over the next few years, whether through graduation, death, a new job, or retirement.   Several companies are turning to expert systems in order to capture the problem-solving expertise of their most valuable people. This preserves the knowledge and makes it available in easily accessible ways to those who must assume the responsibilities of the departing experts.

Considering commercial applications of the technology, expert systems can create value through giving a company a competitive edge. This consideration means that the first companies to exploit this technology to build useful products will obviously be some steps ahead of those that do not.

Some Areas of Application. I next summarize some areas of application where expert systems exist or are being developed, usually by several laboratories. Some of these areas are covered in detail in other presentations as part of this symposium. I want to emphasize that this is a partial list primarily of scientific and engineering applications. A similar list could easily be generated for operations research, economics, law, and so forth. Some of the areas are outside strict definitions of the fields of chemistry and chemical engineering, but I have included them to illustrate the breadth of potential applications in related disciplines.

- Medical diagnosis and treatment

- Chemical synthesis and analysis

- Molecular biology and genetic engineering

- Manufacturing: planning and configuration

- Signal processing: several industries

- Equipment fault diagnosis: several industries

- Mineral exploration

- Intelligent CAD

- Instrumentation: set-up, monitoring, data analysis

- Process control: several industries

Many readers will have read about medical applications, the MYCIN and INTERNIST programs. There are many systems being developed to diagnose equipment failures. Layout and planning of manufacturing facilities are obvious applications. Chemistry and molecular biology systems were among the earliest examples of expert systems and are now embodied in commercial systems.

There is a suite of related applications involving signal processing. Whether the data are from images, oil well-logging devices, or military sensor systems, the problems are the same; vast amounts of data, only some of which are amenable to numerical analysis. Yet experts derive valid interpretations from the data. Systems have already been built to capture this expertise.

There are many diagnosis and/or advisory systems under development, applied to geology, nuclear reactors, software debugging and use, manufacturing and related financial services.

There are several applications to scientific and engineering instrumentation which especially relevant to chemistry and chemical engineering. These include building into instruments expertise in instrument control and data interpretation, to attempt to minimize the amount of staff time required to perform routine analyses and to optimize the performance of a system. There are several efforts underway in process control, focused currently in the electrical power and chemical industries.

Before looking at some applications in more detail, let me briefly describe why the number and scope of applications is increasing so dramatically.

The Technology is Maturing Rapidly. The work that computers are being required to do is increasingly knowledge intensive. For example, instrument manufacturers are producing more powerful computer systems that are integral to their product lines. These systems are expected to perform more complex tasks all the time, i.e., to be in some sense "smarter". Two developments are proceeding in parallel with this requirement for "smarter" systems. The software technology for building expert systems is maturing rapidly. At the same time, workstations that support AI system development are making a strong entry into the computer market. For the first time, the hardware and software technology are at a point where development of systems can take place rapidly.

Beginning in 1970, programming languages such as LISP became available. Such languages made representation and manipulation of symbolic knowledge much simpler than use of conventional languages. Around 1975, programming *environments* became available. In the case of LISP, its interactive environment, INTERLISP, made system construction, organization and debugging much more efficient. In 1980, research work led to systems built on top of LISP that removed many of the requirements for programming, allowing system developers to focus on problem solving rather than writing code. Some of these research systems have now evolved to become commercial products that dramatically simplify development of expert systems. Such products, often referred to as *tools*, are specifically designed to aid in the construction of expert systems and are engineered to be usable by experts who may not be programmers.

Supporting evidence for the effects of these developments is found by examining the approximate system development time for some well known expert systems. Systems begun in the mid-1960's, DENDRAL and MACSYMA required of the order of 40-80 man-years to develop. Later systems of similar scope required less and less development time, of the order of several man years, as programming languages and system building tools matured. With current, commercially available tools, developers can expect to build a prototype of a system, with some assistance, in the order of one month. The prototype that results already performs at a significant level of expertise and may represent the core of a subsequent, much larger system (examples are shown below). Such development times were simply impossible to achieve with the limited tools that existed before mid-1984.

<u>Developing Expert Systems</u>.   How has such rapid progress been achieved?   The improvement in hardware and software technologies is obviously important.   Another important factor is that people are becoming more experienced in actually building systems. There has emerged, from the construction of many systems designed for diverse applications, a strong model for the basic steps required in constructing an expert system. The four major steps are as follows:

- Select an appropriate application

- Prototype a "narrow vertical slice"

- Develop the full system

- Field the system, including maintenance and updates

First, one must select an appropriate application.   There are applications that are so simple, that require so little expertise, that it is not worth the time and money to emulate human performance in a machine.   At the other end of the spectrum, there are many problems whose methods of solution are poorly understood.   For several reasons, these are not good candidates either.   In between, there are many good candidates, and in the next section I summarize some of the rules for choosing them.

Second, a prototype of a final system is built.   This prototype is specifically designed to have limited, but representative, functionality.   During development of the prototype, many important issues are resolved, for example, the details of the knowledge representation, the man-machine interface, and the complexity of the rules required for high performance.   *Rapid prototyping* is already creeping into the jargon of the community. The latest expert system building tools are sufficiently powerful that one can sit down and try various ideas on how to approach the problem, find out what seems logical and what doesn't, reconstruct the knowledge base into an entirely different form, step through execution of each rule and correct the rules interactively.   This approach differs substantially from traditional methods of software engineering.

The third step, however, reminds us that we do have to pay attention to good software development practices if a generally used, and useful system is to result from the prototype.   Development of a full system, based at least in part on the prototype, proceeds with detailed specifications as the system architects define and construct its final form.

The last step is just as crucial as its predecessors.   The system must be tested in the field, and the usual requirements in the software industry for maintenance and updates pertain.

The primary differences, then, between development of expert systems and more traditional software engineering are found in steps one and two, above.   First, the problems chosen will involve symbolic reasoning, and will require the transfer of expertise from experts to a knowledge base.   Second, rapid prototyping, the "try it and see how it works, then fix it or throw it away" approach will play an important role in system development.

The only phase of development of expert systems that I will say any more about is the first, and in many ways the most crucial, step for those who are contemplating building expert systems for the first time. How do you go about selecting an appropriate application? Here are the basic criteria:

- Symbolic reasoning

- Availability and commitment of expert

- Importance of problem

- Scope of problem

- General agreement among specialists

- Data and test cases available

- Incremental progress possible

First, the application should involve symbolic reasoning. There is no point in trying to develop an expert system to perform numerical calculations, for example, Fourier transforms.

Second, there should be experts available that can solve the problems involved in the selected application and they must be committed to spend their time working with the system and other experts in developing the knowledge base. If such experts are not available, or will not commit to the effort, forget the application.

Third, the problem must be important. It must be a problem whose computer-aided solution creates value by some measure. Such problems may require substantial expertise, or they may be simple, repetitive, and labor intensive. test. No one will invest in a system if the problems are infrequently encountered and can be solved quickly by persons of normal intelligence.

Fourth, the scope of the application must be bounded. There must be some specification of the functionality of the expert system and characteristics of the problems it is expected to solve. Trying to build an expert system to solve the world's economic problems is not a good application to choose. However, selecting a product mix from an oil refinery based on the current state of supply and demand in the world's energy markets might be a good application.

Fifth, there must be general agreement among experts on how to solve the problem, on what constitutes the facts in the domain, and what are judgemental rules. Without such agreement, the values mentioned previously of extending the knowledge base beyond any single individual's contribution, and fusion of expertise across several domains will not be realized. More practically, without general agreement, other experts will criticize the performance of the system.

Sixth, there must be ample data and test cases available to convince the system builders that some defined level of performance has been achieved. Although this may seem obvious, some systems have been built and tuned to perform well on a single test case. Needless to say, such systems usually fail when confronted with a second test case.

Seventh, it must be possible to build the system incrementally. It must be easy to extend the knowledge base and modify its contents, because as you all know, rules often change as new evidence is gathered. The progress of science and technology are always working to make our knowledge inadequate or obsolete. We must learn new things; we must be able to instruct the expert system accordingly.

Selected Applications

Biological Reactors. In this section I discuss some applications that are at least indirectly related to chemical science and engineering. The first example, illustrated in Figure 1, is derived from a simulation and diagnosis of a biological reactor that we put together for a demonstration.

Because the expert system was not connected to a real reactor, we built a small table-driven simulation to model the growth of cells in suspension. The graphical interface includes images representing the reactor itself, several feed bins and associated valves. Also shown in Figure 1 are several types of gauges, including a strip chart, monitors of various states and alarm conditions, temperature, and the on/off state of heaters and coolers.

The simulation runs through a startup phase, then through an exponential growth phase which is inhibited by one of several conditions. The expertise captured in the rules in the knowledge base is designed to diagnose one of several possible faults in the system and to take action to correct the condition. Growth inhibition may be caused by incorrect temperature, depletion of nutrients, incorrect pH or contamination. The system is able to diagnose the fault and to take action to adjust temperatures, the pH, add nutrients or recommend the batch be discarded due to contamination. A simple example, but one that illustrates several points mentioned earlier. The graphical interface is essential for non-experts. The system was developed rapidly as a prototype. As such, it does useful things, it can be examined, criticized, refined, and can represent the beginnings of a larger system. Combinations of relatively simple rules can diagnose problems and take specific actions.

Communication Satellites. The next example illustrates an expert system similar to those under development in process control and instrumentation companies. These systems are designed to diagnose faults and suggest corrective actions.

An aerospace company in California monitors telecommunication satellites in geosynchronous orbit, 23,000 miles away in space. When something goes wrong on that satellite, $50 to $100 million are dependent on taking the right corrective action. This company is using expert systems to capture the knowledge of the developers of the satellites in diagnosing and correcting problems, and to make this knowledge available to all operators responsible for monitoring the condition of on-board systems.

Like many modern instruments, their instrument, the satellite, is connected to their computer systems through an interface, in this case an antenna dish that transfers data from the satellite to computers at a ground control center.

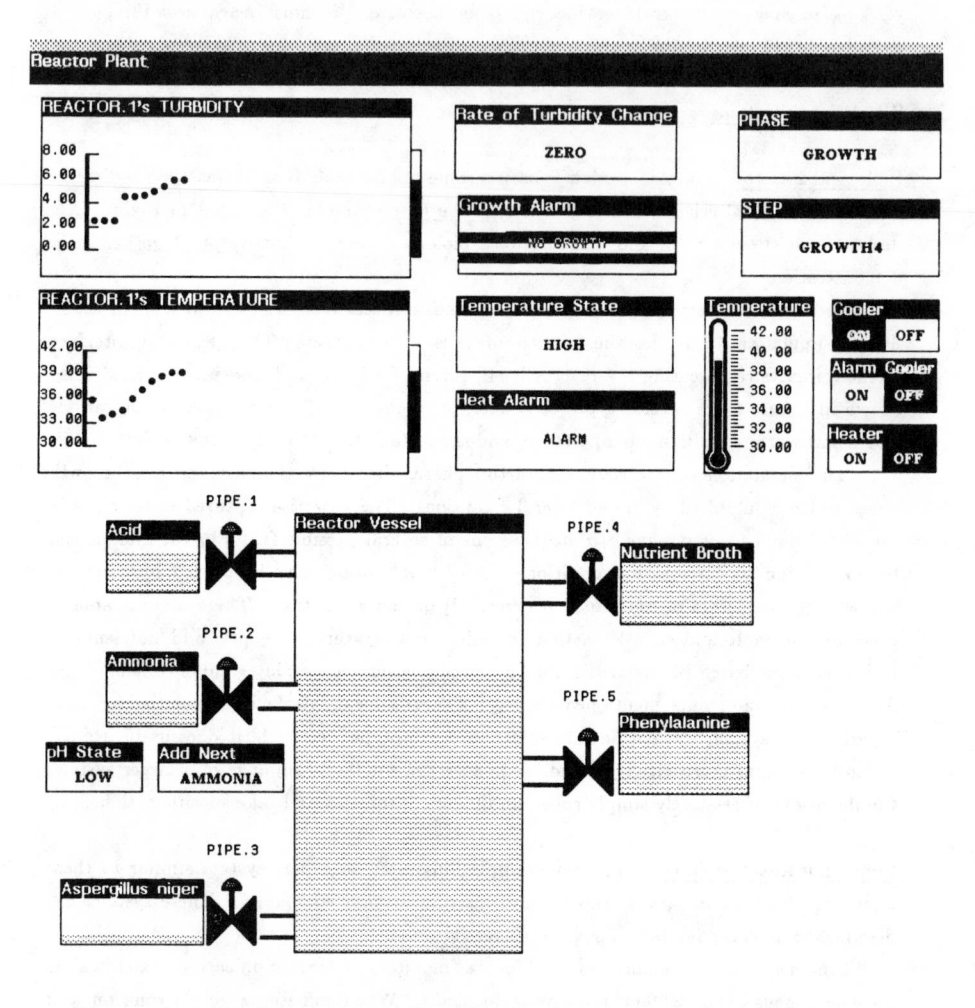

Figure 1. Graphics screen for the prototype expert system for
diagnosing faults in a biological reactor. The screen shows a
schematic of the reactor, together with gauges, strip charts, and
"traffic lights" indicating the state of the reactor obtained from
sensor readings. (Reproduced with permission. Copyright 1983
IntelliCorp.)

What is especially interesting about their problem of diagnosis of failures and advice ·on corrective measures is their treatment of the alarm conditions that trigger the execution of the expert system. The first goal of their rules is to focus on the single, or small set of, alarm(s) that are of highest priority, thereby ignoring what may be many lower priority alarms for a single problem. This usually allows isolation of the problem to a specific subsystem, such as the energy storage and heating system shown schematically in Figure 2. When the problem is localized, the system provides advice on what actions to take, then examines the other alarms to determine if they are of secondary importance or represent concurrent, major problems. Here, the graphical presentations, for example, Figure 2, provide information to the operator on which systems are being examined and where the faults may occur.

Space Stations. The final example I have selected results from work done by the National Aeronautics and Space Administration (NASA) in preparation for flying the space station. NASA's general problem is that many space station systems must be repairable in orbit by astronauts who will not be familiar with the details of all the systems. Therefore, NASA is looking to the technology of expert systems to diagnose problems and provide advice to the astronauts on how to repair the problems.

The problem they chose for their prototype is part of the life support system, specifically the portion that removes CO_2 from the cabin atmosphere. This system already has been constructed, and NASA engineers are already familiar with its operation and how it can fail. Using this information they were able to build as part of their knowledge base a simple simulation for the modes of failure of each of the components in the system. The life support system is modular, in that portions of it can be replaced, once a problem has been isolated. The graphical representation chosen for the instrument schematic and panel is shown in Figure 3.

On the left of Figure 3 is a schematic of the system, with hydrogen gas (the consumable resource) flowing through a valve to the six-stage fuel cell. Cabin atmosphere enters from the right, excess hydrogen plus CO_2 exits at the H_2 Sink, and atmosphere depleted in CO_2 exits at the Air Sink. There is a variety of pressure, flow, temperature and humidity sensors on the system. The lower subsystem is a coolant loop that maintains temperature and humidity in the fuel cell. On the right of Figure 3 is a schematic of an instrument panel that contains many of the instruments the astronauts will actually see.

Each component in the schematic is *active*. Pointing to any component with a mouse yields a menu of possible modes of failure for that component. Selection of a failure results in setting parameters in the underlying knowledge base, which are of course reflected in the settings of the meters and gauges on the instrument panel.

Simply pointing to the *IDENTIFY* button runs the rule system, which diagnoses the problem and provides advice on action to take to fix the life support system. The remainder of the screen is devoted to various switches and output windows that are used to build and debug the knowledge base.

As an indication of how rapidly the technology of expert systems has matured, this

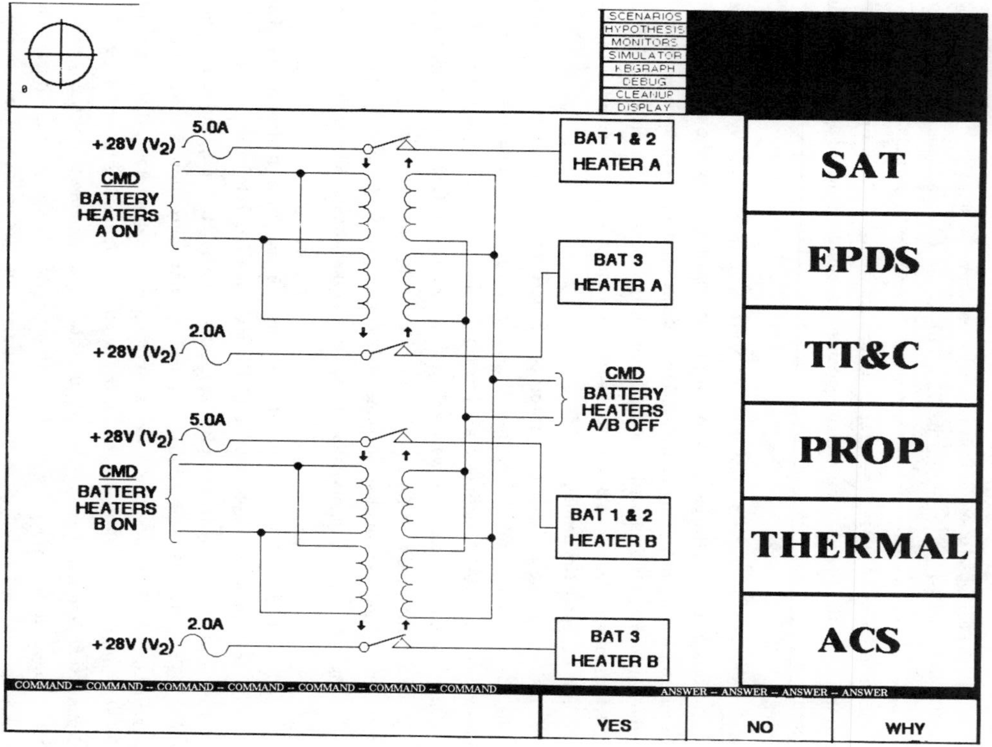

Figure 2. Graphics screen for a portion of the expert system developed for an alarm advisory system for communications satellites. This screen displays a portion of the battery and heater subsystem used to maintain thermal balance on the satellite.

prototype was built in our offices by two people from NASA, one a programmer who knew nothing about LISP, the other an expert on the life support system who knew nothing about programming. Neither had seen KEE^{TM}, our system building tool, before receiving training and beginning work on the prototype. The system, including all the graphics, the simulation and the rules, was built in four weeks. It is capable of diagnosing many of the important modes of failure of this portion of the life support system. Much work remains to be done before a final version of the expert system is completed, but this prototype provides an important starting point.

Concluding Remarks

I have used this paper as an introduction to what amount to revolutionary change in the software technologies of expert systems. At the same time, a revolution is occurring in hardware technology as well. At the moment, tools for building high performance expert systems run on special purpose hardware, LISP machines. These machines have been quite expensive, making entry into this area difficult for many laboratory groups. Several things are happening that are changing this situation dramatically. First, applications developed on LISP machines can now be ported to midi and minicomputers, making replication of a developed system much less expensive. Second, hardware vendors such as Xerox have recently announced LISP machines at modest prices, just under $10,000 for one such machine. Third, Texas Instruments has a contract to produce a VLSI implementation of its LISP machine on a chip. Successful development of this chip will further reduce the cost of a machine. Fourth, better programming environments are becoming available on midi and minicomputers, and in the short run some of these systems will mature to the point where significant work can be done, albeit at performances substantially below the LISP machines.

In the longer term, better hardware for symbolic computation will become available. These machines will support large knowledge bases, and be able to perform rapid retrievals of data from them. Logical inferences will be performed at much higher rates, approaching those now achieved by arithmetic operations. Parallel architectures will further improve the speed of symbolic computations, just as they have done for numeric computations. The keyboard is already becoming obsolete in expert systems products. Menus accessed by pointing devices, or special purpose, programmable touchpads are much easier for most people to use. Speech and picture input is already achievable in simple systems; the improvement of this technology will continue to revolutionize human-machine interactions.

An important characteristic of expert systems technology is that it can be added on to existing technologies. Such systems are already compatible with modern distributed computing environments, and can be networked easily with existing systems. Thus, investments in hardware and software are protected, and machines of more conventional architectures can be used as they are used now, for example, to support large data bases or to perform numerical calculations. An expert system can make use of these machines, passing requests for retrievals or calculations over a network, and gathering results to be used in the problem solving activities.

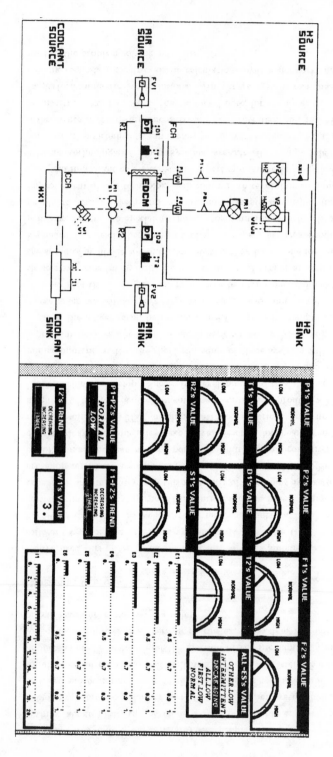

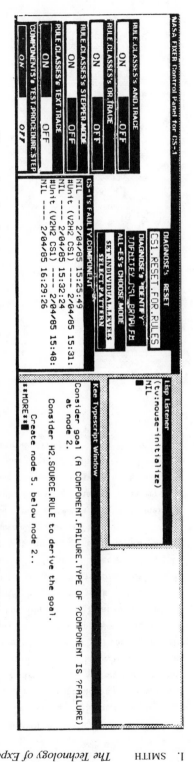

Figure 3. Graphics screen for the prototype expert system developed by NASA for diagnosis and repair of the life-support system. This portion of the system strips cabin atmosphere of CO_2. (Reproduced with permission. Copyright 1983 IntelliCorp.)

In my opinion, these technologies will have substantial impact on the practice of chemistry and chemical engineering. Everyone is familiar already with the extent to which computers have taken over routine tasks of data acquisition, reduction and presentation. Machines for data interpretation are now being constructed. Robotics is another discipline of AI that is now being used in simple systems to perform repetitive laboratory operations. The fusion of vision and expert systems technologies with robotics will make the latter much more flexible and adaptable to changing conditions. These changes, and many others brought on by the new technologies, will probably not diminish the total number of jobs available in the physical sciences, but it certainly will change what work is done in these jobs. There is already a history of jobs requiring limited skills being displaced by computers and automation. Expert systems will create additional displacements. At the same time, more jobs related to building and maintaining such systems will become available, but these jobs will require substantially more education and skills.

For jobs that already require substantial skills, expert systems will serve to make the people holding these jobs more productive. An analogy has been made to engineers who used to calculate trajectories by hand, but now use computers to perform these routine tasks, thereby freeing their time for more intellectual pursuits. Chemists and chemical engineers will see similar improvements to their own productivity.

RECEIVED January 24, 1986

EXPERT SYSTEMS

2

A Knowledge-Engineering Facility for Building Scientific Expert Systems

Charles E. Riese and J. D. Stuart

Radian Corporation, Austin, TX 78766-0948

RuleMaster is a general-purpose software package for building and delivering expert systems. Its features include 1) knowledge acquisition by inductive learning, 2) specialized artificial intelligence programming skills are not required, and 3) it runs on a wide range of micro-computers and mini-computers. RuleMaster was developed to enable scientists and engineers to incorporate human-like decision making as part of their computer applications. One such application is TOGA, an expert system to diagnose faults in large transformers based on gas chromatographic analysis of the insulating oil.

An expert system is a computer program which contains the captured knowledge of an expert in some specific domain. The program is able to give advice within the domain in much the same manner as the human expert would, asking for information as it is needed, volunteering partial diagnoses as they are reached, and functioning with incomplete or possibly erroneous information. The expert system is able to provide an explanation of the line of reasoning upon demand.

Until recently, most expert system building took place in the research departments of universities and a few major corporations. The primary emphasis was investigation of artificial intelligence principles, and the application was of secondary importance. The expert systems tools used reflect this interest. They are typically stand-alone AI computer systems, using special hardware and software environments (usually Lisp-based) not commonly found in scientific and engineering organizations.

But applications usually need a different type of computing environment. The reasoning task, accomplished by AI techniques, often constitutes ten percent or less of the code of an application. The majority of the code is for conventional programming tasks, such as data acquisition, data base access, numerical calculations, and graphics. In each application domain, computer hardware and software has been selected to match the needs of its tasks. In

0097-6156/86/0306-0018$06.00/0

established fields like chemistry, computer solutions have been implemented and in use for years. It is not reasonable for the AI component, a relatively small addition to the total system, to dictate major changes to the computing environment.

While the original expert system approaches were suitable for AI research, several types of problems are encountered when the emphasis is shifted to scientific expert system applications. In the original approaches, expert system building is slow and expensive due to the amount of expert and knowledge engineer time required to express and test rules. The cost of AI hardware and special AI programmers makes small applications prohibitively expensive. The expert systems are stand-alone programs, and it is difficult or impossible to integrate their reasoning with existing scientific software. Sometimes, finished expert systems can not be used in the field because they are too slow, or require inappropriately expensive hardware.

Because of the current high demand for expert system applications, software packages which are optimized for application building, rather than for AI technique research, have been developed. One of these is RuleMaster (1), which is designed to extract expert reasoning and to incorporate it into a wide range of scientific and engineering applications. In contrast with many other AI approaches, RuleMaster is based on contemporary structured programming principles. Conventional micro- and mini-computers may be used by any computer professional to build expert systems integrated with existing computer programs. A knowledge acquisition system based on inductive learning speeds up the rule generation and testing process. A procedural representation of the rule base is automatically generated, providing consistency and completeness checking and efficient run-time behavior. Embedding expert system reasoning into existing systems is supported by two features: access to external user programs from the RuleMaster rule language, and the automatic generation of a C code representation of the expert system.

RuleMaster Description

History. Radian Corporation is a technical consulting company, employing about 1000 people. About half of Radian's business is in the chemistry and chemical engineering fields. In 1981, Radian management realized that expert systems capability could enhance and complement existing consulting activities. Radian entered into an agreement with Donald Michie, of Edinburgh University and Intelligent Terminals Limited (ITL). For a number of years, he had done research in inductive learning and in other expert system techniques, and often used conventional structured programming languages like Pascal. He noted that the special AI environments were primarily useful for research into AI techniques, and were not necessary for an expert systems package oriented toward building applications. RuleMaster was designed and developed by ITL and Radian during 1982 and 1983. Since then, both companies have continued enhancing RuleMaster, and several dozen expert system applications are under construction or completed.

<u>Components</u>. The two principle components of RuleMaster are:

Radial: a procedural, block structured language for
             expressing decision rules, and

RuleMaker: the knowledge acquisition system; induces decision
             trees from examples of expert decision-making, and
             expresses these decisions trees as executable
             Radial code.

RuleMaster expert systems are represented as Radial programs. To
build an expert system, domain knowledge is normally entered in two
parts: a module structure and the bodies of the modules. The
structure defines the hierarchical organization of decisions used to
solve the problem. The code within each module defines the details
of one of these decisions.

      RuleMaker is a knowledge extraction utility for building and
testing the decision logic contained within Radial modules. The
logic is specified as a table of examples of correct expert
decisions for each module. RuleMaker transforms each example set
into an equivalent decision tree, and automatically generates the
body of the module in the form of Radial code. System builders may
also choose to enter Radial code directly, although they usually
prefer to work with example tables.

      Consultation of an expert system is accomplished by using its
Radial code representation as input to the Radial interpreter. The
interpreter first performs completeness and consistency checks, and
then provides interactive run-time support.

<u>Inductive Learning (RuleMaker)</u>. Experts are best able to explain
complex concepts to human apprentices implicitly by using examples
of the expert's decision-making, rather than by explicitly stating
fundamental theoretical principles. The apprentice quickly
generalizes these example decisions to form working rules, which he
applies when similar situations are encountered.

      RuleMaster's knowledge acquisition tool, RuleMaker, employs a
learning process similar to that of the apprentice. To teach a
concept to RuleMaker, the expert provides a set of examples (called
a training set) of correct decisions within some context. Each
training set contains a list of the attributes which are factors for
determining the choice of action. Each example contains a value for
each of the attributes, together with the specified actions to be
taken when that combination of attribute values is encountered. The
RuleMaker utility checks each training set for completeness and
consistency, and then generates a procedural representation of the
knowledge embodied in the example.

      To illustrate this, the example set of Figure 1 shows how a
simple corona detection decision (likely, possible, or unlikely) in
TOGA (Transformer Oil Gas Analysis) might be specified. TOGA is an
expert system that diagnoses faults in large electrical transformers
and will be described in detail later in this paper. The corona
decision is based on four attributes: H_2, thermal, H_2/C_2H_2, and
temperature. The attribute "H2" is the concentration of hydrogen
gas; it may be low, medium, or high, according to numerical ranges

set by the expert in another Radial module. "Thermal" refers to thermally generated hydrocarbon gases, which may be absent, slight, or definitely present. The other two attributes are the hydrogen-to-acetylene ratio and the estimate of the temperature at which the hydrocarbon gases were generated. A hierarchy of rules supplied by the expert determines the value of each of these attributes, based eventually on the numerical concentrations received from the gas chromatograph.

The decision for each example is expressed as an "action-next state" pair. The "action" is a reference to executable Radial code, which consists of a sequence of Radial statements. These statements may contain references to external programs in various languages (this will be discussed further later). The "next state" describes the context to which control is to pass after the action is completed. For diagnostic expert systems, such as TOGA, the next state will usually be the "goal" state of the module. This passes control back to the calling module. For procedural expert systems, such as robotics and instrumentation control applications, the control will be transferred between several states within a module to implement looping.

The decision tree for the training set of Figure 1, as generated by RuleMaker, is shown in Figure 2. The generated tree agrees with all decisions represented in the example set, and generalizes to reach decisions for unspecified portions of the space. The rule induction algorithm, called ID3 (2), uses information theoretic techniques to reduce the number of decision nodes in the generated tree.

Rule Language (Radial). RuleMaster expert systems are expressed in Radial, a block structured interpreted language with a syntax similar to Pascal and ADA. Radial is a simple, easy-to-learn language which supports the full range of expert system capabilities.

The building block of Radial, corresponding to the Pascal procedure, is called a "module". The syntax within each module is based on finite automata theory, to provide the control structures needed to support both diagnostic and planning aspects of expert systems applications. Other language features include recursive routine calls, argument passing, scoped variable and functions, abstract data types, and user-defined overloaded operators. Built-in data types include string, integer, floating point, and boolean.

The Radial code for the decision tree of Figure 2 is shown in Figure 3. This code was generated by RuleMaker. Experts have difficulty correctly generating a deeply nested conditional phrase like this, but they are able to inspect it for possible errors or omissions.

TOGA uses the built-in numerical capabilities of Radial to compute functions of concentration values, which are used extensively in the rules. The ratio of hydrogen to acetylene concentration in the corona rule is a simple example of this. User-defined compound data types are used to handle blocks of data as a single named structure. These features are invaluable in building practical expert systems, but are not available with all packages.

Most Radial code is constructed by RuleMaker from training sets

H2	thermal	H2/C2H2	temperature		action	next state
high	–	high	low	=>	(likely,	GOAL)
med	absent	high	low	=>	(likely,	GOAL)
high	–	high	moderate	=>	(possible,	GOAL)
med	absent	high	moderate	=>	(possible,	GOAL)
high	–	high	high	=>	(unlikely,	GOAL)
med	absent	high	high	=>	(unlikely,	GOAL)
med	present	–	moderate	=>	(unlikely,	GOAL)
med	slight	–	moderate	=>	(unlikely,	GOAL)
low	–	–	–	=>	(unlikely,	GOAL)
–	–	low	–	=>	(unlikely,	GOAL)

Figure 1.   Example set for corona rule.

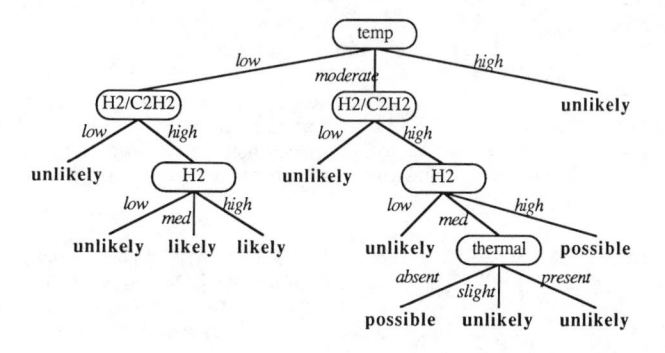

Figure 2.   Decision tree for corona determination.

```
IF (temp) IS
    "low" : IF (H2/C2H2) IS
        "high" : IF (H2) IS
            "low" : ( "unlikely" -> result, GOAL )
            "med" : ( "likely" -> result, GOAL )
            ELSE ("likely" -> result, GOAL )
        ELSE ( "unlikely" -> result, GOAL )
    "moderate" : IF (H2/C2H2) IS
        "high" : IF (H2) IS
            "low" : ( "unlikely" -> result, GOAL )
            "med" : IF (thermal) IS
                "absent" : ( "possible" -> result, GOAL )
                "slight" : ( "unlikely" -> result, GOAL )
                ELSE ( "unlikely" -> result, GOAL )
            ELSE ("possible" -> result, GOAL )
        ELSE ( "unlikely" -> result, GOAL )
    ELSE ( "unlikely" -> result, GOAL )
```

Figure 3.   Corona determination rule induced from Figure 1
examples, as expressed in automatically generated Radial code.

of examples, as described in the previous section. However, Radial code can also be entered directly by the system builders, if they so desire.

Explanation. A user may ask for explanation of the line of reasoning at any time during an expert system consultation. RuleMaster presents explanation as a list of premises and conclusions in English-like text. The explanation describes the execution path which led up to the current conclusion or question. Explanation is presented in proof ordering, which usually differs from the order in which the questions and conclusions were encountered. This is perceived as more relevant and understandable than the time-ordered presentation of fired rules, as is present in most expert system approaches.

A sample explanation for the corona decision is as follows:

> Since the estimated oil temperature is moderate
>   when H2/C2H2 is above 4
>   and the concentration of H2 is medium
>   and overheating of oil is absent
>   it follows that a corona is possible

This text was constructed at run-time by the Radial interpreter from text fragments provided beforehand by the system builders. It displays, in English, the path through the corona decision tree (Figure 2).

When explanation is requested at intermediate points in a session, just the reasoning for the current decision tree is presented. By asking for elaboration, the user can inspect the reasoning underlying the current rule. Elaboration of the corona decision above would yield descriptions of the lines of reasoning which determined the premises: that the oil temperature was moderate, that the concentration of H2 was medium, etc. Elaboration may be repeated until the user is satisfied or until all the steps have been examined.

If explanation is requested at the end of a session, the entire line of reasoning leading up to the latest top-level conclusion is presented in proof order. Intermediate conclusions are derived before they are used in premises.

The number of levels of explanation available depends on the nesting of routine calls at run-time. The hierarchical organization of modules makes it easier to control and understand the run-time behavior of rule execution.

Explanation- driven expert system building leads to robust systems. By testing to ensure that the right conclusions are reached for the right reasons, the probability of the reasoning being correct for unforeseen situations is enhanced. Quality explanation also makes systems more useful as teaching tools.

External Processes. The Radial language supports interfacing to software written in the various computer languages available under UNIX: Fortran, C, Pascal, Lisp, etc. The Radial language takes care of the details of passing arguments to and from external routines. This capability allows Radial to be used just for the reasoning portion of an application. The remainder of the domain-dependent code can be written in whatever language is most suitable.

External code may be used to obtain input (e.g., from data bases, instrumentation, numerical data base routines, other computers), to send output (to data bases, printers, graphic devices), or to perform actions when decisions are reached (e.g., instrument control). The operators for user-defined data types will usually be implemented with external routines.

A RuleMaster program may also be set up to be called from another program. By combining several expert systems in this manner, a large application can be modeled as a set of cooperating experts.

C Code Generation. The primary representation of a RuleMaster expert system is as Radial code, much of which is generated from example tables. The building and testing is carried out by interpreting this Radial program. The advantage of interpreting is speed of development and support for interactive operation.

Once an application is tested and fairly stable, another delivery mechanism is available. The system can generate a C source code representation of the expert system. When compiled and executed, the same behavior as the interpreted Radial version of the expert system will be exhibited.

There are several reasons for using the C version of an expert system. Although interpreted Radial is already faster than many expert system approaches, compiled C is faster still. For expert systems which get inputs from instrumentation (rather than from a person at a keyboard) and need to respond in real time, this speed may be essential. Another advantage is portability. The C code may be compiled on computers other than the one on which the system was developed. The third advantage is the small size of the compiled code. For larger applications, the compiled object code is about one eighth the size of the corresponding resident code for the interpreted version. This allows expert systems to be delivered in systems with limited computing resources, such as embedded in chemical instrumentation.

Efficiency. Much of the computer resource requirement of traditional production rule expert systems is used to decide which rules are legal for firing at each step of a consultation. With RuleMaster, this part of the inference engine job is accomplished at expert system building time, rather than at execution time. For every knowledge base and inference engine combination, there is an equivalent procedural representation. Radial is designed so that this procedural representation can be determined at system building time, and pointers between conditional branching statements are set up at that time. During a session, it is only necessary to traverse these rule pointers. The execution speed improvement resulting from rule compilation increases with the size of the knowledge base.

A side effect of this approach is the consistency and completeness checking which is performed at building time when the rule pointers are being set up. Errors and oversights are caught at this stage and corrected before the iterative development cycle is continued. Most expert system approaches do not support error detection of this type and inconsistency and redundancy in those knowledge bases are difficult to detect.

For even faster operation, C code versions of the expert system may be used. This will result in at least an order of magnitude faster response over the interpreted version.

Portability. RuleMaster is written in the C language, making it portable to a wide range of micro- and mini-computers with the UNIX, VMS, or PC-DOS operating systems. By late 1985, RuleMaster had been installed on more than twenty brands of computers, ranging in size from IBM PCs to large mini-computers.

TOGA: An Expert System for Transformer Fault Diagnosis

Hartford Steam Boiler Inspection and Insurance Company (HSB) insures distribution transformers for power generation utility companies. The cost to HSB when an insured transformer fails often exceeds a million dollars. The possibility of losses of this magnitude has given HSB the incentive to develop a transformer fault early detection and diagnosis program, based on chemical analysis of the transformer insulating oil.

Diagnostic Approach. Possible causes of transformer failure include general insulation deterioration, overheating due to overload, shorting at failed joints, corona activity near insulation, arcing, and grounded core. Each failure mode causes heating of the oil, which may be local and intense or widespread and moderate. The oil decomposes when subjected to heat, and some of the decomposition products are gases which dissolve in the oil: hydrocarbon, carbon monoxide, carbon dioxide, and hydrocarbons. The relative concentrations of the various gases depends on the heating history, and is thereby related to the cause of failure. The concentrations of these gases can be accurately measured with gas chromatographs, and this information used to diagnose the cause of an incipient breakdown prior to catastrophic failure.

Diagnosing a transformer's condition from chemical analysis of its oil is an expert skill which has been developed over the past 20 years. It is relatively easy to find skilled chemists who can provide the chemical analysis, but experts who can diagnose a transformer's condition from this data are rare. The diagnosis is typically based on a mixture of science and heuristic rules developed from years of experience.

Function of TOGA. An HSB employee, Richard I. Lowe, is one of the handful of transformer diagnosis experts in the U.S. His rules have been incorporated in an expert system called TOGA, which was built with the RuleMaster expert system building package.

The function of TOGA is to transform the results of chemical analysis, together with descriptive information about a transformer, into a diagnosis of transformer condition and a recommended action. The rules were created by a process of successive refinement, using the HSB data base of past transformer histories as a source of test cases.

Motivation. TOGA contains only a small portion of the knowledge of the expert and its potential performance is limited to something less than that of the expert. However, there are still a number of reasons for building the system.

Document Expert Techniques. Richard Lowe became an expert by making thousands of diagnostic decisions over more than twenty years. Most rules used in this diagnosis are heuristic (rather than based solely on theory) and they had not been written down very well by anyone. Building TOGA was an effective method for eliciting a consistent, complete, and tested description of the diagnostic rules. Value resides in the written expression of the rules, and not just in the computer program which executes them.

Training. By using a diagnostic system built by an acknowledged expert, novices can quickly learn to diagnose transformers by observing decisions which are reached and lines of reasoning.

Distribute Expertise. TOGA allows novices to perform as experts at chemistry laboratories and utility sites, especially for the simpler and more prevalent situations covered by the rules.

Consistency. TOGA can be used to insure that the same diagnosis and recommendation is made for the same transformer data at all locations and times. Thus, it can be a tool for both informing of implementing standard diagnostic procedures.

Automate Decision-making Process. For daily operation, TOGA is run automatically from gas chromatograph output and data bases (as opposed to interactively) to generate expert interpretation of data. this speeds up the data analysis task and removes the element of human error from routine diagnoses.

Aid Expert With Complex Decisions. TOGA helps prevent the judgment mistakes which can occur when rare transformer conditions are encountered or when experts are forced to make a hurried diagnosis.

Validation. TOGA was validated by comparing its diagnoses to those previously made by the expert who supplied the rules. A set of 859 test cases from a historical data base were used. The data base contained the gas analysis results, transformer descriptive information, and the expert's detailed diagnoses (which had been prepared several years before TOGA was built). None of the cases were used in rule construction.

The comparison (Table I) shows that the expert system is an excellent representation of the expert's decision-making process.

Table I. TOGA Validation Results

Transformer	TOGA and Expert:	
Condition	Agreed	Disagreed
No Problem	651	0
Problem	204	4

One would also like to compare the diagnoses with the actual
transformer condition, and not just with the expert's previous
assessment of the condition. Unfortunately, this is usually not
possible. it is expensive to remove a transformer from service,
open it up, and determine its condition. However, this was done for
ten of the 208 "problem" cases. Engineers overhauled these
transformers and determined the nature and cause of their problems.
For all ten of these cases, both the expert system and the expert
had made the correct diagnosis.

Operational Use. TOGA is used daily by chemists in Radian's
analytical laboratory to screen the analytical results for
indications of possibly faulty transformers. This helps insure that
HSB can take quick action when it is necessary, and also helps
Radian's chemists learn the relationship between various hydrocarbon
gas concentrations and the transformer condition. At HSB, TOGA is
also used to diagnose transformers and prepare reports, which are
sent to the transformer owner after being verified by the expert.

Using RuleMaster

Knowledge Extraction. Expert systems are usually used to solve
"hard" problems for which the solution methodology is not
documented. An expert is a person who can provide the highest
quality answers or advice for a specific problem domain. Unless the
expert routinely teaches the problem-solving method, he or she will
probably have difficulty in clearly describing the method.
RuleMaster provides an example-based knowledge input mechanism
which practicing experts find comfortable to use.
For TOGA, a top-down procedure was used to begin the knowledge
extraction process. The expert was interviewed to determine the
terminology and the coarse framework of the solution method. He
selected a set of 40 transformer test cases to jog the memory during
the expert system building process. Then a list of possible top-
level decisions or actions was generated to define the scope of the
expert system. This consisted of the list of possible diagnoses (no
problem, corona, arcing, . . .) and the list of recommendations (no
action, resample at specified time, remove transformer from service,
. . .)
Then the expert was asked about the factors used to arrive at
each decision. Sometimes the factors were raw data available from
the gas chromatograph, and sometimes the factors were intermediate

attributes yet to be defined (like presence or absence of thermally generated hydrocarbon gases). Whenever new quantities were introduced, the expert was asked about the factors used to determine it. This process was repeated recursively until eventually the entire solution was described in terms of chromatograph data.

At this point, there was enough information together to build a first prototype. Each intermediate or final conclusion defined a decision module. These modules were organized into a hierarchical structure. Within each module, example table structures were created. Based on the interviewing records, a first cut at the example sets was entered. At this point, a running prototype expert system existed.

The value of this approach is that a running expert system is rapidly created, without forcing the expert to articulate a general problem-solving procedure. The prototype system is available for the iterative knowledge refinement process, which draws out more details of the decision-making procedure from the expert to gradually build a complete and tested expert system.

Knowledge Refinement. The first prototype is only a rough approximation of the expert's decision strategy. Many details are missing. Refinement of the prototype is accomplished by a continuation of example-based learning steps.

For TOGA, the 40 test cases formed the basis of knowledge refinement. The prototype was exercised for each case. Wrong advice, or correct advice reached for the wrong reasons, indicated the need for changes to the knowledge base. Whenever one of these errors was encountered, that test case was stepped slowly through the prototype expert system again. The point where the prototype reasoning differed with the expert's reasoning specified exactly where the knowledge base needed to be changed.

With the problem localized in the module hierarchy, the fix is easy. Usually, it required adding a single example (matching the test case) or correcting an existing example. Sometimes the error pointed out the need for more detail, as when two different conclusions could be reached for the same example vectors. In these cases, the expert was asked to provide a new attribute, which could distinguish between the two conclusions. On rare occasions, the expert and knowledge engineer noticed that the module hierarchy no longer seems suitable. This suggests a possible re-organization of the module structure.

Learning from examples is especially effective because the knowledge representation (in the form of example tables) is close to the way that experts normally think about their field. The translation from the expert's notation to a more abstract rule language is done by the inductive learning algorithm. Not only can knowledge be generated and tested effectively in the form of example sets, but colleagues in the field of expertise will be able to easily and thoroughly understand the reasoning incorporated in the system.

Programming Skills. One of the first steps in building RuleMaster expert system is creating the module hierarchy for the prototype. This requires skill in top-down design and structured programming.

People without some course work and experience in these computer science disciplines tend to make mistakes and flounder at this stage.

For the majority of the iterative refinement process, however, only minimal computer skills are required. The modules are small enough so their logic can be easily understood by anyone familiar with the application. Changes are usually limited to editing examples, and the example ordering is not important. The inductive learning algorithm automatically takes care of control flow. Most of knowledge refinement can be done by anyone who knows a little editing and file management. This is often the expert himself.

Therefore, additional programmers with highly specialized skills are not required to add an expert reasoning capability to an existing computer program. The programmers already on the project can also build the expert system. Not only does this save money, but these people understand the problem and are likely to do a better job than someone whose primary interest lies elsewhere.

<u>Conclusions</u>. TOGA is an expert system built with RuleMaster which has been validated and is in daily use. The primary benefit from building TOGA is that the transformer diagnostic knowledge now exists in a form which can be used to pass the skill on to a new generation of engineers. HSB will not lose its transformer diagnosis capability when the current expert retires. Other employees can use the expert system to diagnose transformers, or they can learn the technique by studying a written version of the knowledge base.

Other applications built with RuleMaster demonstrate additional reasons for building expert systems.

WILLARD (<u>3</u>) is a severe storms forecasting expert system which can obtain all input data from National Weather Service data lines. When severe storm situations occur, forecasters become very busy and do not have time to utilize all the data which is available. The expert system can take over the routine portion of the forecasting, leaving the experts free to focus on the more difficult and critical portions of the analysis.

TURBOMAC (<u>4</u>) diagnoses faults in large rotating machinery, such as power generation turbines. This expert system allows field engineers to incorporate the reasoning of one of the top experts in vibration diagnosis in their maintenance and operational decisions.

GloveAID (<u>5</u>) predicts the most effective glove materials to choose for protection against hazardous chemicals. There are no established experts in this field, because much of the protection effectiveness measurements are just now being performed. The inductive learning aspect of RuleMaster is used to help organize the date which is available and to suggest which measurements should be performed next.

The objective of QualAID (<u>5</u>) is to provide advice on how much and what type of quality assurance (QA) and quality control (QC) is needed for various types of environmental analyses. The purpose of this system is to provide consistently good advice to chemists whose primary field of expertise is other than QA/QC.

Literature Cited

1.   Michie, D.; Muggleton, S.; Riese, C. E.; Zubrick, S. M.
      Proc. of the First Conference on Artificial Intelligence
      Applications, IEEE Computer Society; Washington, D.C., 1984,
      pp.591-7.
2.   Quinlan, J. R., In Expert Systems in the Micro-electronic
      Age, (D.Michie, ed.), Edinburgh Univ. Press, Edinburgh,
      U. K.; pp 168-201.
3.   Zubrick, S. M.; Riese, C. E.   Proc. 14th Conf. on Severe
      Local Storms, American Meteorology Society; Boston, MA;
      1985; pp 117-122..
4.   Stuart, J. D.; Vinson, J. W.   Proc. 1985 ASME International
      Computers in Engineering Conference, American Society of
      Mechanical Engineers: New York, N. Y.; Vol. II, pp 319-328.
5.   Keith, L. H; Stuart, J. D.   In "Artificial Intelligence
      Applications in Chemistry"; Hohne, B. Pierce, T., Ed.; ACS
      Symposium Series (in print), American Chemical Society:
      Washington, D.C., 1985.

RECEIVED January 17, 1986

A Rule-Induction Program
for Quality Assurance–Quality Control and Selection
of Protective Materials

L. H. Keith and J. D. Stuart

Radian Corporation, Austin, TX 78766-0948

This chapter describes two prototype expert systems for chemical applications being developed using Rule-Master.(1) The first, QualAId, is a traditional type of system where knowledge on how much and what type of quality assurance (QA) and quality control (QC) is needed for various types of environmental analyses. The second, GloveAId, is being developed to help select the best glove material(s) for protection against a wide variety of hazardous chemicals. However, unlike the former example, the knowledge base for selecting the best glove materials is not yet known. Therefore, experimental data is being subjected to the rule-induction process of RuleMaster and the resulting correlations are examined and tested to help formulate the rules which are, in turn, used to build the expert system.

QualAId

The prototype of QualAId currently in existence is one small part of the total framework needed for a useful expert system. The objective of QualAId is to provide advice on how much and what type of QA/QC is needed for various types of environmental analyses. The rules for determining these needs have been derived from the American Chemical Society (ACS) publication, "Principles of Environmental Analysis," (2) and from various protocols and recommendations of the U.S. Environmental Protection Agency (EPA).

This particular demonstration module only incorporates decisions involving analysis of volatile and semivolatile organic compounds from water. These compounds are, by definition, volatile enough to be separated by gas chromatography (GC). The complete expert system will incorporate decisions based upon any type of chemical in any type of matrix and will also be capable of providing advice specifically for selected EPA methods commonly in use, i.e., EPA Methods 624, 625, 1624, 1625, the various non-mass spectrometric 600 Methods, etc. (Figure 1).

0097–6156/86/0306–0031$06.00/0
© 1986 American Chemical Society

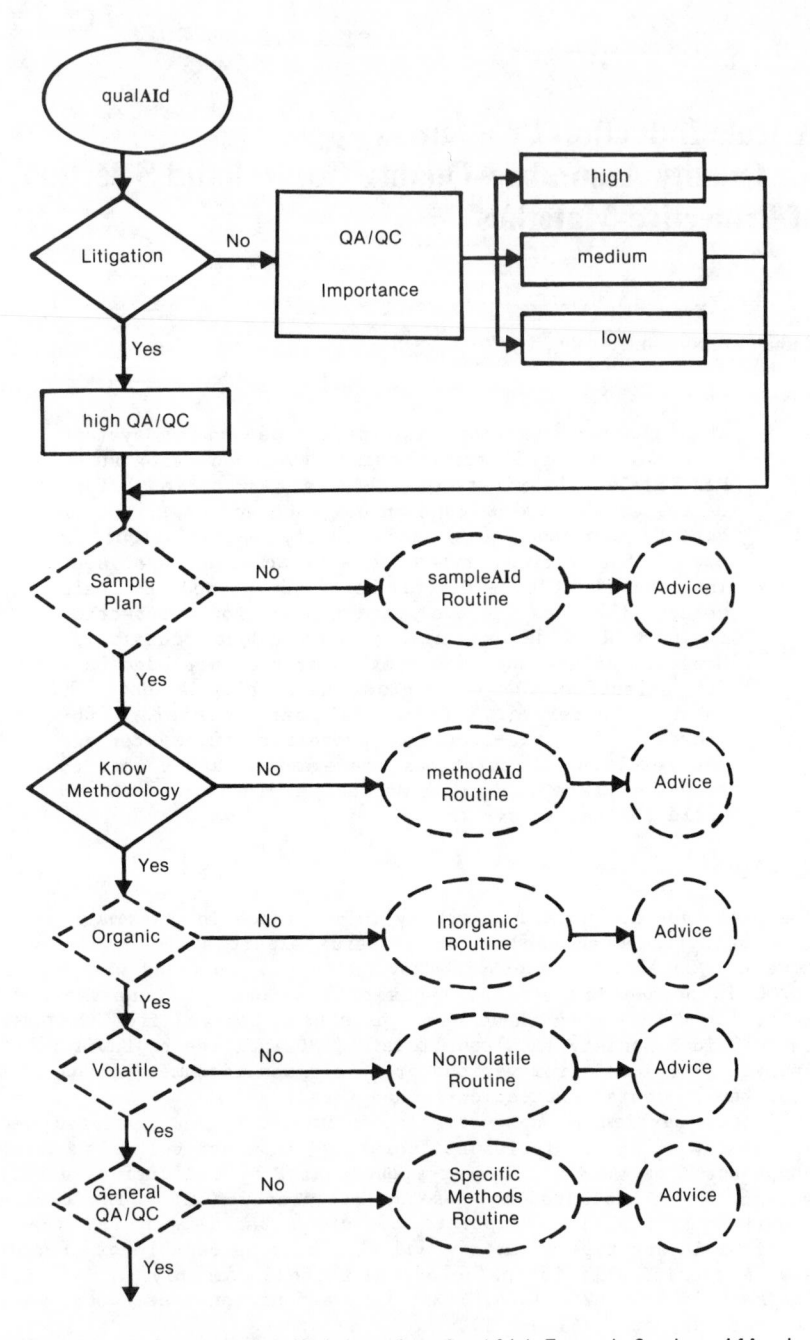

Figure 1a.   Diagram of Modules for QualAId Expert System (first half).

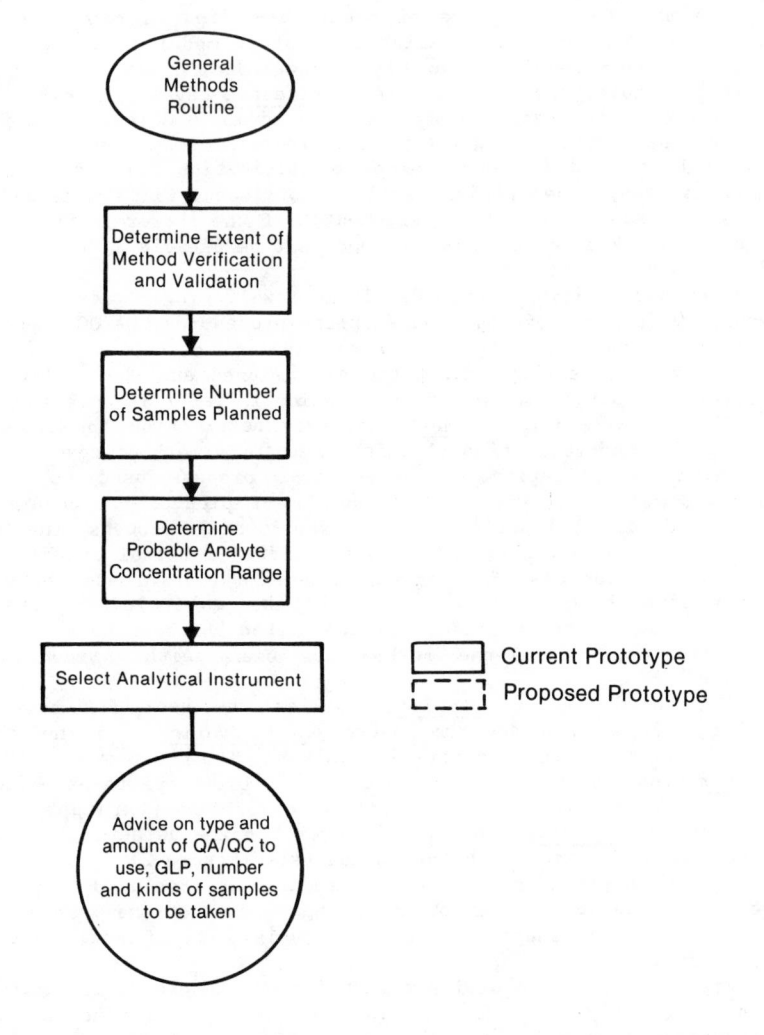

Figure 1b.   Diagram of Modules for QualAid Expert System (second half).

The purpose of this expert system is to provide consistently good advice in both the types and amounts of QA/QC to use. There are many decisions to make and errors are very expensive in terms of time and money.

The expert system is comprised of a series of modules encompassing the many varied aspects of decision-making. Information from each of these modules is available to other modules to make decisions where they require interrelated knowledge.

For example, the first module, Confidence Level, is key to many of the decisions that will be made in other modules. The first query by the computer asks the user whether the resulting analytical data will be used for enforcement or litigation actions. If the answer is "yes," then a high level of confidence will be needed and the user is advised of this assignment. If the answer is "no," then the user is asked to specify how important he/she views the accuracy and precision of the data.

Routines awaiting future development will provide advice on the best analytical methodology and sampling procedures, QA/QC needs for inorganic, nonvolatile organic, and selected methodologies (Figure 1). For the present system, these are skipped and the routine for general QA/QC advice for volatile organics in water is entered.

The second module, Method, involves determining the level of verification and validation to which the user's methodology has been subjected. Verification is the general process used to decide whether a method in question is capable of producing accurate and reliable data. Validation is an experimental process involving external corroboration by other laboratories (internal or external) of methods or the use of reference materials to evaluate the suitability of methodology (1). A menu of choices includes: (1) the method has only been verified, (2) the method has been both verified and validated, or (3) the method has been neither verified or validated.

The third module, Samples, queries the user for how many samples will be taken and the fourth, Concentration, for the expected range of probable concentration values. The choices of probable concentration values are: (1) high [> 10,000 parts-per-billion (ppb)]; (2) Medium [10-10,000 ppb]; or (3) Low [< 10 ppb]. The fifth module, Detector, queries the user for the detector that will be used in conjunction with the GC analysis (Figure 2).

The information from these five modules is then used to provide a series of advisory statements relating to whether the user will or will not meet the stated confidence levels and, if not, what the options are.

Figure 3 is the resulting advice for an example of a good QA/QC match with the user's needs. In this example, a high level of confidence was established, the methodology was both verified and validated, two samples were to be taken and analyzed by gas chromatography-mass spectrometry (GC-MS) at levels below 10 parts-per-billion (ppb). These conditions might be typical of analyses for 2,3,7,8-Tetrachloro-p-dioxin (TCDD) in polluted water.

We need to establish what instrument you plan to use for the analysis.

Since the compound(s) you are analyzing are sufficiently volatile to be separated by gas chromatography, I am assuming that you will use a GC for your separations. Here are the detector choices we have to consider:

    a   =   Mass spectrometer (general purpose)

    b   =   Hall detector (in the halogen, nitrogen or sulfur mode)

    c   =   Flame photometric detector (for phosphorous)

    d   =   Photoionization detector (for olefins and aromatics)

    e   =   Electron capture detector (for pesticides, halogens, etc.)

    f   =   Flame ionization detector (general purpose)

Choose one of these by typing the corresponding menu letter: [a,b,c,d,e,f] a

Figure 2.   Queries for the Module Detector.

Advice:

You have your methodology both verified and validated as required for measurements needing a high level of confidence. But, you must also assure that your analyst is experienced in performing the type of analysis you need, that you have standards for the analytes available, and that you have a written quality assurance plan that documents good laboratory practice.

You indicated that you need a high level of confidence in your data and GC-MS is an instrument combination that will provide you with the capability of generating that degree of quality. You should also use capillary GC columns if at all possible in order to obtain the best chromatographic resolution. When you have analytes that are completely, or almost completely resolved from other compounds, the quality of the mass spectra generated will be better. You will also need documentation that the mass spectrometer is operating and is tuned up correctly and that the resolution obtained from the gas chromatograph meets your requirements. These requirements should be met with QC standards for each instrument prior to analyzing your samples.

Less than 10 ppb is the most difficult range of analytes to quantify with a high degree of confidence. Quantitative measurements which do not fall within the "Region of Quantitation" cannot be reported. This region is defined as resulting from signal measurements which are 10 or more standard deviation units above the background noise. In order to meet this requirement at levels below 10 ppb, very good cleanup procedures are usually needed in addition to capillary GC columns and experienced analysts. All measurements reported must also be within the range of your calibration curves. If you cannot meet these stringent requirements, then you will probably have to lower your desired level of confidence in the data.

For a high level of confidence you will need to have both "field" and "method" blanks. Field blanks are blanks from a similar source that do not contain the analytes of interest. Control sites (uncontaminated sites) are used to obtain field blanks and if field blanks are not available, every effort should be made to obtain blank samples that best simulate a sample that does not contain the analyte (such as a simulated or synthetic field blank). Your method blanks will consist of all solvents, resins, etc. that you will use for extracting, concentrating and cleaning up the samples prior to analysis. You may want about half of these unspiked and the remainder spiked with known levels of your analyte standards. Similarly you may want to spike about half of your field blanks with known levels of your analyte standards. This plan would provide you with:

- 25% unspiked field blanks for control samples,
- 25% spiked field blanks for monitoring matrix effects,
- 25% unspiked method blanks for workup/instrument QC, and
- 25% spiked method blanks for workup/instrument QC.

This plan would provide you with: analyte standards so that any matrix effects will be identified during the analysis.

The total number of blanks you would need, based on the number of samples you plan to take, is: 4.

Advice: This will give you the required quality control

(RETURN continues)

[Note: 4 blanks were recommended even though only 2 samples were planned.]

Figure 3. Example of Advice Provided for a Good QA/QC Match with User Needs.

GloveAId

GloveAId is an expert system being developed for the National Toxicology Program. It has been programmed to choose from seven glove materials the one most likely to provide the greatest protection at the cheapest cost against a variety of chemicals. Chemical input is selected by choosing one of seventeen chemical classes. Glove tactility needs and the desired amount of protection (in units of minutes) are also input. The computer provides advice as to the probable best glove to select and, if none meet requested criteria, it advises the best choice it has available and explains the limitations of that choice with respect to the users request. Factors used in making the decisions include: chemical class, molecular weight, volatility (boiling point), reaction with glove materials (weight change), tactility and glove cost.

The prototype GloveAId system was developed using a data base generated from chemical permeation measurements performed at Radian. Experimental data from these tests were entered into a LOTUS-1-2-3 spreadsheet and sorted by all classifiable respects in order to make visual correlations with the protective character of seven different glove materials. The data base consisted of 90 chemicals with associated physical properties (molecular weight, boiling point and linearity of the molecule), chemical class and measurements of breakthrough times, steady-state permeation rate and degradation characteristics. The latter consisted of percent weight change when a piece of the material was immersed in the test chemical for four hours. Each of the chemicals was tested against all seven glove materials for weight change but only against four of the glove materials for breakthrough and permeation rate data so that 1,300 measured values and 540 associated pieces of information were available. Visual correlation of this data produced the protective rating approximations listed in Table I.

It is time consuming and difficult for humans to make visual comparisons of a numerical data set and draw the simplest possible correlations between them; the larger the data set, the more difficult this is to do. A lot of time and effort was expended to make the approximate evaluations listed in Table I. When the data set is a dynamic one, i.e., it is changing due to the addition of new data, it simply adds to this problem. However, one strength of computer usage is that such tasks can be performed with ease and, when this capability is coupled to the ability to induce correlations or "rules" from a data set, an extremely powerful tool for evaluating data is created. This second way of evaluating the data is currently being pursued and is described in more detail in the next section.

The ratings in Table I are based only on the safety aspects of the glove materials; i.e., protection from exposure to chemicals as indicated by the majority of breakthrough times observed within the members of a chemical class. However, tactility is often an additional important ergonomic factor; it is impossible to perform delicate tasks with thick, bulky gloves. Tactility of the gloves was rated subjectively using a dime. If the features of a dime could be readily felt through the glove, it was assigned a rating of "very good." If the features were not very distinguishable through

TABLE I. INITIAL GLOVE MATERIAL RATINGS SORTED BY CHEMICAL CLASS

Chemical Type	Viton	Butyl Rubber	PVA	Neoprene	Nitrile	PVC	Latex
Hydrocarbons							
1. Aliphatic - not halogenated	B	P	P	P	F	?	?
2. Aliphatic - halogenated	B	F	B	F	VP	VP	?
3. Aromatic - not halogenated	B	F	B	F	F	VP	P
4. Aromatic - halogenated	B	F	B	?	P	?	?
Oxygenated Compounds							
5. Alcohol	B	B	F	B	F	VP	VP
6. Aldehyde	VP	B	VP	P	P	VP	VP
7. Acid	B	B	?	G	P	VP	VP
8. Acid halide	P	B	?	?	F	VP	VP
9. Epoxide	F	B	P	?	?	P	VP
10. Ester	B	B	B	?	B	?	?
11. Ether	F	B	F	?	P	P	VP
12. Ketone	B*	B	P	P	P	P	VP
Nitrogen Compounds							
13. Nitro compound	F	B	B	F	P	VP	?
14. Cyanate	G**	F	B	VP	G	?	VP
15. Amine	G	G	B	F	F	VP	P
16. Amide	P	B	?	F	P	?	?
17. Nitrile	?	VP	B	F	?	VP	VP

B = Best   G = Good   F = Fair   P = Poor   VP = Very Poor   ? = Unknown due to insufficient data.

* Only if molecular weight (MW) is greater than 50.

** Only if molecular weight (MW) is greater than 100.

the glove, but the dime could be easily picked up from a flat surface, it was assigned a "moderate" rating. If the dime was picked up with difficulty from a flat surface, the rating of the glove was "poor."

Finally, the approximate costs of gloves from the seven materials were also considered. All of this information was provided to the computer with rules for prioritizing the choices, i.e., safety first, tactility second, and cost third.

An introductory screen is printed after log-on into GloveAId. This is followed by a menu from which the chemical class is chosen. Multiple functional groups cannot be handled by the system yet. The user is then queried for the amount of time that the glove needs to provide protection. This is followed by a request for the tactility requirements of the user.

The final screen summarizes the answers given to the computer and provides the best advice possible from the information and rules supplied (Figure 4). In this example, there were no gloves that met the user's needs, so the computer provided the next best choices. The recommended materials are a moderately tactile (nitrile) glove with probable short protection time or a thick (butyl rubber) glove with poor tactility but probable good protective properties. When safety and tactility requirements can be met, then the most cost-effective choice is provided.

This prototype expert system is currently being tested by comparing the GloveAId predictions before a chemical is tested with the best gloves after their performance has been documented. To date, 62 additional chemicals have been tested. Fifty-six (56) of these (90%) had one or more gloves correctly predicted by the expert system. Although this is good for a prototype system, we are striving to improve the percentage of total choices. Often, more than one glove material will have very good breakthrough protection. For example, with the 62 chemicals, there were a total of 132 gloves with good performance. The expert system correctly advised only 60 of these (45%).

Rule Induction

The easiest way to describe the rule induction capabilities of Rule-Master is to demonstrate its use with a relevant set of data. This data consists of information on a series of nonhalogenated aromatic compounds which were tested with five different glove materials. An arbitrary protective rating was assigned to each test based on the following breakthrough times:

Protective Rating	Breakthrough Time
Very Poor	<5 min.
Poor	5 - <15 min.
Fair	15 - <100 min.
Good	100 - <200 min.
Best	> = 200 min.

Readily available information for each of the compounds consisted of molecular weight and boiling point. In addition, the compounds were

```
        The specific glove and protection requirements are:
        Chemical type is aldehyde
        Protection time requirement in minutes is 200
        Tactility requirement is moderate tactile

        There are no glove materials in the data base
        meeting the requirements that you specified.  The
        closest are:

        Nitrile
        Approximate cost is $3.00 per pair of gloves.
        Protection time is probably greater than 5 minutes
        Tactility is moderately tactile

        Butyl Rubber
        Approximate cost is $10.00 per pair of gloves.
        Protection time is probably greater than 200 minutes
        Tactility is not tactile
```

Figure 4.   Example of Summary and Advice from GloveAId When User
Needs Are Not Met.

assigned a linear (1) or non-linear (0) shape. In this example all compounds were assigned (0) for shape designation since aromatic compounds are not linear. Other measured data included the steady state permeation rate and percent weight gain or loss. The permeation rate usually exhibits the reverse trend as the breakthrough time (i.e. as breakthrough time increases the permeation rate usually decreases). Therefore, permeation rates were not included in the data set since they seldom result in a different relative protective rating than would be derived from breakthrough times. However, weight gain or loss is a good indication of a chemical either reacting with the protective material or being absorbed into it.

RuleMaker, a subsystem of RuleMaster, induces rules for all situations from examples that may cover only some of the cases. At the heart of the induction process is the creation of an induction file, which in part includes examples indicating what the expert system should do under different circumstances. Now, in the example above, THE RULES FOR CORRELATING VARIOUS CHEMICAL AND PHYSICAL PARAMETERS OF THE HAZARDOUS CHEMICALS TESTED WITH THE PROTECTIVE ABILITY OF THE SELECTED GLOVE MATERIALS ARE NOT KNOWN -- THEY WILL HAVE TO BE INDUCED FROM THE ANALYTICAL DATA.

The RuleMaster induction file produced from the example data set is shown in Figure 5. The name given to this induction file module is "Class10". The STATE in a module is essentially the name of sub-modules that will carry out actions within a module. In this simple example there are no sub-modules so the name given to the state is "only".

The CONDITIONS section of the module is comprised of descriptions of the various parameters upon which a decision will be based. Each line in the conditions section is made up of three parts:

- the name of the decision parameter (for example, glove, molecular weight, etc.)

- the specified method of determining the parameters value (for example the statement "integer.read What is the molecular weight?" means the computer will display that question and will expect a numerical answer from the user); this part is denoted using square brackets, and

- the allowable values for the parameter. However, in this case we don't know what the allowable values for the parameter are so any value is allowed by typing the word "integer". Later, after rules have been defined and the allowable values are known, they can be used to replace any integer. This will be an important part of the second phase when the expert system is refined to include this knowledge.

The experimental data in this illustration comprise the actual rule base for RuleMaker. The first column of data in the EXAMPLE section of Figure 5 consists of the glove material tested. The second column of data consists of the molecular weights, and the third column consists of the boiling points in degrees centigrade.

```
/*
CLASS: AROMATIC NOT HALOGENATED
*/

MODULE: class10

STATE: only
    CONDITIONS:
        glove       [ask "What is the glove type?"
                    "Butyl_Rubber,Neoprene,Nitrile,PVA,PVC,Viton"]
                    [Butyl_Rubber Neoprene Nitrile PVA PVC Viton]
        molwt       [integer.read "What is the molecular weight?"] integer
        boilpt      [integer.read "What is the boiling point?"] integer
        shape       [integer.read "What is the shape?"] integer
        change      [integer.read "What is the percent change?"] integer
    EXAMPLES:
        Butyl_Rubber    130    195    0     56    =>(good,GOAL)
        Butyl_Rubber    106    144    0    190    =>(fair,GOAL)
        Butyl_Rubber    106    138    0    181    =>(fair,GOAL)
        Butyl_Rubber    106    136    0     90    =>(fair,GOAL)
        Butyl_Rubber    106    139    0    189    =>(fair,GOAL)
        Neoprene        148    193    0     64    =>(fair,GOAL)
        Nitrile         148    193    0     11    =>(best,GOAL)
        Nitrile         106    136    0     95    =>(fair,GOAL)
        Nitrile         106    144    0     60    =>(poor,GOAL)
        Nitrile          78     80    0     58    =>(fair,GOAL)
        Nitrile         106    139    0     77    =>(fair,GOAL)
        Nitrile         106    138    0     82    =>(fair,GOAL)
        Nitrile         130    195    0     63    =>(fair,GOAL)
        PVA             130    195    0     62    =>(best,GOAL)
        PVA             148    193    0      0    =>(best,GOAL)
        PVA             106    138    0      3    =>(best,GOAL)
        PVA             106    144    0      0    =>(best,GOAL)
        PVA             106    136    0      0    =>(fair,GOAL)
        PVA              78     80    0      0    =>(fair,GOAL)
        PVA             106    139    0      0    =>(best,GOAL)
        PVC              78     80    0     40    =>(very_poor,GOAL)
        PVC             106    138    0      8    =>(very_poor,GOAL)
        Viton           106    138    0      1    =>(best,GOAL)
        Viton           106    139    0      1    =>(best,GOAL)
        Viton           148    193    0      0    =>(best,GOAL)
        Viton            78     80    0      3    =>(best,GOAL)
        Viton           130    195    0      0    =>(best,GOAL)
        Viton           106    136    0      0    =>(best,GOAL)
        Viton           106    144    0      1    =>(best,GOAL)
    ACTIONS:
        best        [advise "This glove has a *best* rating."]
        good        [advise "This glove has a *good* rating."]
        fair        [advise "This glove has a *fair* rating."]
        poor        [advise "This glove has a *poor* rating."]
        very_poor   [advise "This glove has a *very poor* rating."]
```

Figure 5.   Induction Module for Nonhalogenated Aromatic Compounds.   The symbol => means "then".

The fourth column pertains to the designation of a non-linear shape (0), and the last column of data lists the percent change in weight gain or loss when the material is soaked in the test chemical for 4 hours. The data within a row is associated with a specific compound but the compounds were listed in random order within a glove material group in order to emphasize an important feature of RuleMaker -- that information (data) can be entered as it is thought of. This is an extremely important (and powerful) difference between RuleMaster and other artificial intelligence programs which are written in a highly structured interrelative fashion. The powerful inductive logic of RuleMaker enables this limitation to be ignored and this frees the user to add, change, or delete example data which influence the rulemaking logic easily and at will. This feature is very important when working with a growing/ changing data base.

The part of the example to the right of the arrow (=>), is an action-next-state-pair. It indicates what will happen when the specified combination of condition values occur. In this example the action is the designation of the relative protection of the material (good, fair, etc.) and the word "GOAL" which indicates that the goal of the module will have been reached when the action section of the module has been carried out and the computer can exit this particular module. Since there is only one module in this simple example, the program would then end.

The ACTIONS section of the module is comprised of two parts:

- the action keyword corresponding to the third part of the EXAMPLE section, and

- the action that is to be carried out (for example to advise the user by a print on the screen and/or a printer that "This glove has a *best* glove rating".

After the information in the induction module is entered, the program is assembled by the computer. During this phase, two actions take place automatically with no further input from the user:

1. Rules are induced from the examples given the computer, and

2. The actual program for running the computer is COMPILED AND WRITTEN by the computer itself!

These two actions by the computer are key to the success of this project. This is because it will be impossible for a human to consider all the possibilities of a large data set and to deduce the best (most simple and therefore cost effective) rules to use in order to choose the best protective materials to use. And when the data base is dynamically growing it would be impossible to use a highly structured artificial intelligence system where the user had to rewrite the program modifications himself every time there was a change in the information.

The rules induced by the computer are shown in Figure 6. The program which the computer wrote for itself (in "Radial" which is similar to a C-type language) is shown in Figure 7. Both of these abbreviated notations say the same thing which, in English is as follows:

"The rules induced from the example data given are:

1.   If the glove material is PVC, the rating is VERY POOR.

2.   If the glove material is nitrile, and compounds have a molecular weight <118 and boiling points >= 142°C, the rating is POOR.

3.   If the glove material is neoprene, the rating is FAIR.

4.   If the glove material is PVA, and the compounds have a molecular weight <92, the rating is FAIR.

5.   If the glove material is butyl rubber, and compounds have a molecular weight <118, the rating is FAIR.

6.   If the glove material is nitrile, and the compounds have a molecular weight between 118–139 or if the molecular weight is <118 and the boiling point is <142°C, the rating is FAIR.

7.   If the glove material is butyl rubber, and the compounds have a molecular weight >118, the rating is GOOD.

8.   If the glove material is Viton, the rating is BEST.

9.   If the glove material is nitrile and the molecular weight is >139, the rating is BEST.

10.   If the glove material is PVA and the molecular weight is >118 or if the molecular weight is 92–118 and the boiling point is >137°C, the rating is BEST."

It is interesting to correlate these rules with the first rules that were estimated with no help from RuleMaster. These were the rules used to construct the first prototype expert system, GloveAId for non-halogenated aromatic compounds:

1.   If the glove material is PVC, the rating is VERY POOR.

2.   If the glove material is nitrile, the rating is POOR.

3.   If the glove material is butyl rubber, the rating is FAIR.

4.   If the glove material is PVA, the rating is FAIR.

5.   If the glove material is neoprene, the rating is FAIR.

6.   If the glove material is Viton, the rating is BEST.

```
<class10>

    0    (all states)
    1    only
[glove]
        Butyl_Rubber : [molwt]
                    <118 :  => ( fair, GOAL )
                    >=118 :  => ( good, GOAL )
            Neoprene :  => ( fair, GOAL )
            Nitrile : [molwt]
                    <92 :  => ( fair, GOAL )
                    >=92 : [molwt]
                        <118 : [boilpt]
                            <137 :  => ( fair, GOAL )
                            >=137 : [boilpt]
                                <139 :  => ( fair, GOAL )
                                >=139 : [boilpt]
                                    <142 :  => ( fair, GOAL )
                                    >=142 :  => ( poor, GOAL )
                        >=118 : [molwt]
                            <139 :  => ( fair, GOAL )
                            >=139 :  => ( best, GOAL )
                PVA : [molwt]
                    <92 :  => ( fair, GOAL )
                    >=92 : [molwt]
                        <118 : [boilpt]
                            <137 :  => ( fair, GOAL )
                            >=137 :  => ( best, GOAL )
                        >=118 :  => ( best, GOAL )
                PVC :  => ( very_poor, GOAL )
                Viton :  => ( best, GOAL )
```

The induced rule has 11 test nodes and 16 leaf nodes.

Figure 6.   The Induced Rules for Nonhalogenated Aromatic
Compounds.   The following are meanings assigned to symbols:
[...] means "If ... is"; => means "then"; and a colon means
"and".

```
MODULE: class10

STATE: only
IF [ask "What is the glove type?"
         "Butyl_Rubber,Neoprene,Nitrile,PVA,PVC,Viton"] IS

"Butyl_Rubber" : IF [ integer.read "What is the molecular weight?" < "118" ] IS
      "T" : [ advise "This glove has a *fair* rating.", GOAL ]
      ELSE [ advise "This glove has a *good* rating.", GOAL ]
"Neoprene"
"Nitrile" : IF [ integer.read "What is the molecular weight?" < "92" ] IS
      "T" : [ advise "This glove has a *fair* rating.", GOAL ]
      ELSE IF [ integer.read "What is the boiling point?" < "137" ] IS
      "T" : [ advise "This glove has a *fair* rating.", GOAL ]
      ELSE [ advise "This glove has a *fair* rating.", GOAL ]
      ELSE IF [ integer.read "What is the boiling point?" < "139" ] IS
      "T" : [ advise "This glove has a *fair* rating.", GOAL ]
      ELSE IF [ integer.read "What is the boiling point?" < "142" ] IS
      "T" : [ advise "This glove has a *poor* rating.", GOAL ]
      ELSE [ advise "This glove has a *fair* rating.", GOAL ]
"PVA" : IF [ integer.read "What is the molecular weight?" < "92" ] IS
      "T" : [ advise "This glove has a *fair* rating.", GOAL ]
      ELSE [ advise "This glove has a *best* rating.", GOAL ]
      "T" : IF [ integer.read "What is the molecular weight?" < "118" ] IS
      "T" : [ advise "This glove has a *fair* rating.", GOAL ]
      ELSE [ advise "This glove has a *fair* rating.", GOAL ]
      "T" : IF [ integer.read "What is the boiling point?" < "137" ] IS
      "T" : [ advise "This glove has a *best* rating.", GOAL ]
      ELSE [ advise "This glove has a *best* rating.", GOAL ]
"PVC" : [ advise "This glove has a *very poor* rating.", GOAL ]
      ELSE [ advise "This glove has a *best* rating.", GOAL ]
ELSE [ advise "This glove has a *best* rating.", GOAL ]

GOAL OF class10
```

Figure 7. The Computer-Generated Program for Using the Rules Induced by RuleMaker in an Expert System for Advising Glove Materials To Be Used for Protection Against Nonhalogenated Aromatic Compounds.

As can be seen by a comparison of the two rule sets, the one induced by RuleMaster has significantly more refinement to it and will come much closer to making accurate predictions than the human induced rule set.

It is useful to display these rules as a series of bar charts in order to be able to view them in relation to one another. This is presented in Figure 8 so that the human induced ranges can be compared to the ranges induced by RuleMaster. It is readily seen that there is good agreement between the two ranges in that all of the initial human assignments are still present in the RuleMaster assignments. The notable difference is that there is considerably more refinement to the possible choices in the RuleMaster chart. The significance is that based on the simpler human induced rules if long term protection (more than 1 hour) was needed for working with nonhalogenated aromatics, Viton was the only good choice. However, Viton gloves are not only very expensive ($30 a pair), but they have poor tactility, so work involving much dexterity is precluded when wearing them. With the RuleMaster information new possibilities are now available for consideration:

- If the compounds have molecular weights >138 then Nitrile may be used; nitrile gloves offer greater tactility and they are much less expensive than Viton.

- If the molecular weight of the compounds is <118 or >93 with boiling points greater than 137°C, then PVA may be used; PVA gloves have no better tactility properties than Viton gloves but they are cheaper so the expenses could be lowered.

Thus, the rules induced by RuleMaster offer possibilities for reducing cost and allowing more dextrous work to be performed than would have been available using the human induced rules.

The important caveat to remember, however, is that the computer has produced the best rules possible from the data it was given and has extended those rules to cover examples past that data set where possible. Thus, until proven with a sufficient number of examples any set of rules must always be viewed simply as the best ADVICE available. There can always be "outliers" caused by additional factors that have not yet been discovered.

Once the computer has induced the rules governing a particular set of complex data then it is easy for a human to check and see if they are true. This can be done in two ways:

1. a simple Rule Table can be constructed, and

2. additional known examples can be analyzed to challenge the rules and see if they hold true; if they don't then additional data is given the computer so that modified rules can be induced.

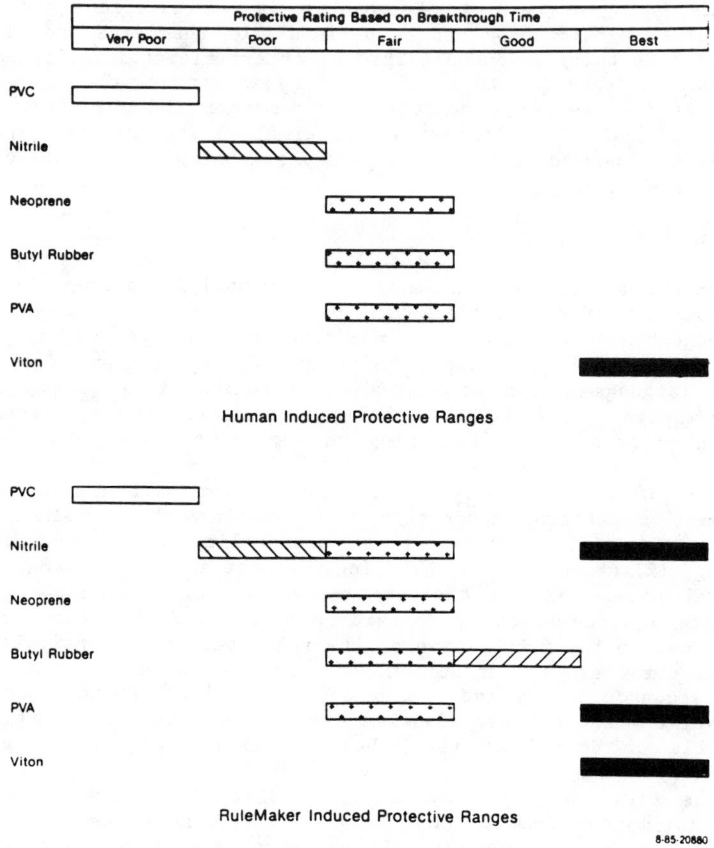

Figure 8. Protective Ranges of Six Glove Materials Against Non-halogenated Compounds.

An example of the Rule Table that can be constructed from this data set is Table II.

Now, once the Rule Table is constructed it is easy to check the data again and visualize these relationships; that is, to verify that they are true. But, remember the lack of obvious relationships when the example data was first examined.

The use of artificial intelligence, and specifically a rule inductive program such as RuleMaster is an excellent way that meaningful relationships can be derived from the large and diverse mass of data being produced. The use of artificial intelligence in this way is referred to as "knowledge manufacturing". Thus, the strongest features of a computer (to remember and correlate large numbers of data) and humans (to be creative and to use reasoning capabilities beyond that of a computer) are being used to solve very complex problems.

Summary

In summary, RuleMaster is an expert system building package intended to solve many of the problems involved in the construction of large knowledge based programs. Its inductive learning system (RuleMaker) allows rapid and effective acquisition of expert knowledge. The Radial language allows structured organization of large quantities of knowledge. Radial also provides a facility for presenting ordered explanation of reasoning to any level of elaboration required.

Use of an expert system in conjunction with a statistical program for pattern recognition such as Ein*Sight or SIMCA is a concept that offers an excellent probability of success in (1) finding, (2) ordering, and (3) using the most selective chemical and physical parameters for choosing the best protective materials to use with a wide variety of hazardous chemicals. No other program can be used both to help develop the rules needed for analysis of a complex data base (by induction) and then to use these rules in a logic sequence to provide a diagnostic decision. Furthermore, the basis of any and all decisions made by the computer are completely available on demand so that they can easily be checked and/or verified.

The first prototype system used rules which were derived as "best estimates" from a data base of about 1300 tests using 90 different chemicals. However, the prototype system is being revised using computer-generated rules. Thus, it is becoming "smarter" and better as it's data base and the resulting rules derived from it is expanded. Using a computer to evaluate large masses of data is not novel, but using it to help generate rules by an inductive logic process from large masses of data is an important new achievement. One of the significant advantages of this expert system will be a consistent unbiased interpretation of the data in a rapid manner once the expert system has been developed. And lastly, RuleMaster is structured so that it is easy to add, change, or delete data from the expert system so that it can continue to grow and improve with use and experience. These features will be invaluable as the data base continues to grow and change.

TABLE II. RULE TABLE FOR NONHALOGENATED AROMATIC COMPOUNDS

Glove Material Rating				
Best	Good	Fair	Poor	V Poor
	BuR if MW >= 118	BuR if MW < 118		
		Neoprene		
Nitrile if MW >= 139		Nitrile if MW < 118 and bp = <142 - or - MW >= 118 -< 139	Nitrile if MW < 118 and bp >= 142	
	PVA if MW >= 118 - or - MW >= 92 -< 118 and bp >= 137	PVA if MW < 92 - or - MW >= 92 -< 118 and bp < 137		
	Viton			PVC

MW = Molecular Weight
bp = Boiling Point
BuR = Butyl rubber
PVA = Polyvinyl acetate
PVC = Polyvinyl chloride

Literature Cited

1. D. Michie, S. Muggleton, C. Riese and S. Zubrick, "RuleMaster –
    A Second Generation Knowledge Engineering Facility," from
    Proceedings of the First Conference on Artificial Intelligence
    Applications, Denver, Colorado, 5–7 December 1984.
2. L.H. Keith, W. Crummett, J. Deegan, Sr., R.A. Libby, J.K.
    Taylor and G. Wentler, "Principles of Environmental Analysis,"
    Anal. Chem., 55, p. 2210–18, 1983.

RECEIVED January 15, 1986

4

A Chemistry Diagnostic System for Steam Power Plants

James C. Bellows

Westinghouse Electric Corporation, Orlando, FL 32817

A diagnostic system for the steam system chemistry of utility power plants is described. It is an expert system which accepts data from a monitoring system and generates recommendations for action to improve the chemistry of the plant. The monitors collect data from important points in the steam cycle. Data is transferred to a central data center for transmission to a centralized diagnostic center. At the diagnostic center, the monitors readings are validated before being used in the diagnosis of the power plant. Recommendations are transmitted to the data center for display. The removal of a malfunctioning sensor from consideration is given as an example of the operation of the system.

Downtime at a steam power plant can be valued at as much as $1 million/day. The actual value depends upon the size of the plant and the cost of replacement power. For 1000 MW nuclear plants, such as those that supply approximately 50% of Chicago's electricity, the $1 million/day is fairly accurate. One of the major causes of downtime, especially unscheduled downtime, is corrosion due to improper steam and water chemistry. Replacement of corroded turbine blading often requires downtime of a month or more. Replacement of corroded nuclear steam generators has required on the order of 9 months. The chemistry of power plants will be briefly reviewed. The goals of the chemistry diagnostic system will be stated. The supporting monitoring system will be briefly described, and capabilities of the current diagnostic system described. The scheme for diagnosing monitors and removing erroneous data from plant diagnosis will be outlined, and an example of a sensor malfunction diagnosis will be given.

Power Plant Chemistry

A power plant may be viewed as a chemical plant which has taken by-product sale to the limit. It recycles the product and sells only by-product. By most standards, it is a chemical plant, full of reactor vessels, piping, pumps, and tanks. Since the principal product is not a chemical, however, people tend to forget that a power plant is a chemical plant. Chemistry has often been the entry level position, and people were promoted to janitor. The materials are generally chosen to optimize heat transfer and mechanical strength and are not optimized for compatibility. Figure 1 shows a simplified schematic of a power plant.

The condenser may consist of copper bearing alloys, such as aluminum bronze, admiralty brass, and Muntz metal. Titanium is also used, as are stainless and carbon steels. Its purpose is to act as a sink for approximately 2/3 of the heat produced in the boiler. The feedwater heaters are steam to liquid heat exchangers and have have steel or copper alloy tubing and usually carbon steel shells. Restricting the discussion to fossil plants for simplicity, the boiler has carbon and alloy steels which are chosen for resistance to the $1000\text{-}1250^0$ F thermal conditions more than for corrosion resistance. The high pressure and intermediate pressure turbines must be designed to operate with inlet temperatures of the same range. The low pressure turbine operates between approximately 700^0 F and 100^0 F. The final stages of the low pressure turbine are supersonic. A large fossil turbine will be over 10 feet in diameter and weigh on the order of 250,000 lb. It rotates at 3600 rpm. The centrifugal stresses in the last stages of the low pressure turbine dictate high strength alloys in the same region that concentrated salt solutions can form.

There are two fundamental types of boilers: once through and recirculating. In the case of once through boilers, all the feedwater is converted to steam as it passes through the boiler in essentially a plug flow regime. Most once through boilers are supercritical pressure (3500 to 4500 psi), so the distinction between liquid and vapor is lost. In a recirculating boiler, pressures are limited to about 2800 psi, and steam is separated from liquid in a steam drum. The liquid is recirculated back to the bottom of the boiler and the steam is superheated. In both boilers, the steam is reheated after it is passed through the high pressure turbine. Occasionally a second reheat after the intermediate pressure turbine is found.

Which type of boiler is present in the system has a significant influence on the fundamental chemistry used in the plant. In once through boilers, no solids can be used so All Volatile Treatment (AVT) is employed. AVT consists of extremely pure water with the addition of ammonia, or other volatile amine for pH control, and hydrazine for oxygen scavenging. The exact concentration of ammonia is chosen to minimize corrosion of the feedwater heaters and depends upon the alloys used in their construction. The hydrazine feed rate is determined by the amount of oxygen in the feedwater. In recirculating boilers at pressures over 1500 psi, the AVT regime is used, but a solid conditioning

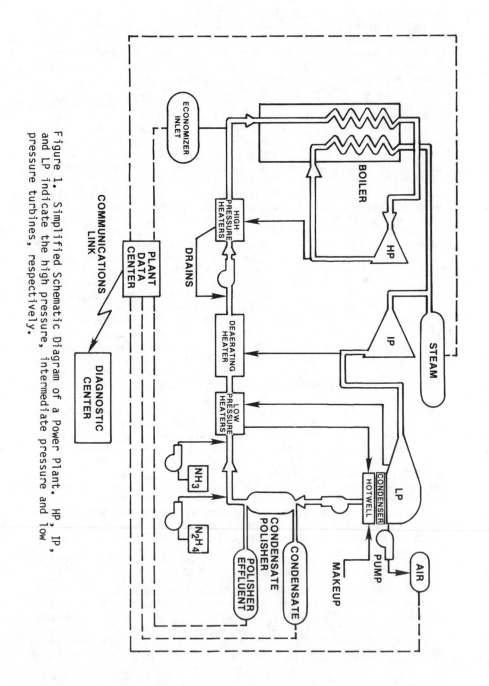

Figure 1. Simplified Schematic Diagram of a Power Plant. HP, IP, and LP indicate the high pressure, intermediate pressure and low pressure turbines, respectively.

agent may be added to the recirculating water in the boiler. The purpose of this conditioning agent is to control pH and to precipitate impurities in compounds which do not adhere to the boiler surfaces. In the United States, this is usually a mixture of disodium and trisodium phosphate. In other countries, sodium hydroxide may be used. In both types of boilers the fundamental chemistry problems are to avoid oxygen and to avoid deposits of chemicals on the heat transfer surfaces. Failure to avoid either of these conditions leads to corrosion, and ultimately to rupture of boiler tubing.

Considerable dissolution of salts may occur in the high pressure steam. As the steam density decreases through the turbine, the solubility of the salts decreases, and the salts deposit on the turbine. Two categories of deposition exist. Alkali metal hydroxides form stable water solutions at all pressure and temperature conditions within the turbine. Sodium hydroxide concentrations can be as high as 90%. These concentrated hydroxide solutions lead to rapid stress corrosion cracking of turbine materials and must be rigorously avoided. The second case is represented by sodium chloride, which deposits as a solid throughout most of the turbine. However, salts elevate the boiling point of water enough that near the transition from superheated to saturated steam, a region exists in which salt solutions of 30% are stable. Sodium chloride solutions of this concentration at temperatures of 100° C are quite corrosive and lead to stress corrosion and corrosion fatigue of turbine alloys.

The problem of power plant chemistry becomes one of determining which sources of chemicals are active at any given time and whether the purification systems are working properly. The condenser is a common source of impurities. On one side is the steam, which must be maintained pure to a few parts per billion; on the other is cooling water, which may be sea water. The condenser will commonly consist of tens of thousands of tubes, each of which is sealed to two tube sheets. The sum of all the leaks must be on the order of pints per day. The condenser and some other parts of the system operate below atmospheric pressure. Oxygen and carbon dioxide from air leaking into the system represent significant contaminants. Condensate polishers are large ion exchange units which remove trace impurities in the feedwater. They must be operated properly, or they may add more impurities than they remove. Deaerating heaters remove dissolved gases from the feedwater. Finally, a drum boiler is a still, and the efficiency of liquid-vapor separation is critical to the purity of the steam.

Definitions

At least in the power industry, the terms "monitoring" and "diagnostics" are often used interchangeably or without careful definition. Much confusion can arise when these terms are used. For purposes of this paper, these terms and the terms "expert system" and "malfunction" will be defined here.

Monitoring. Collection and manipulation of data for control and diagnostic purposes. Monitoring systems will have sensors to measure temperatures, pressures, flow rates, and the concentrations of chemicals in streams. They may store this information in a computer or a data logger. They may perform transformations on the data, such as conversion of voltages to engineering units, computation of averages, and comparison of two values. They may provide alarms when variables are beyond acceptable limits. Monitoring systems may even plot graphs.

Diagnosis (Diagnostics). Determination of condition and specific cause of this condition. A diagnostic system determines that all conditions are as they would be expected, or that there is some malfunction of a component that is causing an undesired condition. A diagnostic system may also generate recommendations for correction of an undesired condition.

Expert System. A computer reasoning system based on rules generated by questioning experts in a given field. Expert systems generally consist of three parts: a rule entering and editing program, a rule base, and an inference engine which takes data and applies the rules to it to reach conclusions about the system which generated the data.

Malfunction. Any condition in which a piece of equipment or system is imperfect for any reason. Examples might be exhausted condensate polishers, leaky condenser tubes, and sensors for which the power has been unintentionally turned off. This definition of malfunction includes deterioration of equipment due to normal wear. By this definition, worn out car brakes are a malfunction.

Goals of the Diagnostic System

The goal of an artificial intelligence diagnostic system is to provide the available expert advice to the user, in a time that is probably faster than the human could deliver it. A number of decisions have been made about the scope of the system which should be stated here. One goal was to use only on-line monitors, since only then could the system be responsible for the quality of the data. The quality of manual analyses varies in unpredictable ways, and we chose not to depend upon it. If one is to work with on-line sensors as the primary source of data, then the validity of sensors must be determined within the system. Since on-line monitors are expensive there is a corollary goal of a minimum number of sensors consistent with diagnosis of important equipment malfunctions and sensor condition. The diagnosis must be done centrally so that experience gained from one plant can be immediately available to other plants by rapid revision of the diagnostic rules. Chemistry upsets in power plants generally require several hours to develop, so transmission once or twice per day to the Diagnostic Center would usually be adequate to detect upsets which were developing slowly. To handle the upsets that were faster than the regular transmission would detect, it was decided that the data gathering computer at the plant should be sophisticated enough to determine that something is happening and make a special transmission of data at that time. The monitor set has been chosen to allow high reliability diagnosis of common power plant conditions, but it will support some unusual conditions as well. Those unusual conditions are included in the diagnostic system simply because the supporting data are present. Finally, it was decided that no information which might be relevant, including manual analysis data, should be rejected completely, and manual entry points have been included for that data. Manual entry of data requires validation of the data before entry.

Monitoring System

In order for any diagnostic system to draw valid conclusions about the condition of a plant, it must have an appropriate monitoring system for gathering data. The monitoring system chosen is shown schematically in Figure 1. Sensors are placed on the influent and effluent streams of each chemically active component of the plant. Thus, by looking at changes in concentrations from condensate to condensate polisher effluent, as well as the concentrations in the polisher effluent, one determines the effectiveness of the polishers. For the chemical feed, the polisher effluent is the influent to the zone, and the final feed (economizer inlet) is the effluent. The sensor set is kept as small as is reasonable, consistent with high certainty of sensing malfunctions of the plant and of the sensors. The sensors used are given in Table I.

The output of the sensors is transmitted to a data center in the plant, which stores the data. Normally the data are transmitted to the central diagnostic center at least once per day, and the diagnosis is returned to the data center for display. The data center also computes rates of change of variables. The data and rates of change are compared with alarm limits and a more sensitive limit, which we call a diagnosis activation limit. If the diagnosis activation limit is reached, a special transmission of data is made immediately so that a diagnosis may be made immediately. It is believed that this is a suitable compromise among the expense of continuously on-line diagnostics, the need for immediate diagnosis of an upset, and the need to keep the diagnostic system centralized to allow rapid improvement in diagnosis as experience with the automated system develops.

Diagnostic System

Diagnosis is accomplished by the expert system. The central part of the expert system is the rule base. The rule base consists of ideas, called nodes, and rules which interconnect them as shown in Figure 2. The upper node is the evidence; the lower node is the conclusion. The rule between them will state that if the evidence is known to be true with absolute certainty, then the conclusion will be known to be true (or false) with a specific confidence.

Table I. Sensors for a once-through boiler

Sensor Description	Condensate Pump Discharge	Condensate Polisher Effluent	Economizer Inlet	Hot Reheat	Makeup
Cation Conductivity	X	X	X	X	X
Specific Conductivity	X	X	X	X	X
Sodium	X	X	X	X	
Chloride	X	X	X	X	
Dissolved Oxygen	X		X	X	
Hydrazine			X		
pH	X	X	X		
Silica	X	X			
Air Exhaust	X				
Makeup Flow					X
Electrical Load			X		

This aspect of the rule is known as sufficiency. The rule will also state that if the evidence is known to be false with absolute certainty, that the conclusion will be known to be false (or true) with another specific confidence. This aspect of the rule is known as necessity. The sufficiency and necessity need not be equal. There are many times when something may indicate the presence of a condition but not be a necessary consequence of that condition. The increases in monitor readings that occur at the start of malfunctions are good examples of indicators which will signal the presence of a malfunction, but when the malfunction becomes stable at some severity, the increase will no longer be present. Of course a high value for the monitor reading will then be present. Evidence may be sensors or the conclusions from other rules. Several rules may support a single conclusion and the same evidence may be used for several rules.

When the system is used to diagnose the power plant chemistry, the inference engine will activate all the rules for which evidence exists. Thus all possible conclusions are examined

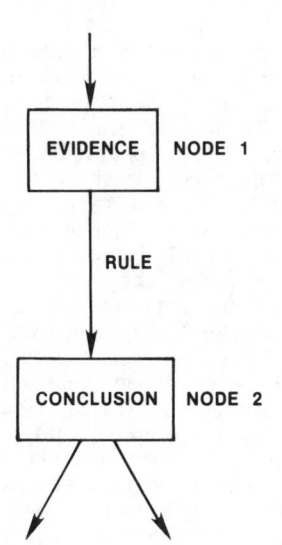

Figure 2. Basic Step in an Expert System Rule Base.

simultaneously. All the evidence for and against all conclusions is always considered. Diagnosis of simultaneous malfunctions occurs simply because evidence for those malfunctions exists. The structural details of our expert system have been published elsewhere (1).

The nodes and rules for an expert system are based on expert judgements. Usually, definitive statistics are unavailable for the relationships between ideas, but experts have a good feel for the relationships. We have found that when the information in the rule set is broken down into small enough steps, experts tend to have substantial agreement concerning the sufficiency and necessity of evidence to a given conclusion. It is quite common to find that what an experienced engineer considers to be one step is in fact several. When the rule base is constructed, the small steps are used. The use of small steps promotes clarity in the rule base and, at times, provides experts with new insights. Since the diagnostic process must be broken down into small steps, the process of building the rule base is much like that of training an able, but rather ignorant person.

It has been arbitrarily decided to say that any malfunction for which there is less than 30% confidence is probably not present with enough severity to cause concern. Between 30% and 50% confidence, one should be concerned that the malfunction may be developing. This represents an early warning, but with increased possibility of error. Between 50% and 70% confidence, action is appropriate to confirm or disconfirm the presence of the malfunction by collecting additional information, if necessary. Above 75% confidence, a plant malfunction is present with enough confidence that action ought to be taken to correct the malfunction. Action on a sensor malfunction indication should take place above 50% confidence, since by that time the system has lost substantial sensitivity to the plant malfunctions supported by the sensor.

Results for a Fossil Once-Through Steam System

There are currently over 50 malfunctions of a fossil once-through steam system that can be diagnosed. Some of these malfunctions are listed in Table II. It will be noted that some of these malfunctions occur as sets of related malfunctions. In some cases the members of the set are mutually exclusive, as in the ammonia feed malfunctions. In other cases, such as contaminated makeup, there is a malfunction which can be broken into smaller, more detailed malfunctions. The system can diagnose a variety of sensor malfunctions as well. The diagnosable malfunctions related to each sensor are shown in Table III. To accomplish these diagnoses, the system contains over 1300 rules. To test the system we have made use of whatever monitoring data we have had accessible. This has consisted of Steam Purity Analyzer System (2) data which is single location data, Total Plant Survey (3) data which is system wide but grab sample, and such plant data as has been accumulated from record reviews and diagnostic missions. None of these data sets conforms exactly to the monitor system that is envisioned as input to the diagnostic system, so estimates of values of other data which were deemed necessary to test the system have been used. A number of diagnoses use the rates of change of variables. Since the only available continuous (one minute interval) data sets were for the Steam Purity Analyzer System, these data sets were sometimes moved to other locations to test the sensitivity of plant malfunction confidences to sensor malfunctions which were known to be in the data. Where intermediate values between grab samples or discrete readings were necessary, they were either linearly interpolated with time or made proportional to a sample for which continuous data were available.

One of the important tasks of the system is to diagnose the sensor malfunctions and remove the malfunctioning sensors from consideration in the plant diagnosis scheme. Chemical sensors are high maintenance and high malfunction rate devices. If they were not removed from consideration when they malfunction, they could generate spurious plant malfunction diagnoses and discredit the diagnostic system. The task of removing a malfunctioning sensor from consideration is accomplished by taking the confidence that

Table II. Representative Malfunction Groups for a Once-Through Boiler System

Malfunction Numbers	Description
1.	Condenser cooling water leak
2-8.	Contaminated makeup
9-12.	Air in leakage
13-17.	Polisher malfunctions
18-31.	Ammonia feed Malfunctions
32-45.	Hydrazine feed Malfunctions
46-47.	Contaminated feed Chemicals
48-51.	Organic contamination
52.	Contaminated boiler

Table III. Number of Malfunctions Diagnosed for each Sensor

Sensor	Number of Distinct Malfunctions
Cation Conductivity	5
Specific Conductivity	4
pH	4
Dissolved Oxygen	3
Sodium	2

there is a malfunction in the sensor and using it to drive a rule that changes the sufficiency and necessity of rules coming from that sensor so that the plant diagnostic system is less sensitive to the information coming from that sensor. This is shown schematically in Figure 3. Data from the sensor and other sensors are used to drive rules that diagnose the sensor of interest (4). The results of these rules are accumulated in a sensor malfunction node. The confidence in this node is used to drive rules which alter the sufficiency and necessity of other rules. The scheme is analogous to setting a valve point based on the values of a number of sensors.

An example of sensor validation and removal from consideration illustrates the working of the diagnostic system. The malfunctioning sensor is an acid cation exchanged conductivity monitor, commonly called "cation conductivity." It consists of a cation exchange resin in the hydrogen form followed by a conductivity meter. The cation exchange resin removes ammonia from the sample stream and the resulting conductivity provides a good estimate of total ionic content, except for hydroxide. The monitor is very sensitive to most of the impurities that are important to power plants. However, when the cation exchange resin is exhausted, the monitor reverts to a specific conductivity monitor and the output is dominated by the ammonia concentration.

Figure 4 shows a test of the diagnostic system for an incident of resin exhaustion for a cation conductivity sensor. The data are a combination of real and synthesized plant data and are given in Table IV. The condensate values for the condensate sensor are those recorded during the actual exhaustion of the cation resin column at a plant installation. The steam and polisher effluent values would be reasonable based on the starting value of the real sensors. All of the sensor values other than the condensate cation conductivity were held constant to clearly

Table IV. Sensors related to a cation conductivity resin exhaustion incident

Sensor	Value	Data Source
Condensate cation conductivity	See Fig. 4	Real
Condensate specific conductivity	7.87	Real
Condensate sodium	2.2	Real
Makeup addition	Off	Estimated
Steam cation conductivity	0.17	Estimated
Steam specific conductivity	7.8	Estimated
Steam sodium	2.1	Estimated
Polisher effluent specific conductivity	0.16	Estimated

Note: These values were held constant to show the effect of the variation in the single variable.

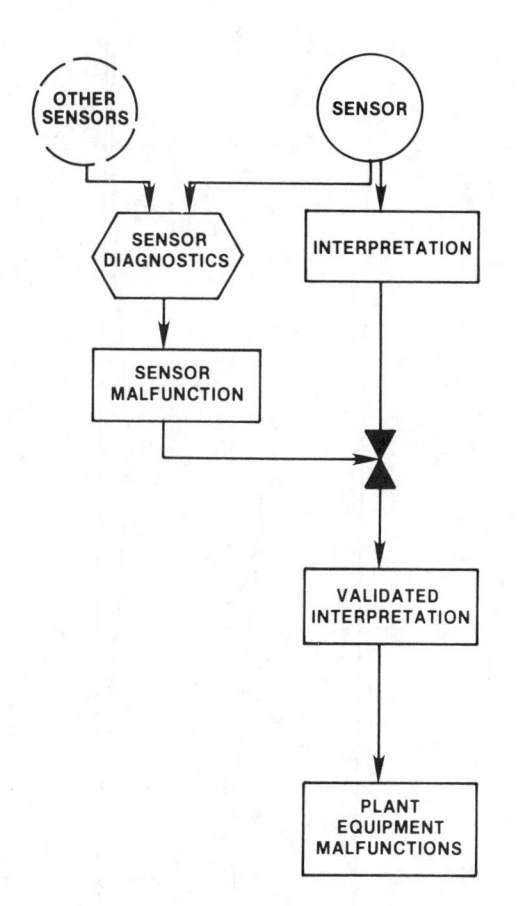

Figure 3. Block Diagram of Sensor Validation.

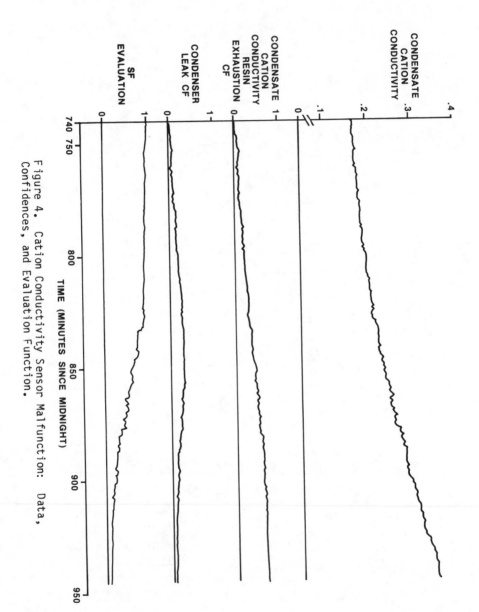

Figure 4. Cation Conductivity Sensor Malfunction: Data, Confidences, and Evaluation Function.

show the effect of the single malfunction on the confidence in other malfunctions. Of particular interest is the condenser tube leak, which has great sensitivity to the value of the cation conductivity of the condensate. One sees that the confidence in the sensor malfunction, the exhaustion of the resin in the cation conductivity sensor, parallels the increase in the cation conductivity reading. At first, the confidence in the condenser tube leak also parallels the increase in the cation conductivity. However, as the confidence in the malfunction of the cation conductivity sensor increases, the confidence in the tube leak peaks at 30% and declines with further increase in the confidence in the sensor malfunction. The value of the evaluation function, which is used to reduce the sensitivity of the plant diagnosis to the malfunctioning sensor, is shown at the bottom of Figure 4. It starts at 100% sensitivity and declines as the sensor malfunction becomes more certain. By the time the sensor malfunction confidence has reached 70%, the plant diagnosis practically ignores the sensor. Of course, if the malfunction had been a condenser leak, the condensate sodium would have increased at the same time as the cation conductivity. The rule base would have recognized this occurrence, the confidence in a malfunction of the cation conductivity sensor would have been substantially reduced, and the confidence in the condenser leak would have increased due to the increases in both the cation conductivity and the sodium concentration.

Data Center Displays

The data center displays the diagnosis and a number of different utility screens. Figure 5 is a picture of the RECOMMENDATION SUMMARY screen. It shows the actions which are most important to improving the chemistry of the unit at the current time. They are listed in priority based on confidence in the existence of the malfunction and on the seriousness of the consequences of the malfunction at its current severity. On the data center screen, the recommendations are color coded, with red recommendations having a confidence in the underlying malfunction of at least 70%. Yellow indicates 30-70% confidence, and green indicates 0-30% confidence. The rectangles on the right hand edge and along the bottom are touch buttons to allow access to other screens. They blink if new information is available on those screens. Their color is determined by the color of the most urgent information on the screen. The RECOMMENDATION screen shown in Figure 6 displays the action, a cryptic reason for taking the action, and the consequences of not taking action. The consequences are as specific as the current state of knowledge will allow.

Experience with a Generator Diagnostic System

Although the subject is hardly chemistry, it would be appropriate to make mention of a companion project in the diagnosis of conditions in electrical generators. Such a system is in

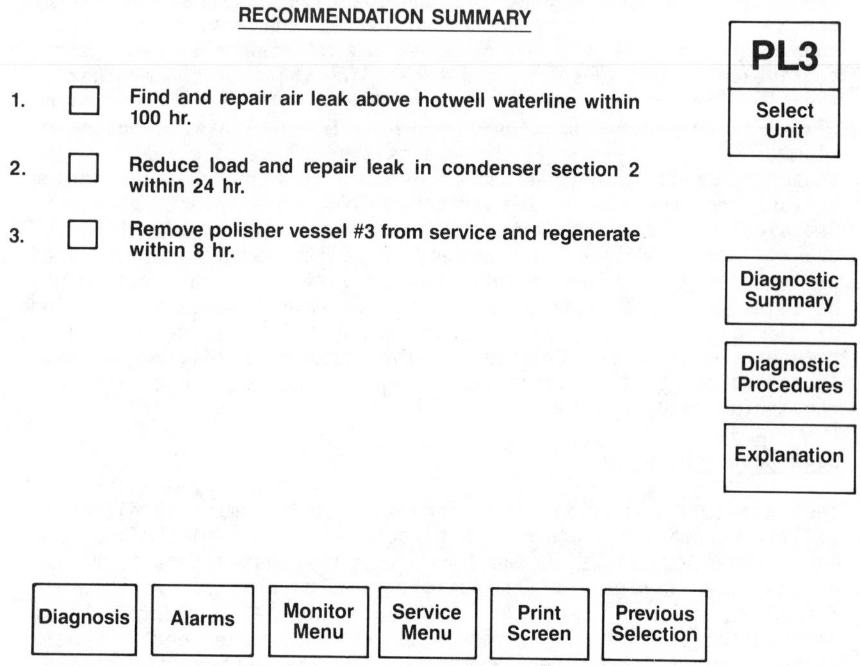

RECOMMENDATION SUMMARY

	PL3
	Select Unit

1. ☐ Find and repair air leak above hotwell waterline within 100 hr.

2. ☐ Reduce load and repair leak in condenser section 2 within 24 hr.

3. ☐ Remove polisher vessel #3 from service and regenerate within 8 hr.

Diagnostic Summary

Diagnostic Procedures

Explanation

Diagnosis	Alarms	Monitor Menu	Service Menu	Print Screen	Previous Selection

Figure 5.   Recommendation Summary Screen.

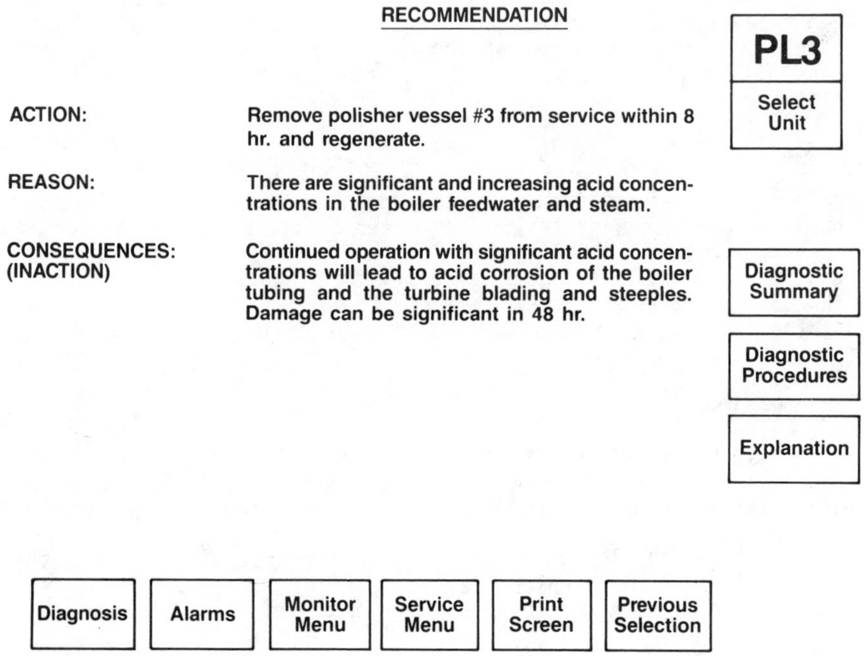

RECOMMENDATION

PL3

Select
Unit

ACTION: Remove polisher vessel #3 from service within 8
               hr. and regenerate.

REASON: There are significant and increasing acid concen-
               trations in the boiler feedwater and steam.

CONSEQUENCES: Continued operation with significant acid concen-
(INACTION)   trations will lead to acid corrosion of the boiler
               tubing and the turbine blading and steeples.
               Damage can be significant in 48 hr.

Diagnostic
Summary

Diagnostic
Procedures

Explanation

| Diagnosis | Alarms | Monitor Menu | Service Menu | Print Screen | Previous Selection |

Figure 6.   Recommendation Screen.

operation for seven power plants in Texas. While yet in the
prototype stage, it detected a developing generator malfunction
several hours before any alarms sounded. Since the malfunction
was known, appropriate resources could be mobilized before the
generator was taken out of service, and the problem was repaired
in four days. This particular malfunction normally requires two
to three weeks for repair when it is allowed to progress to the
point where the automatic generator control systems take the
generator out of service. Working with a customer in the final
stages of the development of the generator system has influenced
many decisions in both the generator and the chemistry diagnostic
systems.

Summary

An artificial intelligence system for the chemistry of a fossil
once-through steam system has been constructed. It is based on
on-line monitors. It diagnoses both sensor and plant malfunction
and removes malfunctioning sensors from diagnosis of plant malfunctions. The system has been tested off-line using real and
synthesized power plant data and is now ready for testing in a
plant.

Acknowledgments

The assistance of C. T. Kemper and S. Lowenfeld with the
artificial intelligence and of numerous Westinghouse Engineers in
providing data and information for the rule base is gratefully
acknowledged.
    Portions of this paper were previously presented at the 45th
International Water Conference, Pittsburgh, Pa., October 22-24,
1984 and permission to republish them is gratefully acknowledged.

Literature Cited

1. Fox, M. S. Kleinosky, P. Lowenfeld, S. Proc. 8th Internat.
    Joint Conf. Artificial Intelligence, 1983, p. 158.

2. Bellows, J.C.; Carlson, G.L.; Pensenstadler, D.F.   J.
    Materials Energy Systems 1983, 5, 43.

3. Peterson, S.H.; Bellows, J.C.; Pensenstadler, D.F.; Hickam,
    W.M. Proc. 40th Internat. Water Conf. 1979, p. 201.

4. Gonzalez, A.J.; Osborne, R.L.; Bellows, J.C. "On-Line
    Diagnosis   of   Instrumentation   through   Artificial
    Intelligence," presented at ISA Power Industry Symposium, New
    Orleans, La., May 1985.

RECEIVED January 10, 1986

A Real-Time Expert System for Process Control

Lowell B. Hawkinson, Carl G. Knickerbocker, and Robert L. Moore

LISP Machine Inc., Los Angeles, CA 90045

Expert systems technology can provide improvements in
analysis of process information, intelligent alarming,
process diagnosis, control and optimization of processes. However, to realize these benefits, a real-time
expert system capability is required. A program
design is described which supports forward and
backward chaining inference in a real-time environment, with dynamic measurement data. The knowledge
base for the program is implemented in structured
natural language form for application to a broad range
of process expert systems. Plant test results are
described.

In the real-time application of expert systems, a number of design
considerations, beyond those usually considered in expert systems,
become important. Execution efficiency is a prime consideration.
In conventional expert systems, the facts and knowledge upon which
the inference is based are static. In the industrial application,
the facts or process measurements are dynamic. In an industrial
application there may be several thousand measurements and alarms
which may significantly change in value or status in a few minutes.
The problem posed by an operator advisor, to give expert
diagnosis of plant condition and to recommend emergency actions or
economic optimization adjustments, illustrates these real-time
requirements. Some of the plant conditions which can occur
include:

1. Critical measurement failure. In this case, the information
   presented to the operator is incorrect. An expert system would
   use a process knowledge base to detect inconsistencies and to
   alert the operator.
2. Process upset. In this case, the expert system would identify
   underlying process problems, distinguishing causes from
   effects, and would advise the operator accordingly. Heuristic

0097–6156/86/0306–0069$06.00/0

rules of optimization would be applied by the expert system to give control advice.

In these examples, the expert system is simply applying the expertise used in its development. The potential advantage of the operator advisor is that this expertise is available quickly, on any shift, for providing organized advice to the operator.

To meet these requirements, several design considerations must be addressed:

1. Data access. An efficient real-time data interface must be established with the distributed measurement system.
2. Inference paradigms. The basic inference mechanisms of forward-chaining and backward-chaining must be integrated into a real-time execution environment.
3. Computational efficiency. The efficiency of inference is enhanced by program and knowledge-base structure and by machine speed. Also, heuristic procedures, as used by experts, can augment the deductive procedures of conventional inference.

The program developed by LMI in response to these design requirements is called Process Intelligent Control (PICON). The individual design considerations are addressed in the following discussion.

Process Intelligent Control

The expert system package is designed to operate on a LISP machine interfaced with a conventional distributed control system. The design assumes that up to 20,000 measurement points and alarms may be accessed. The Lambda machine from LMI was utilized. The real-time data interface is via an integral Multibus connected to a computer gateway in the distributed system.

Data transfers, in floating point engineering units or in status states, are requested by the expert system. Thus the distributed system does not transmit all measurements and alarms on a fixed scan basis, but rather the process data are accessed as required for inference. In a sense, the expert system is acting like an expert operator, who focuses attention or scans the process operation selectively, using expertise to determine specific areas of attention.

The basic inference paradigms supported by the expert system are forward-chaining and backward-chaining. Within the context of an alarm advisor, there are requirements for both of these para- digms. An expert process operator, during normal plant operation, will scan key process information. This is for purposes of moni- toring control performance and detecting problems which may not cause explicit alarms. The programming paradigm which reflects this approach is a scanned forward-chaining inference. The heuristic rules which determine possibly-significant-events are scanned, and rule condition matching triggers an alert to the expert system monitor program. Conventional alarms also may trigger an alert, if they are heuristically ranked as possibly- significant-events.

An expert process operator, once alerted, will focus attention
on the problem. This may involve invoking procedure rules for
safety or other reasons, and it may involve assembling information
and primary analyses to allow inference about the problem. Logic
rules and procedures are used when required for the diagnostic
inference. The expert system mimics the expert process operator
in this regard: Logic rules and procedures are invoked
specifically when they are required for diagnosis of a process
problem, or as requested for a specific step in inference.

In working through process control examples, we found that many
calculations, data checks, rate checks and other computationally
intensive tasks are done at the first level of inference.
Considerations of computational efficiency led to a design
utilizing two parallel processors with a shared memory (Figure 1).
One of the processors is a 68010 programmed in C code. This
processor performs computationally intensive, low level tasks
which are directed by the expert system in the LISP processor.

The processing of data applies a level of intelligence. Instead
of mere measurement values, the expert may base inference on
trends or patterns of measurements. Thus the system must be able
to access primitive functions of data, such as averages and trends
of values, and quality information, such as the presence of noise
or discontinuous values. Such functions are conveniently
calculated in the parallel 68010 processor, coded in C language
for execution efficiency.

An expert, given time to do so, may utilize calculations to
develop inference results. For example, a material balance
calculation around a process unit may indicate a measurement
inconsistency. To mimic this expertise, general mathematical
operations on combinations of measurements or functions of
measurements are implemented in the parallel processor also.

Higher levels of inference depend on the truth conditions of
the first level antecedent conditions, and thus higher levels of
inference involve pattern matching and chained-inference logic.
Higher level inference is done in the LISP processor, using
various expert system paradigms, while the first level antecedents, which are computationally intensive, are evaluated in the
parallel 68010 processor.

The expert system package is designed so that an algorithm of
reasonably arbitrary structure can be dynamically loaded into the
68010 from the LISP processor. This allows, for example, the
expert system to implement process-monitoring functionality in a
dynamic fashion, the equivalent of:

"look closely at the energy balance around the specific process
unit for the next few minutes."

The expert system design includes the ability to change the time
period of measurement and algorithm processing in individual
cases. Thus, in effect, the system can "focus attention" to a
specific area of the process plant, and put all associated
measurements and rules for that area on frequent scan. This can
be done under control of the LISP program. Thus, for example:

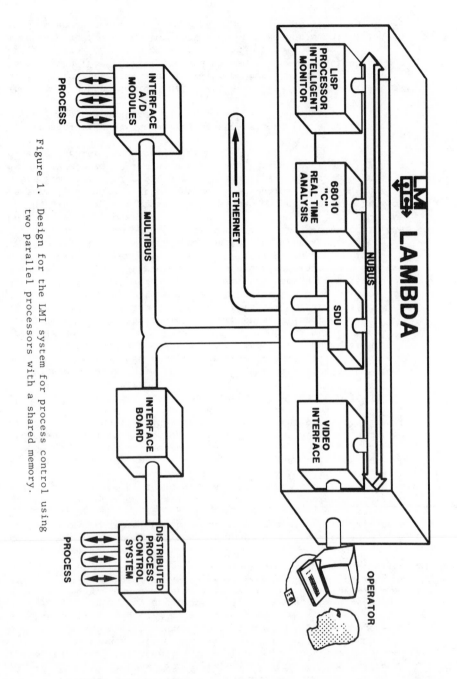

Figure 1. Design for the LMI system for process control using two parallel processors with a shared memory.

A back-chaining diagnostic expert system could reach a point
where an inference test is required.
The LISP program would tell the 68010 processor to "focus" on
the measurements and low-level inferences required around a
process unit.
The inference could then be tested.

Another use of this "focus" facility is to scan the plant in a
background mode, focusing attention on parts of the plant to
evaluate unit process performance and detect subtle problems,
utilizing both the programmed knowledge of the the expert process
operator and the expert process engineer. It is not practical to
examine an entire plant continuously with this intensity, but the
individual parts of the plant could be scanned in a background
mode. This is equivalent to the way a process engineer would
analyze plant performance during normal plant operation.
It should be noted that the ability to focus not only emulates
the way a human expert works, but also it avoids the problem
associated with overloading the distributed process system with
requests for information. While the expert system knows about all
20,000 measurement and alarm points in the process environment,
only those of interest to the expert system need be accessed.
The LISP environment contains the higher-level functionality of
the expert system. A truth-maintenance design structure is used.
The design assumption is that lower-level intelligent processing,
done in the 68010, will signal potentially significant process
events. Thus, only a table of truth condition triggers needs to
be checked by the LISP programs.

Some general examples of inference using the system:

- detecting process problems, particularly on complex
  combinations of conditions which require expertise for proper
  interpretation.
- focus inference, in which rules of all priorities are activated
  for a unit process. In the typical use, a
  possibly-significant-event (detected by a high priority
  procedure rule) would trigger a focus on the process unit, thus
  initiating the gathering of information required for inference
  around the process unit.
- diagnosis, a backward chaining inference procedure, which would
  be triggered by a possibly-significant-event or by operator
  request. Diagnosis uses the focus mechanism. An explanation
  is then given of the diagnostic conclusion.

Summary and Future Extensions

Virtually all tasks which require the routine application of human
expertise, in an organized way, are candidates for expert systems.
The computer implementation of expertise has such advantages as
speed, around-the-clock availability, and ease of expansion of the
knowledge base. As such, expert systems represent the next
generation of higher level software, performing tasks presently
done by human operators.

Expert systems have been investigated for 20 years. The implementation of expert systems is now being undertaken on a widespread basis, due to the availability of hardware and software tools which alleviate the "knowledge-engineer bottleneck", allowing cost effective implementation. In a similar way, real-time applications of expert systems require tools to allow straightforward implementation. We have presented a software/hardware structure which supports knowledge-base capture and real-time inference for process applications.

In general, the LMI package (Figure 2) provides a knowledge-base structure, facilities for acquiring the knowledge base in an organized manner, and real-time collection of data with some parallel processing of inference, and higher-level inference tools. The individual applications require specific knowledge engineering, which is facilitated using the tools we have described. The system is currently installed at Texaco and Exxon facilities and is in pilot plant or laboratory testing at seven additional sites.

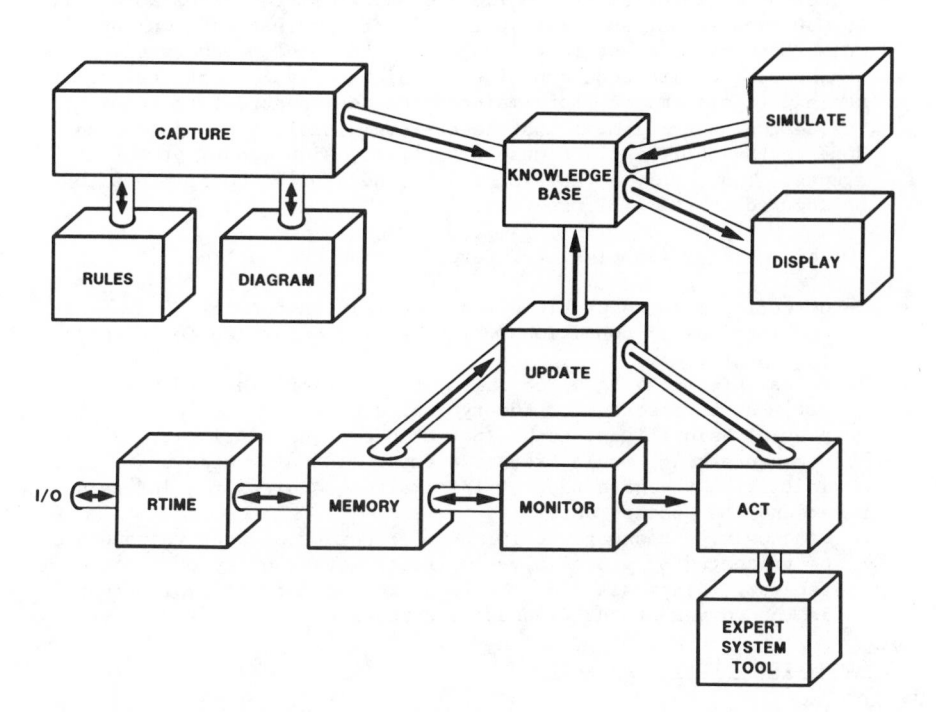

Figure 2.   General structure of the LMI package.

RECEIVED December 17, 1985

Interpretation and Design of Chemically Based Experiments with Expert Systems

David Garfinkel[1], Lillian Garfinkel[1], Von-Wun Soo[2], and Casimir A. Kulikowski[2]

[1] University of Pennsylvania, Philadelphia, PA 19104
[2] Rutgers University, New Brunswick, NJ 08903

Expert system building programs, e.g., EXPERT, can now supervise numerical calculations in addition to performing qualitative reasoning and choosing among possible alternatives. This capability can be used to interpret experiments, calculate optimal designs for them, and automate model construction and manipulation, as well as to resolve associated problems due to differing conceptual frameworks and definitions. Three hierarchically arranged applications are suggested to (a) determine and manage free Mg^{2+} levels; (b) construct an expert system to derive enzyme kinetic models (including Mg^{2+}) and fit them to data; (c) design experiments (including enzyme kinetics) using minimal numbers of animals to prove drugs safe and effective.

Expert systems, and artificial intelligence in general, are new fields whose breadth of application, and indeed, whose exact definitions, are not yet completely settled. It is sometimes claimed that no two experts on artificial intelligence agree exactly on what its definition is. Definitions of expert systems at least agree on the necessity for expertise, but even here there are differences in emphasis and in priority.

Expert systems, which evolved from many sources, were recognized as a distinct system type because of a large body of work on medical consultation problems. The resulting systems, such as MYCIN, CASNET, and INTERNIST/CADUCEUS, essentially solved what are considered classification problems, by choosing among a set of possible diagnostic or treatment alternatives. Such systems have usually obtained information by asking the user questions. They have usually performed qualitative reasoning with "knowledge" rules of the type: if conditions A are true then conclude hypothesis B with probability X. There exist other types of expert systems, such as DENDRAL, which produces interpretations of quantitative experimental evidence, and MOLGEN, which formulates plans for the design of experiments. Most expert systems have been written in some variant of LISP or a re-

0097-6156/86/0306-0075$06.00/0

lated language, which were originally not as well suited for calculation as for logical manipulation. More recently it has been
possible to get an expert system to supervise calculations, digest
considerable masses of observational data, and draw conclusions which
are not strictly computational, as in the case of ELAS and the oil-
well drilling programs. These involve the EXPERT system builder
(1), which has the following advantages: it is written in FORTRAN
and can therefore easily communicate with FORTRAN programs; a PROLOG
version has also recently been prepared; it has data base capabilities; and it is good at explaining what it is doing and why. Interaction between artificial intelligence and modeling has evolved to
where modeling societies routinely program artificial intelligence
sessions at meetings, and are forming technical committees on this
subject.

This paper reflects the past activities of some of its authors
in computer modeling of the chemical aspects of biological systems.
This activity requires expertise in both model-building and in the
relevant biology. It also involves examination of the actions of and
results obtained by experts, like that routinely done in building expert systems. It also involves keeping track of and coherently
explaining sequences of decisions, which expert systems are equipped
to do.

In this paper we are concerned with a set of relatively similar
possible applications involving management of calculations and of
modeling. These involve actions (calculation, information retrieval,
and "intelligent" reasoning) at more than one hierarchical level.
Particular attention will be given to the design and interpretation
of experiments in enzyme kinetics. Designing an experiment may involve computation of optimal conditions, and its interpretation may
involve fitting of optimal parameters of a model, but non-numerical
reasoning procedures are also involved. Attention is therefore required to the kinds of reasoning employed in designing experiments
and to the critiquing of the reasoning and techniques involved in
such experiments. A high-level description of an experimental design
cycle can be given in such steps as: definition of the problem (what
questions are to be addressed? what hypotheses are to be tested?);
quantitative modeling; design and then performance of the necessary
experiments; analysis of the results; and then model reinterpretation
and possible problem redefinition (2).

A Problem of Definition

The process of building expert systems usually involves determining
the conceptual framework and pattern of decision making of experts
(often one outstanding expert). These are often not written down
and may not be clearly explainable because there is heavy reliance on
heuristics and even hunches. However, we would like to suggest that
this may not be the only way to apply expertise. We have encountered
workers in different fields handling the same subject matter differently because they have different conceptual frameworks and different
jargon as well as different heuristics and priorities. We offer the
following example involving a relatively simple multiple equilibrium
calculation.

Although there is no controversy about the basic definition of stability constants, physical chemists and biochemists handle the concepts involved and the resulting calculations differently. Physical chemists think in terms of reactive species and biochemists in terms of total concentrations of components. A further source of confusion is the differing definitions of "apparent constant". To a physical chemist the stability constant for $MgATP^{2-}$ formation

$$Mg^{2+} + ATP^{4-} = MgATP^{2-}$$

is defined as

$$K_{MgATP} = \frac{[MgATP^{2-}]}{[Mg^{2+}][ATP^{4-}]}$$

For a given temperature the standard state is at zero ionic strength. The constant observed experimentally at finite ionic strengths would be considered "apparent". A biochemist would call such a constant "intrinsic". The presence of interfering ions (H^+ and K^+) which form H and K chelates of ATP by binding to ATP^{4-} would be handled by calculations involving the corresponding equilibria.

Biochemists handle these calculations differently, and define apparent constants in terms of total components. Thus an apparent constant for $MgATP^{2-}$ at low pH in the presence of K^+ would be defined as

$$K' = \frac{K_{MgATP}}{(1 + [H^+] K_{HATP} + [K^+] K_{KATP})}$$

While it is relatively easy to show that the two calculations are equivalent in simple systems, it is not so easy with more complex in vivo systems, as when these equilibria are studied with ^{31}P NMR spectra from perfused or intact organs. We recently (3) became involved in a controversy where a 4-fold difference in magnesium ion level was calculated from substantially identical NMR spectra as a result of such differences in definition. Our experience indicates that an intelligent program to supervise such calculations would be quite useful.

In such a situation an intelligent program may function as an "intelligent interface", a program which can translate information from one conceptual framework to another. Even though there are many experts in the subject matter involved, programs of this type would be useful for the many others who are not expert in the subject matter or the calculations involved or who have difficulties in communication. The advent of software for small expert systems on microcomputers would add the advantage of convenience as well.

Applications

We describe here three possible applications of expert systems to
supervise calculations and design experiments which are largely chemically based, although they have biological content as well.   These
are arranged in a hierarchically increasing order of complexity
(i.e., each level needs the capabilities of the preceding one).   The
simplest of these applications is to supervise complex equilibrium
calculations.   The example described is of a type which often occur
in studying biological systems where it is necessary to control concentrations of reactive species.   Such calculations are often not
properly handled.

Calculations Involving Magnesium Ions

Enough examples of poorly designed experiments and poor calculations
involving magnesium ions exist in the biochemical literature to indicate a need for a better method.   This also applies to other
equilibria of comparable complexity, as with other metal ions.   Experiments involving enzyme kinetics are particularly affected.
Magnesium ions affect many enzymes by binding strongly both to the
enzymes and to important reactants such as ATP.   In a review on the
kinetics of magnesium-dependent enzymes, Morrison (4) stated "it is
unfortunate that studies on many metal-activated enzymes . . . have
been undertaken using conditions that preclude interpretation of the
data."
        The relevant calculations are commonly handled poorly, because
the equilibrium equations involved are difficult to solve manually
(but not with computers).   The few calculations that are actually
reported in the biochemical literature use simplified methods of
limited and frequently unknown validity.   Large excesses of magnesium
ion are frequently used in experiments, perhaps in an attempt to
avoid such calculations.   The relevant theory is well worked out and
there are excellent reviews.   The limitation appears to involve dif-
fusion to the (mathematically) inexpert user, which is one of the
motivations of building expert systems.
        The computational and other (e.g., data base and design) capa-
bilities to meet these needs can be specified.   We may need to determine how much magnesium ion (or other substance of interest in an
equilibrium system) is present in a cell interior or a solution emu-
lating the cell interior.   Here a complex series of equilibria may
be affected by conditions such as temperature or ionic strength.   Or
it may be necessary to work through a pattern of concentrations of
some particular molecular or ionic species to determine an ultimate
effect, or to keep particular species or particular side effects
within certain limits while changing others.   Computations may have
to start from any of the participating substances which are either to
be controlled or are observable.
        Computation of amounts of species present in straightforward
equilibria can usually be done without much difficulty, e.g., (5).
Some of the other requirements mentioned above are demanding enough
to define a minimal interesting problem in artificial intelligence
("toy problem").   Included are conversions among sets of conditions
(i.e., different temperature or ionic strength), which have caused

considerable difficulty, and which could be handled by providing an
expert system with the necessary conversion algorithms and data.
Such a system would include a program similar to that of Storer and
Cornish-Bowden to do equilibrium calculations. A communication-
control subprogram would be linked to an expert model by using the
EXPERT knowledge-base shall (or system-builder) which is advantageous
here because it can interact with procedures such as those written in
FORTRAN for numerical computation. Additional programs and a small
data base, which EXPERT can handle, would keep track of which chemical was what array element, and other requirements mentioned above.

The system could be used to answer questions such as:

How could I add to a solution combinations of ATP and magnesium
ion so their chelate is constant and free ATP^{4-} varies systematically
so as to define families of curves with (different) constant magnesium
ion?

This type of capability could be extended to magnesium-control-
led enzymes without substantial expertise regarding their kinetics
by adding another limited expert system to manage simple calculations
involving modifications to their kinetics. This would require adding
a small data base of the binding and inhibition constants of magnesium ion with important enzymes. We have assembled this information
for some of the enzymes we have worked with (6). This would permit
answering questions like:

How much magnesium ion can I add to solution X without inhibit-
ing enzyme Y by more than 10%?

Calculations Involving Enzyme Kinetics

At the next hierarchical step would be an expert system for the design of experiments in enzyme kinetics (and mathematically similar
systems like transport kinetics). Such a system would largely arise
from our experience in modeling enzyme kinetics. It could systematically perform, correctly, routine operations that are either not
done or done incorrectly, because they are too tedious or require
particular expertise. (For this subject there exists a sizeable body
of well-worked out theory, and yet considerable work is done as if
this theory did not exist). Such an expert system could offer the
expert user better modeling strategy and completeness and the inex-
perienced user the advantage of "friendliness". As an extreme
example, we modeled (7) what is probably the best data set in the
phosphofructokinase literature, and improved on its interpretation
(which included modeling) by the original experimenters, who are
highly expert in this subject. It was found that one important
effect was not determined by their data, but could have been with a
few additional measurements. If an expert system such as that
described below were available, this could have been done before
their experiments were concluded--and left permanently incomplete.
Also, the entire interpretation task would have required considerably less time and effort.

In working with enzyme and transport kinetics we already have a
program of considerable sophistication, PENNZYME (8) to fit experimental data to rate laws by optimization methods and to display the
results of the fitting process. This program would require extension
to perform experimental design functions (such as calculating design

and information matrices). For most applications it would be best
for the scientist to remain in the loop. An interface between
PENNZYME and EXPERT such that EXPERT could direct PENNZYME'S calculation functions would be very similar to the interface between
EXPERT and several oil-well logging programs (9-10). To help in
assessing and documenting modeling applications it would be desirable
to have EXPERT produce a record of its actions, decisions, and
reasoning, in addition to the chemical or biological output. This
would require only a straightforward extension of EXPERT'S very good
existing capabilities for explaining its actions to a user on-line.
        The major operations that would have to be performed by such an
expert system are:

Selection of a conceptual model. As the first step in modeling, it is
necessary to decide what kind of a conceptual model to try. For an
enzyme this includes a choice of mechanism and an indication of the
numerical values that go with it (determination of the best values
comes later). Probably this will be better done by an expert human
than by a program for some time. Examples of rules (domain knowledge)
for enzyme kinetics which are applicable (regardless of the methods
of calculation used) are:
        1. Kinases usually have Km's for ATP considerably lower than
tissue levels of ATP.
        2. Most other Km's approximate the usual tissue level of the
substrate involved.
        3. Certain classes of enzymes tend to have characteristic
mechanisms. (Examples: transaminases often have ping-pong mechanisms, kinases usually do not).
        4. The commonly used linearized plots of kinetic data are a
usable initial guide to determining the mechanism.

Selection of a computational model. Once a conceptual model has been
selected, it is necessary to encode it in a form usable for calculation, i.e., a rate law giving the velocity of the enzyme as a function of the relevant chemical concentrations. An expert model would
include determination of the situations where a given rate law should
be tried together with control information that determines how this
is to be done. The expert model obtains such information by querying the user or by deduction from its knowledge rules using results
from past calculations. Means for deriving such rate laws exist,
e.g., the KINAL program of Cornish-Bowden (11), which we have modified (PROKINAL) to facilitate interfacing with EXPERT to derive rate
laws automatically. The operations involved included:
        1. Obtaining the proper rate law from an existing library or
generating a new one, as with KINAL;
        2. Matching the generalized designations for reactants in the
rate law (reactant A, reactant B, . . .) with the real ones in the
system being studied;
        3. Asking the user for corrections if there is a problem;
        4. Obtaining starting values of parameters, as from information
on analogous enzymes;
        5. Fitting the rate law to the data and obtaining the optimal
parameters;
        6. Making appropriate modifications to the rate law.

Fitting of models to data. Fitting rate laws representing models to
the experimental data is the lowest-level and most frequent operation
this expert system would do. PENNZYME does a two-step optimization,
first using this simplex method which is robust and independent of
starting guesses, and then the more accurate Fletcher-Powell method,
which requires better starting estimates. Examples of heuristic
rules on how to operate PENNZYME (problem-solving knowledge) are:

1. Do at least one simplex optimization before doing a Fletcher-
Powell optimization.

2. Always get an optimization report. If the percentage reduction of the least-squares error is 0.00%, do not repeat the last
type of optimization performed.

3. If a simplex optimization has not converged for a model with
at least two parameters after many interactions, and the least-
squares error reduction is 0.00%, then something is wrong with the
rate law equation or the rate law file.

Having EXPERT operate PENNZYME under particularly favorable
conditions is expected to be straightforward to the point of being
uninteresting. Obtaining the percentage reduction resulting from a
given optimization and deciding from its value and place in pattern
of operations and resulting values what operation to ask for next
would be conceptually similar to the usual one-query-at-a-time
directed to a human user of the usual expert system. Determining
what is wrong with a rate law file (which is not a common problem)
would normally require user intervention. At the other extreme, this
program combination will not be able to extract from a poor data set
information that is not there to begin with. The most useful application is to the intermediate situation, where there is useful but
limited or noisy data, or where the experimental design is not optimal.

Experimental Design. It is now possible, but inconvenient, to use
PENNZYME in an inverse mode, by determining the parameters in a rate
law and then manipulating the chemical concentrations so as to find
the point in concentration space that maximizes a given effect. An
immediate application is to maximize the difference between two rate
laws by means of a discrimination function (2). This amounts to designing a critical experiment to distinguish between them. The user
who has as an appropriate set of experimental data of sufficiently
good quality and two alternative rate laws that might fit it could
have the EXPERT-PENNZYME combination:

(a) Find the optimal parameters for these rate laws;

(b) Determine the point(s) or region(s) in concentration space
where they differ most (following (2));

(c) Recommend one or more experimental measurements at those
points.

In designing sequential experimental measurements or groups of
them, other functions that might be performed with appropriate
calculations are:

(1) Minimize the confidence limit or variance of a given parameter, such as a Michaelis constant. This requires picking a point
or points in concentration space where the value of the parameter is
maximally sensitive to the experimental result obtained, i.e., a
kinetic constant basically representing the binding of a small

molecule is insensitive to measurements where its binding is very
small or very large, and more sensitive to measurements where its
binding is near half maximal.

(2) Maximize or minimize information or design matrices.

While performing the Fletcher-Powell optimization, PENNZYME
calculates the variance-covariance matrix of the parameters. This
can be used to test model acceptability: the parameters of a good
model should be relatively (although never completely) independent
of each other; if they are not, there is something questionable
about it. More important, this matrix is also usable for design
calculations. Endrenyi (12) points out "optimal designs aim at
minimizing the volume of the joint confidence region of the parameters. In the linear least-square approximation, this is proportional to the determinant of the parameter variance-covariance
matrix V . . ." The important D-optimality criterion maximizes the
determinant of the information matrix which is proportional to its
inverse. (Other optimality criteria may be more robust or better in
special situations.) The necessary matrix manipulations can be co-
ordinated with the PENNZYME program using existing matrix manipulation software packages. Appropriate expert rules to use such
computations to design experiments would then have to be derived.
These would have to consider the probable accuracy or difficulty of
a given measurement. A small net signal above a large background
noise will probably be inaccurate. An experimenter might prefer two
measurements under convenient conditions to one measurement under
inconvenient (or scarce-material consuming) ones. Considerations of
minimizing experimenter's effort, number of animals used, etc. can
either be included in a body of rules, or by adding some kind of
penalty function to the calculations.

Special consideration of metal ions. The effects of metal ions such
as magnesium ion could be calculated by effectively incorporating
into this system the software described   previously. Qualitative
considerations could then be included by assembling a set of knowledge rules applicable to magnesium ion behavior with regard to
enzymes, e.g.

Magnesium ion is usually involved (for "charge neutralization")
where "high-energy phosphate" is moved from one molecule to another
by an enzyme, i.e., the metabolically active form of ATP is usually
the magnesium chelate.

The ensemble of EXPERT plus data knowledge bases and calculation
routines would then be used to solve problems such as determining a
change in enzyme activity on changing metal ion level--or determining
whether there is an effective change in mechanism as well.

Pharmacokinetics and Drug Dosage Regimen Design--A Possible Application Requiring Construction and Manipulation of a Complex Model and
Data Base with an Expert System

A major part of the slow and expensive drug development process consists of testing to determine that a given potential drug is both
safe and effective. The number of drug (or cosmetic) toxicity tests
performed annually in the United States is very large, involving
perhaps 15 million animals and considerably more dollars. The
expense of testing and qualification may be prohibitive for use in

animals: it may cost more to prove a drug safe and effective in a given species than could ever be earned by sales for use in that species. The techniques used, especially to test toxicity, are now being strongly criticized, especially because of the large number of animal-based experiments. Computer-based methods of predicting toxicity from the chemical structure are being developed in response to this problem.

An suitable expert system which can manage pharmacokinetic simulation could substantially improve the speed and efficiency of this process. Such a system would contain information about drugs, drug metabolism, excretion, etc. and the relevant physiological parameters. It would supervise construction of models from quantitative measurements of the behavior of the drug under test in animals. The expert system would be needed because large-scale biological modeling has thus far been slow. Also, pharmacokinetic modeling has emphasized simple systems and given little attention to qualitative data or to extrapolation from one species to another.

Modeling Considerations. Drugs for internal use must enter the body in some way, reach the blood stream, be transported to the relevant organs and active sites, exert their action, perhaps be metabolized or modified, with subsequent departure from the body. These processes involve the action of enzymes and of kinetically similar transport mechanisms, so the techniques and software described above (an expert system involving EXPERT and PENNZYME) are applicable here. The major variable which this type of analysis would try to predict and manage is the (free) plasma level of a drug. This level is likely to be identical to the drug level at the site of action. It has been shown to be directly related to therapeutic effect for many drugs--but less closely related to the dose administered. Important theoretically predictable perturbing factors here include disease conditions such as renal failure, old age, and physiological factors; an important but unpredictable one is failure to take a drug as directed. The effects of such factors on the behavior and apparent toxicity of a given drug would require systematic exploration with appropriate models, which is best supervised by an expert system because it would otherwise take too long.

Compartments between which drugs do not mix, or mix only slowly, commonly exist in the body. Metabolism within them is carried out and controlled by enzymes in the usual way. These compartments can be detected by time-curve analysis of the blood levels of drugs. Compartments determined in this way have the limitations that:

1. They are difficult to predict a priori;
2. Their structure may depend on particular numerical values associated with a system under study as well as its structure or organization;
3. Sometimes different competent workers disagree as to the compartmental structure of the same system.

An important methodology of extrapolating pharmacokinetic or drug properties from one species to another which is relatively independent of such compartmental modeling has been developed by Bischoff and collaborators (13). It is instead based on known anatomical and physiological functions, such as blood flow to organs which either metabolize drugs or are affected by them, the size and

metabolic rate of the animal, etc. To some extent this approach
("physiological pharmacokinetics") is a chemical engineer's formulation of pharmacokinetic problem. The rate at which a given drug is
delivered to a metabolizing or target organ by the plasma (with its
level of drug) is calculated along with the rates of metabolism or
detoxification by such organs, as well as the rate of removal of the
drug (or its metabolites) from the body. From this information the
total and free (after binding to proteins, etc.) organ content of the
drug and the level at the active site is calculated. This method is
based on the orderly change of many anatomical and physiological
properties with body weight. Anatomical dimensions increase nearly
linearly with weight, while physiological rates vary as the .7 to .8
power (14). Physiological processes are therefore slower in larger
animals; the cardiac output of a mouse per body weight is about an
order of magnitude higher than that of a man. This trend is coherent:
the disposition half-life of hexobarbital approximates 1,680 gut-
beats in a wide variety of mammals (14). Dedrick (15) has described
a formalism for animal scale-up. General application of this method
would require assembly at a data base with sizes of and blood flows
to the most important organs, excretory capacity and renal functions,
etc., and even prediction of potential compartmental sizes where
possible.

The comparative behavior of a few drugs has been thoroughly
studied by these workers, with the largest effort directed to metho-
trexate (16). This drug constitutes a good test case because its
mechanism of action is well-known and simple, the amount of information about it is very large, and it now appears applicable to two
unrelated therapeutic situations requiring different dosage levels.
Bischoff et al were able to fit substantially the same model to data
for mouse, rat, dog (including dogs of different sizes), monkey, and
man. They were then able to successfully extrapolate from these
mammalian studies all the way to the sting ray, which is zoologically
a variety of shark (17).

A second level of sophistication is possible here. To quote
from Bischoff (18), "Williams notes that foreign organic compounds
tend to be metabolized in two phases. Phase one reactions lead to
oxidation-reduction and hydrolysis products. Phase two reactions
lead to synthetic or conjugation products that are relatively polar
and are thus more easily excreted". Species variations of phase one
reactions are very common but hard to predict; phase two reactions
are much more limited in number and more predictable. A more powerful expert system could probably make useful predictions of the
quantitative behavior of toxic metabolites of drugs and perhaps help
get an indication of what presently unkown species-dependent toxici-
ties might be. For this purpose the admittedly incomplete information on which pathways of detoxification and other metabolism are
present in which organ and how active they are, would have to be
collected (this includes heuristics as well as hard data including
the types of information mentioned above). Some of the unpredicta-
bilities as to which toxic products might be formed by the liver of
what species might be compensated for by appropriately designed experiments with such livers or tissue cultures derived from them).

One could set up an expert system by interfacing a suitable
simulation program with EXPERT. Good optimization capabilities and

ability to handle design optimality problems like those mentioned above are important in the simulation program, in addition to the good data-base and explanation capabilities of EXPERT. Such an expert system could then build multi-species pharmacokinetic models by the method of Bischoff and Dedrick. After repeating their work as the test case, this expert system could be used for the other drugs whose kinetics have been sufficiently studied (including sampling in several tissues) as required for such analysis. Subsequent extension to include additional methodologies is possible (e.g. detailed representation of enzyme kinetics). Model construction with only part of the original data could then be repeated to determine the need for completeness of (experimentally determined) information, i.e., which and how many animal experiments are really necessary. Such considerations are important in drug testing, and an expert system would help both by doing the modeling faster than a human, and also more systematically.

A well-established specialized expert system with which the proposed expert system could be compared is the digitalis advisor of Szolovitz and Long (19) which represents a well-understood clinical situation. It performs clinical functions beyond the scope of this proposed system, but it does do some things, like maintaining the blood level of the drug involved, and monitoring its toxicity, that this proposed system is concerned with and should perform adequately.

Starting with appropriate knowledge of the behavior of a prospective drug in one species one could then extrapolate to other species, ultimately including humans. This capability could be used in testing a proposed drug to determine proper dosage and regimen under what conditions it (and possibly its metabolites) is toxic, and how sensitive its behavior might be to perturbing conditions, which presently have to be re-performed for each species involved by empirically and heuristically guided experiments. It is reasonable to hope for significantly improved efficiency in performing these expensive operations.

Conclusion

We have described a set of applications of a conventional expert system which extend the usual functions of such systems from primarily logical reasoning and solution of classification problems to include supervision of calculations and of modeling, i.e., systems management. A hierarchy of applications arising from biochemical research have been discussed. These follow biological systems in being primarily chemical at the lowest level but acquire more biological character at the higher levels. At the lowest level, these permit the convenient performance of calculation which is not being done or done properly. At the intermediate level, they provide a better research tool, especially for experimental design. At the most complex level, they would permit a complex, slow, and expensive process to be carried out with less resource expenditure (calendar time, money, and animal experiments).

Acknowledgments

Supported by NIH grants HL15622, AM33016, RR643, and RR2230.

Literature Cited

1. Weiss, S.; Kulikowski, C. Proc. 6th International Joint Conference on Artificial Intelligence, 1979, p. 942.
2. Mannervick, B. "Kinetic Data Analysis";Endrenyi, L.,Ed.;Plenum Press: New York, 1981; p. 235.
3. Garfinkel, L.; Garfinkel, D. Biochemistry 1984, 23, 3547.
4. Morrison, J. F. Methods in Enzymology 1979, 63, 257.
5. Storer, A. C.; Cornish-Bowden, A. Biochem. J. 1977, 165, 61.
6. Garfinkel, L.; Garfinkel, D. Magnesium 1985, 4, 60.
7. Waser, M. R.; Garfinkel, L.; Kohn, M. C.; Garfinkel, D. J. Theoret. Biol. 1983, 103, 295.
8. Kohn, M. C.; Menten, L. E.; Garfinkel, D. Comput. Biomed. Res. 1979, 12, 461.
9. Weiss, S.; Apte, C. IEEE Transactions on Pattern Analysis and Machine Intelligence 1985, PAMI7, 586.
10. Weiss, S.; Kulikowski, C.; Apte, C.; Uschold, M.; Patchett, J.; Briggham, R. M.; Spitzer, B. Proc. 2nd Annual Nat'l. Conf. on Artificial Intelligence, Pittsburgh, PA,1982, 322.
11. Cornish-Bowden, A. Biochem. J. 1977, 165, 55.
12. Endrenyi, L. "Kinetic Data Analysis"; Endrenyi, L., Ed,; Plenum Press: New York, 1981, p. 137.
13. Bischoff, K. B. Cancer Chemotheraphy Reports 1975, 59, Part 1, p. 777.
14. Adolph, E. F. Science 1949, 109, 579.
15. Dedrick, R. L. J. Pharmacokinet. Biopharm. 1, 1978, 435.
16. Bischoff, K. B.; Dedrick, R. L.; Zaharko, D. S.; Longstreth, J. A. J. Pharamaceutical Sciences 1971, 60, 1128.
17. Zaharko, D. S.; Dedrick, R. L.; Oliverio, V. T. Comp. Biochem. Physiol. 1974, 42A, 183.
18. Bischoff, K. B. Fed. Proc. 1980, 39, 2456.
19. Szolovitz, P.; Long, W. J. "Artificial Intelligence in Medicine", Szolovitz, P., Ed.: AAAS Selected Symposium 51; Westover Press: Boulder, Colo.; 1982; p. 79.

RECEIVED December 17, 1985

An Expert System for the Formulation of Agricultural Chemicals

Bruce A. Hohne and Richard D. Houghton

Rohm and Haas Company, Spring House, PA 19477

An expert system has been written which helps the
agricultural chemist develop formulations for new
biologically active chemicals. The decision making
process is segmented into two parts. The first is
which type of formulation to use. The second is how
to make a formulation of that type with the chemical
of interest. The knowledge base currently contains
rules to determine which formulation type to try and
how to make an emulsifiable concentrate. The next
phase will add rules on how to make other types of
formulations. The program also interfaces to several
FORTRAN programs which perform calculations such as
solubilities.

What Is An Agricultural Formulation

An essential part of the development of a new pesticide is
establishing a good, dependable formulation. The product's active
ingredient and physical properties should remain acceptable for two
years or more. These formulations are often subjected to storage
conditions of extreme heat, cold, and humidity. Once sold to the
applicator, the concentrated formulation should dilute easily to
field strength and pass freely through conventional spray equipment.
    Agricultural (Ag) formulations that are commonly diluted and
applied by means of spray equipment include water soluble liquids,
emulsifiable concentrates, wettable powders, and flowable
suspensions. The choice of which formulation to develop normally
depends upon the solubility properties of the technical pesticide.
Scientists often must also consider manufacturing costs, field
efficacy and product toxicity.
    A water soluble liquid formulation (WSL) is prepared from
pesticides that are highly water soluble. This is, by far, the
simplest type of formulation. One distinct advantage of WSL's over
other formulations is that the field spray dilutions are infinitely
stable as true solutions. Pesticides that are hydrophilic and
ionic, such as inorganic or organic metallic salts, often fall into
this category. Unfortunately, only a small portion of all
pesticides are adequately soluble in water.

0097–6156/86/0306–0087$06.00/0
© 1986 American Chemical Society

An emulsifiable concentrate is prepared from pesticides that are soluble in common organic solvents, such as xylene and kerosene. Using emulsifiers in the composition causes the formulation to disperse into small particles, called an emulsion, when diluted in water.

Pesticides that are not soluble or have limited solubility in common solvents are formulated as wettable powders (WP) or flowable concentrates (flowables). A wettable powder has the capacity for high active ingredient content, often between fifty and eighty percent by weight, and is made by blending and grinding dry ingredients. Wettable powders are best prepared from pesticides that are high melting, friable solids. Diluents, such as natural clays and synthetic silicates, are used to improve the powder's physical properties. The disadvantages of a WP are: messy handling properties; potential dust inhalation hazard for field personnel; and the need to measure the powder on a weight basis. In some cases these problems can be overcome by formulating the pesticide into a suspension. Water and other ingredients are added to the composition to suspend and disperse the active compound into a flowable.

Regardless of what type of formulation is employed in the field, the formulation must wet, disperse, and remain homogeneous in the application spray equipment. Careful selection of formulating agents, commonly called inerts, is extremely important. These ingredients have no biological activity of their own, but combined, they function as the delivery system for the pesticide.

In addition to solvents and diluents, formulations may contain emulsifiers, dispersants, chelating agents, thickeners, defoamers, and more. The large number and variety of each type makes selecting the components for a formulation difficult and time consuming.

Why Is This A Good Area for an Expert System

The process of choosing application areas for expert system development has been detailed elsewhere, both for the general case and the corporate environment [1]. There are several specific advantages in the formulations application. Experts on one type of formulation are not necessarily experts on other formulation types. Expertise in Ag formulations tends to be in the form of 'rules of thumb', based on experiences with similar chemical systems. Incremental growth, like this, is ideal for expert system development. Formulation scientists are also likely to be more tolerant of the program's mistakes because their skill is measured by how few bad formulations they make before they make a good one.

Multilevel expert systems offer additional advantages over traditional expert systems. Multilevel expert systems draw on computational computer programs to solve parts of the problem. The Ag formulation expert system does this in the areas of computational chemistry, bookkeeping, and communication.

There are numerous computational programs available to chemists today. These programs are algorithmic by nature, and solve problems that do not lend themselves to expert systems. However, a great deal of expertise may be needed by the chemist to decide which program to use and how to actually use it. Most chemists do not have, and are not willing to gain, this computer expertise. Some would rather use traditional, noncomputational, methods rather than navigate the maze of available computer programs and users manuals. Expert systems can be extremely valuable in providing this expertise to chemists.

The Ag formulations expert system has the ability to execute the appropriate computational programs, giving it an advantage over the formulation chemist. Bookkeeping tasks are generally handled better by a computer than a chemist. For example, time tables must be met for long term storage studies, toxicology data, and government registrations. These tasks are easily handled by the computer.

The expert system fills several potential communication gaps. Molecular modeling calculations which are performed by the synthetic chemists, outside the formulation area, can be accessed by the expert system. Through this interface, the expert system can extract useful, structural information directly. Also, if a structure has not been entered, the formulation chemist can use the modeling program to enter the structure into the computer. In addition, the system safeguards against communication gaps between the chemist and management/marketing by including marketing and production considerations in the rule base. In this way, management can determine which new formulations are possible, and what characteristics they will sacrifice with a particular formulation.

Structure of the Problem

The problem of devoloping a new formulation is highly structured. The structure tends to be hierarchical, although this hierarchy does not resemble a traditional decision tree. Each branch point may have any number of branches. The decision about which 'branch' to take at each level can be viewed as an independent expert system. The ability to break the overall problem into smaller, simpler subproblems is desirable for expert systems.

Many of the facts in the system are shared by several subproblems, and subproblems must be. developed by starting at the top of the hierarchy and working down. Other than these stipulations, they are independent problems. Each branch of the tree can be used independently, and need not be complete to be useful in the formulation study. The expert system's competence on each subproblem can be judged independently. In many cases different experts are used to develop the knowledge bases for different subproblems. Figure 1 shows the structure of the problem, tracing one branch from each level.

Structure of the Expert System

The program was written on an Apollo computer in LISP. Apollo's Domain LISP, a version of Portable Standard LISP, was the dialect available.

The expert system has been written to follow the natural structure of the Ag formulation problem. Figure 2 shows the overall structure of the expert system. One nice feature of the program is that at each branch point the user can override the computer's choice, and can also select as many branches to pursue as desired.

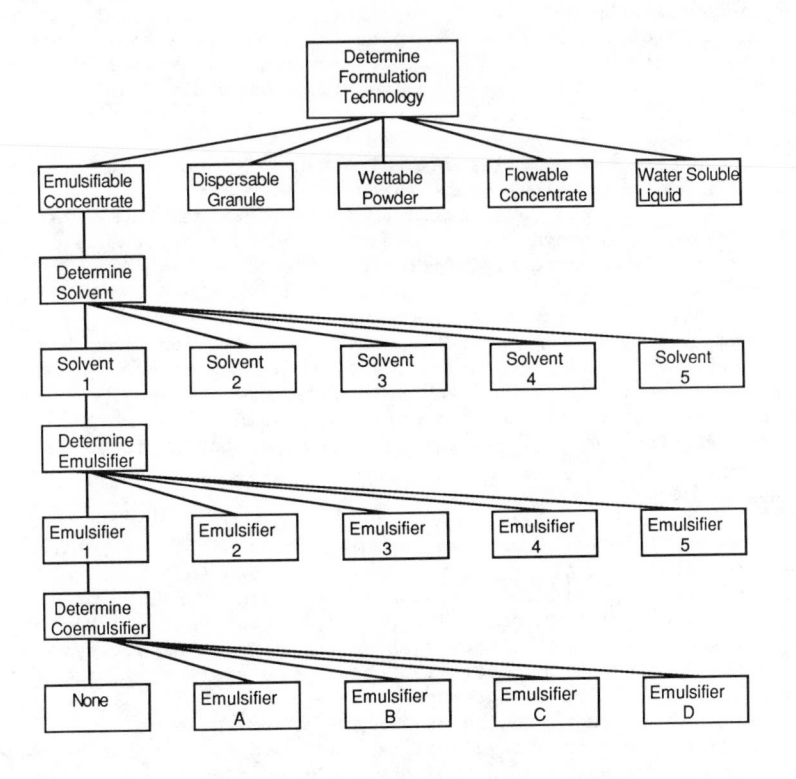

Figure 1. Structure of the Problem

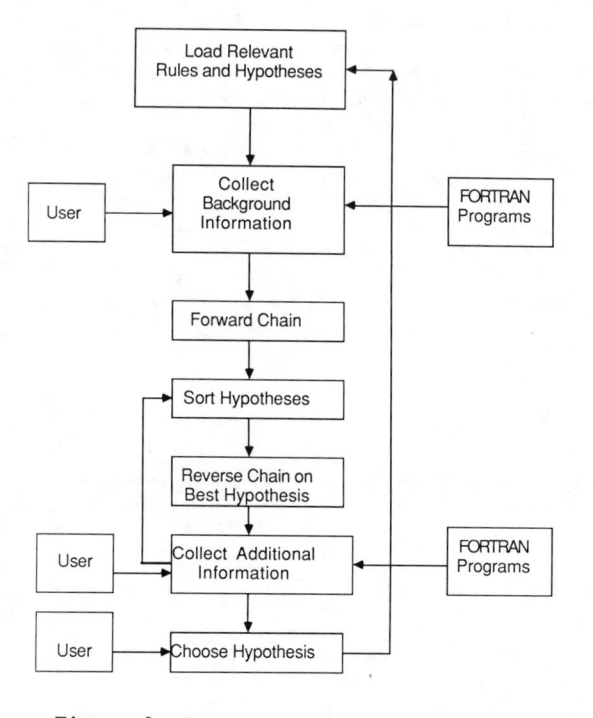

Figure 2. Structure of the Program

The logical deduction portion of the program is based on IF-THEN rules. FACTS, acquired both as the result of logical deductions and by querying the user, are stored in similar data structures. Because the branch points in the problem are also logical deductions, they are stored in a data structure similar to the FACTS. The branch points contain additional flow of control information that relates to the hierarchy of the problem. The difference between FACTS and branch points is transparent to the logical deduction portion of the program.

The top level in the structure of FACTS is the fact name, e.g., ACTIVE_INGREDIENT. Under each fact are various properties relevant to that fact, e.g., H2O_SOLUBILITY. For each property, several pieces of information are stored (see Figure 3). All properties contain a VALUE, which is initialized to a null or missing value. They also contain the method to obtain the VALUE. Currently supported methods are ASKIT, PROVEIT, and CALL.

```
Fact name
     Property 1
          Value
          Where to find it          (Ask, prove, calculate)
          Prompt                    (How to ask user)
          Allowable response        (Checks user's response)
          Explanation               (For prove and calculate)

     Property 2
          Value
          .
          .
          .
```

Figure 3. Structure of Facts

If the method for acquiring a VALUE is ASKIT, then a user PROMPT is stored. In order to guarantee a valid response to the question, a LISP function to check the answer is included with the FACT. Table I lists the currently implemented response checking functions. Whenever the inference engine reaches one of these facts, searching is stopped and the user is prompted for a value.

Table I. User Input

Function	Allowed Response
PercentP	Positive integer between 1 and 100
Yes_NoP	Yes, No, Y, N
Any_Of	Any number of members of the listed possibilities
One_Of	One member of the listed possibilities
PositiveP	Any positive number
ImportanceP	High, Med, Low, H, M, L
IntegerP	Any positive or negative integer
Integer_listP	A list of integers seperated by spaces
Minus1_to_One	Any number between -1 and +1

For values which must be deduced, a TEXT explanation is saved.

This TEXT is used in the various explanation and tracing facilities. Whenever the inference engine reaches one of these FACTS it either continues its search, if possible, or proceeds another level deeper in the reverse search and tries to prove that FACT.

The CALL facility allows the expert system to access software external to the LISP program. Included with the CALL is the name of a LISP function which handles the outside software. In the case of the fact CHEMICAL_NAME, the LISP function executes a FORTRAN program which allows the user to either retrieve the structure of a previously entered compound or enter a new one. The program also breaks the chemical structure into its functional groups. When the FORTRAN program terminates, the LISP function updates the list of facts, and inserts the name into CHEMICAL_NAME and the functional groups into FUNCT_GROUPS. These FACTS are then available to the expert system. In this way, access to outside software is completely data driven.

The structure of the branch points is the same as that of those FACTS which must be deduced, except for the additional control information. The properties correspond to the different branches in the hierarchy at that point. Figure 4 shows the data structure of branch points. For each branch point (property), there is a list of rules which apply. By only considering rules applicable to the specific subproblem, the time required for searching is drastically reduced. A list of FACT-PROPERTY pairs, which are useful background information for the subproblem, is also saved. This background information is collected at the beginning of each subproblem and used in a forward-chaining function. This approach can prevent the reverse chaining portion of the system from appearing as though it is "wandering" at the beginning of each subproblem. The final piece of control information is the name of the next subproblem, and correspondes to the FACT name. These names are stored for each branch point.

```
Conclusion Name
     Branch Point 1
          Value
          Where to find it          (prove)
          Explanation               (Tell the user if it is true)
          Next level name
          Background facts          (Questions always asked)
          Rule names                (List of relevant rules)

     Branch Point 2
          Value
          .
          .
          .
```

Figure 4. Structure of Conclusions

Rules in the expert system are structured to allow flexibility and future expansion. For speed of execution, the IF-THEN clauses are actually executable LISP code. Tables II and III contain examples of how rules are structured. The IF clauses contain functions, called predicates. Predicates have a value of either

true or false when evaluated. If all the IF clauses are true, then
the THEN clauses are executed. The THEN clauses contain ACTIONs
which change the VALUEs of other FACTs. The PREDICATEs and ACTIONs
are the basic building blocks for all the rules in the system.
There is no limit to the number of IF or THEN clauses which a rule
can contain. As more powerful rules are required, additional
building blocks can easily be added by writing new PREDICATEs or
ACTIONs.

Table II. Structure of Rules

```
AgRule_1
    If-1   (Isequal Active_Ingredient Desired_Level Value >40)
    Then-1 (Suggest Form_Type EC -.5)
    Then-2 (Suggest Form_Type WSL -.5)
    Then-3 (Suggest Form_Type Flowable -.5)
Why    EC's, WSL's and Flowables rarely have that high an AI level
Date   11/14/83
Author Houghton


Agrule1
    IF
1. The value of the active ingredient's desired concentration
is >40%
    THEN
1. There is suggestive evidence (-0.5) that the
formulation type should not be emulsifiable concentrate
2. There is suggestive evidence (-0.5) that the
formulation type should not be water soluble liquid
3. There is suggestive evidence (-0.5) that the
formulation type should not be flowable concentrate
    BECAUSE:
EC's, WSL's and Flowables rarely have that high an AI level
```

Table III. Structure of Rules

```
AgRule_13
    If-1   (Isequal Solvent Req_EPA_Clear Value C)
    Then-1 (Avoid NotEqual EC_Solvent EPA_Clear C -1)
Why    It's the law
Date   12/20/83
Author Hohne


Agrule13
    IF
1. The value of the solvent's required EPA clearance
is C
    THEN
1. Avoid (-1) emulsifiable concentrate solvents where EPA clearance
is not equal to C
    BECAUSE
It's the law
```

The rule structure allows simple Boolean functions to be performed. Multiple numbered IF clauses are logically ANDed together. Multiple clauses which are part of the same numbered IF are logically ORed. The logical NOT does not exist, but can be simulated using predicates with the opposite meaning in the IF clause, (i.e. BIGGER is equivalent to NOT SMALLER). Table IV lists the currently available predicates for IF clauses.

Table IV. Relationships (predicates)

Predicate	Meaning
BIGGER	Bigger than
SMALLER	Smaller than
MEMB	Member of the list
NOTMEMB	Not a member of the list
ISEQUAL	Is equal to
NOTEQUAL	Not equal to

The ACTIONs available to the THEN clauses are listed in Table V. These ACTIONs give rise to two types of THEN clauses. The first type affects the VALUE of only one property. The THEN clauses in Table II show the construction of one-property THEN clauses. The second type of THEN clause deals with all of the current branch points. Table III shows the construction of this type of THEN clause.

Table V. Actions

Action	Meaning
SUGGEST	Adjust the property's value using the listed confidence factor
SET_EQUAL	Set the property's value equal to the listed value
ORDER_BY	Order the hypotheses by the value of the listed property
AVOID	Avoid conclusions where the requirement listed

The inference engine was designed to use multivalued logic, i.e., it handles inexact reasoning. Confidence factors (CF) are contained in the THEN clauses of each rule. The equation for combining positive confidences is:

$$CF = old\_value + new\_value - (old\_value \times new\_value)$$

The equation for negative confidences is:

$$CF = old\_value + new\_value + (old\_value \times new\_value)$$

For mixed positive and negative confidences, a simple sum is used. The advantage to these functions is they are bounded by -1 and +1.

The program also handles exact reasoning through the SET_EQUAL

ACTION in the THEN clause. This ACTION can be used to set a
confidence value to +1 (true) or -1 (false), regardless of
previously compiled confidences. SET_EQUAL can also be used to set
FACT values equal to nonnumeric values, where required.

The natural language interpretation of the rules given at the
bottom of Tables II and III was generated by the program. The
natural language generator uses synonyms for FACT names and
properties. The synonyms are simply substituted into one of several
templates to generate a sentence. The template used is determined
by the value of the confidence factor and the combination of ACTIONs
and PREDICATEs.

Current Status of the Project

The project is still in the prototype stage. It is being used, but
not widely. Presently, the knowledge base for the system has less
than 100 rules. This number is misleading because all the work
performed by the FORTRAN programs is not counted in the number of
rules. These programs give the system far more knowledge than would
be expected from the 'small' knowledge base.

The system can help scientists reliably determine what type of
formulation to make. However, the only branch of the decision tree
which has rules is the emulsifiable concentrates (EC) branch. The
system can determine which solvents to try to make an EC. Its
decision relies heavily on rules and solubility calculations. Work
is just beginning on the rules to determine which emulsifiers to
use.

The program has been interfaced to two FORTRAN programs. The
first, MOLY, is a locally developed product for chemical structure
entry, display, and molecular modeling [2]. The expert system only
takes advantage of the chemical structure handling portion of the
program. The other program, UNIFAC [3], performs solubility
calculations for the active ingredient in a group of solvents of
interest to formulation chemists.

The inference engine performs both forward and
reverse-chaining. The reverse-chain algorithm is a depth first
search. Using this algorithm, questions asked by the system are
grouped by subject, making the program appear more logical to the
user. The program handles exact and inexact logic calculations and
explains, in English, why a question was asked and how a conclusion
was reached. The program also allows the scientist to change
answers in case of mistakes, or to investigate "what if" scenarios.

Directions for Future Development

Future developments fall into two classes: additions to the
knowledge base and enhancements to the program. As the program is
used by more people, fine tuning of the rules to select which type
of formulation to try will be needed. Work, from that point, will
continue on the emulsifiable concentrate branch. The solvent
selection portion will require some fine tuning, but the major work
is in adding to the list of solvents. The emulsifier selection
portion of the knowledge base will definitely require additional
rules, to be followed by considerable tuning as it is used. The
remaining four formulation types have yet to be started. They will

require different experts and can be developed concurrently with the EC portion.

The first major enhancement to the program will be the ability to stop sessions at any point and restart at the same point at a later time. This capability will be more than just a convenience, it will be necessary to make the laboratory results requested by the program useful. After this addition, the next major enhancement will be to develop a method of using the rules to trouble-shoot field problems. This enhancement will involve adding some rules, but most of the knowledge should already be in the knowledge base. As the program becomes widely used, the ability to generate reports and data sheets for laboratory results will be a valuable addition. The added ability to remind the scientist about certain deadlines for a project may be easily included, but will not be useful until scientists use the Apollo computer regularly.

The expert system currently has no rule entry or maintenance facilities. Rules are entered and modified using the Apollo computer text editor. This is acceptable for a prototype, but not for a production system. Before these facilities are added, its cost and capabilities will need to be compared to those of commercial expert systems.

Literature Cited

1. Prenau, D.S., "Selection of an Appropriate Domain for an Expert System", AI Magazine, 6(2), 1985
2. Dyott, T., Stuper, A.J., Zander, G.S., "MOLY, an Interactive System for Molecular Analysis", J. Chem. Inf. Comp. Sci., 20(28), 1980
3. Fredenslund, A., Jone, R.L., Prausnitz, J.M., "Group-Contribution Estimation of Activity Coefficients in Nonideal Liquid Mixtures", AIChE Journal, 21(6), 1975

RECEIVED December 17, 1985

COMPUTER ALGEBRA

8

Computer Algebra: Capabilities and Applications to Problems in Engineering and the Sciences

Richard Pavelle

MACSYMA Group, Symbolics, Inc., Cambridge, MA 02142

MACSYMA is a large, interactive computer system
designed to assist engineers, scientists, and
mathematicians in solving mathematical problems. A user
supplies symbolic inputs and MACSYMA yields symbolic,
numeric or graphic results. This paper provides an
introduction to MACSYMA and provides the motivation for
using the system. Many examples are given of MACSYMA's
capabilities with actual computer input and output.

My purpose in this paper is to provide a broad introduction to the
capabilities of MACSYMA. It is my hope that this information will
create new users of Computer Algebra systems by showing what one
might expect to gain by using them and what one will lose by not
using them.

MACSYMA output is used and CPU times are often given. In some
cases I have modified the output slightly to make it more
presentable. The CPU times correspond to a Symbolics 3600 and to the
MACSYMA Consortium machine (MIT-MC) which is a Digital Equipment
KL10. These are about equal in speed and about twice as fast as a
Digital Equipment VAX 11/780 for MACSYMA computations. When CPU
times are not given one may assume the calculation requires at most
10 CPU seconds.

What is MACSYMA. The development of the Computer Algebra system,
MACSYMA, began at MIT in the late 60s, and its history has been
described elsewhere (1). A few facts worth repeating are that a
great deal of effort and expense went into MACSYMA. There are
estimates that 100 man-years of developing and debugging have gone
into the program. While this is a large number, let us consider the
even larger number of man-years using and testing MACSYMA. At MIT,
between 1972 and 1982, we had about 1000 MACSYMA users. If we had 50
serious users using MACSYMA for 50% of their time, 250 casual users
at 10% and 700 infrequent users at 2% then the total is over 600
man-years. MACSYMA has been at 50 sites for four years and is at

400 sites today. Well, we could conclude that at least 1000 man-years have been spent in using MACSYMA. MACSYMA is now very large and consists of about 3000 lisp subroutines or about 300,000 lines of compiled lisp code joined together in one giant package for performing symbolic mathematics.

While this paper is directed towards MACSYMA, the development of MACSYMA and other Computer Algebra systems has really been the result of an international effort (2). There are many systems, world-wide, of various sizes and designs which have been developed over the past fifteen to twenty years (3, 4). Research related to the development of these systems has lead to many new results in mathematics and the construction of new algorithms. These results in turn helped the development of MACSYMA as well as other systems. These systems are now being recognized as important tools allowing researchers to make significant discoveries in many fields of interest (5).

Why MACSYMA is Useful or Necessary. Here are some of the more important reasons for using MACSYMA:

1. The answers one obtains are exact and can often be checked by independent procedures. For example, one can compute an indefinite integral and check the answer by differentiating; the differentiation algorithm is independent of the integration algorithm. Since exact answers are given, the statistical error analysis associated with numerical computation is unnecessary. One obtains answers that are reliable to a high level of confidence.

2. The user can generate FORTRAN expressions that allow numeric computers to run faster and more efficiently. This saves CPU cycles and makes computing more economical. The user can generate FORTRAN expressions from MACSYMA expressions. The FORTRAN capability is an extremely important feature combining symbolic and numeric capabilities. The trend is clear, and in a few years we will have powerful, inexpensive desktop or notebook computers that merge the symbolic, numeric and graphic capabilities in a scientific workstation.

3. The user can explore extremely complex problems that cannot be solved in any other manner. This capability is often thought of as the major use of Computer Algebra systems. However, one should not lose sight of the fact that MACSYMA is often used as an advanced calculator to perform everyday symbolic and numeric problems. It also complements conventional tools such as reference tables or numeric processors.

4. A great deal of knowledge has gone into the MACSYMA knowledge base. Therefore the user has access to mathematical techniques that are not available from any other resources, and the user can solve problems even though he may not know or understand the techniques that the system uses to arrive at an answer.

5. A user can test mathematical conjectures easily and painlessly.

One frequently encounters mathematical results in the literature and questions their validity. Often MACSYMA can be used to check these results using algebraic or numeric techniques or a combination of these. Similarly one can use the system to show that some problems do not have a solution.

6. MACSYMA is easy to use. Individuals without prior computing experience can learn to solve fairly difficult problems with MACSYMA in a few hours or less. While MACSYMA is written in a dialect of LISP, the user need never see this base language. MACSYMA itself is a full programming language, almost mathematical in nature, whose syntax resembles ALGOL.

There are two additional reasons for using MACSYMA that are more important than the others.

7. One can concentrate on the intellectual content of a problem leaving computational details to the computer. This often results in accidental discoveries and, owing to the power of the program, these occur at a far greater rate than when calculations are done by hand.

8. But the most important reason is that, to quote R.W. Hamming, "The purpose of computing is insight, not numbers." This exemplifies the major benefit of using MACSYMA, and I will demonstrate the validity of this statement by showing not only how one gains insight but also how one uses MACSYMA for theory building. However, a second quotation reputed to be by Hamming is correct as well, namely that "The purpose of computing is not yet in sight."

Capabilities and Uses of MACSYMA

Capabilities. It is not possible to fully indicate the capabilities of MACSYMA in a few lines since the reference manual itself occupies more than 500 pages (6). However, some of the more important capabilities include (in addition to the basic arithmetical operations) facilities to provide analytical tools for

Limits	Taylor Series (Several Variables)
Derivatives	Poisson Series
Indefinite Integration	Laplace Transformations
Definite Integration	Indefinite Summation
Ordinary Differential Equations	Matrix Manipulation
Systems of Equations (Non-Linear)	Vector Manipulation
Simplification	Tensor Manipulation
Factorization	Fortran Generation

There are other routines for calculations in number theory, combinatorics, continued fractions, set theory and complex

arithmetic. There is also a share library currently containing
about 80 subroutines. Some of these perform computations such as
asymptotic analysis and optimization while others manipulate many of
the higher transcendental functions. In addition one can evaluate
expressions numerically at most stages of a computation. MACSYMA
also provides extensive graphic capabilities to the user.

To put the capabilities of MACSYMA in perspective we could say
that MACSYMA knows a large percentage of the mathematical techniques
used in engineering and the sciences. I do not mean to imply that
MACSYMA can do everything. It is easy to come up with examples that
MACSYMA cannot handle, and I will present some of these. Perhaps the
following quotation will add the necessary balance. It is an exit
message from some MIT computers that often flashes on our screens
when logging out. It states: "I am a computer. I am dumber than any
human and smarter than any administrator." MACSYMA is remarkable in
both the questions it can and cannot answer. It will be many years
before it evolves into a system that rivals the human in more than a
few areas. But until then, it is the most useful tool that any
engineer or scientist can have at his disposal.

Uses. It is difficult to list the application fields of MACSYMA
because users often do not state the tools that helped them perform
their research. However, from Computer Algebra conferences (7, 8,
9) we do know that MACSYMA has been used in the following fields:

Acoustics	Fluid Dynamics
Algebraic Geometry	General Relativity
Antenna Theory	Number Theory
Celestial Mechanics	Numerical Analysis
Computer-Aided Design	Particle Physics
Control Theory	Plasma Physics
Deformation Analysis	Solid-State Physics
Econometrics	Structural Mechanics
Experimental Mathematics	Thermodynamics

Researchers have reported using MACSYMA to explore problems in:

Airfoil Design	Nuclear Magnetic Resonance
Atomic Scattering Cross Sections	Optimal Control Theory
Ballistic Missile Defense Systems	Polymer Modeling
Decision Analysis in Medicine	Propeller Design
Electron Microscope Design	Robotics
Emulsion Chemistry	Ship Hull Design
Finite Element Analysis	Spectral Analysis
Helicopter Blade Motion	Underwater Shock Waves
Maximum Likelihood Estimation	
Genetic Studies of Family Resemblance	
Large Scale Integrated Circuit Design	
Resolving Closely Spaced Optical Targets	

Examples of MACSYMA

Polynomial Equations. Here is an elementary example that

demonstrates the ability of MACSYMA to solve equations. In MACSYMA, as with most systems, one has user input lines and computer output lines. Below, in the input line (C1), we have written an expression in an ALGOL like syntax, terminated it with a semi-colon, and in (D1) the computer echos the expression by displaying it in a two dimensional format in a form similar to hand notation. Terminating an input string with $ inhibits the display of the D lines.

(C1) X^3+B*X^2+A^2*X^2-9*A*X^2+A^2*B*X-2*A*B*X-

9*A^3*X+14*A^2*X-2*A^3*B+14*A^4=0;

$$
\begin{array}{l}
 3 \quad\;\; 2 \quad\; 2\ 2 \qquad\;\; 2 \quad\;\; 2 \qquad\qquad\qquad 3 \\
(D1)\ X\ +\ B\ X\ +\ A\ \ X\ -\ 9\ A\ X\ +\ A\ \ B\ X\ -\ 2\ A\ B\ X\ -\ 9\ A\ \ X \\[10pt]
\qquad\qquad\qquad\quad 2 \qquad\quad 3 \qquad\quad 4 \\
\qquad\qquad +\ 14\ A\ \ X\ -\ 2\ A\ \ B\ +\ 14\ A\ \ =\ 0
\end{array}
$$

    In (C2) we now ask MACSYMA to solve the expression (D1) for X and the three roots appear in a list in (D2).

(C2) SOLVE(D1,X);

$$
\begin{array}{l}
 2 \\
(D2) \qquad [X\ =\ 7\ A\ -\ B, \quad X\ =\ -\ A\ , \quad X\ =\ 2\ A]
\end{array}
$$

    Notice that MACSYMA has obtained the roots analytically and that numeric approximations have not been made. This demonstrates a fundamental difference between a Computer Algebra system and an ordinary numeric equation solver, namely the ability to obtain a solution without approximations. I could have given MACSYMA a "numeric" cubic equation in X by specifying numeric values for A and B. MACSYMA then would have solved the equation and given the numeric roots approximately or exactly depending upon the specified command.

    MACSYMA can also solve quadratic, cubic and quartic equations as well as some classes of higher degree equations. However, it obviously cannot solve equations analytically in closed form when methods are not known, e.g. a general fifth degree (or higher) equation.

Differential Calculus. MACSYMA knows about calculus. In (D1) we have an exponentiated function that is often used as an example in a first course in differential calculus.

$$
\begin{array}{l}
 X \\
 X \\
(D1) \qquad\qquad\qquad X
\end{array}
$$

We now ask MACSYMA to differentiate (D1) with respect to X to obtain this classic textbook result of differentiation. Notice how fast, 3/100 CPU seconds, MACSYMA computes this derivative.

(C2) DIFF(D1,X);
Time= 30 msec.

$$
\text{(D2)} \qquad X^{X^X} (X^X \ \text{LOG}(X) \ (\text{LOG}(X) + 1) + X^{X-1})
$$

Below is a more complicated function, the error function of the tangent of the arc-cosine of the natural logarithm of X. Notice that MACSYMA does not display the identical input. This is because the input in (C1) passes through MACSYMA's simplifier. MACSYMA recognizes that the tangent of the arc-cosine of a function satisfies a trigonometric identity, namely TAN(ACOS(X)) = SQRT(1-X^2)/X. It takes this into account before displaying (D1).

(C1) ERF(TAN(ACOS(LOG(X))));

$$
\text{(D1)} \qquad \text{ERF}\left(\frac{\text{SQRT}(1 - \text{LOG}^2 (X))}{\text{LOG}(X)}\right)
$$

Now when MACSYMA is asked to differentiate (D1) with respect to X, it does so in a straightforward manner and simplifies the result using the rational canonical simplifier RATSIMP. This command puts the expression in a numerator-over-denominator form canceling any common divisors. In (D2) the symbols %E and %PI are MACSYMA's representations for the base of the natural logarithms and pi, respectively.

(C2) DIFF(D1,X),RATSIMP;
Time= 1585 msec.

$$
\text{(D2)} \qquad - \ \frac{2 \ \%E^{1 - \frac{1}{\text{LOG}^2 (X)}}}{\text{SQRT}(\%PI) \ X \ \text{LOG}^2 (X) \ \text{SQRT}(1 - \text{LOG}^2 (X))}
$$

Factorization

MACSYMA can factor expressions. Below is a multivariate polynomial in four variables.

$$
\text{(D1)} \ - 36 \ W^2 \ X^7 \ Y^4 \ Z^8 \ + 3 \ W^2 \ X^6 \ Y^3 \ Z^8 \ - 24 \ W^3 \ X^7 \ Y^4 \ Z^6
$$

$$+ 2 W^3 X^6 Y^3 Z^6 + 96 W^2 X^8 Y^6 Z^5 - 168 W^4 X^7 Y^6 Z^5$$

$$+ 12 W^2 X^7 Y^6 Z^5 - 216 W^2 X^{10} Y^5 Z^5 - 8 W^2 X^7 Y^5 Z^5 + 9 X^7 Y^5 Z^5$$

$$+ 14 W^4 X^6 Y^5 Z^5 - W^2 X^6 Y^5 Z^5 + 18 W^2 X^9 Y^4 Z^5 + 87 X^7 Y^3 Z^5$$

$$- 3 W^2 X^6 Y^3 Z^5 + 6 W X^7 Y^5 Z^3 + 58 W X^7 Y^3 Z^3 - 2 W^3 X^6 Y^3 Z^3$$

$$- 24 X^8 Y^7 Z^2 + 42 W^2 X^7 Y^7 Z^2 - 3 X^7 Y^7 Z^2 + 54 X^{10} Y^6 Z^2$$

$$- 232 X^8 Y^5 Z^2 + 414 W^2 X^7 Y^5 Z^2 - 29 X^7 Y^5 Z^2 - 14 W^4 X^6 Y^5 Z^2$$

$$+ W^2 X^6 Y^5 Z^2 + 522 X^{10} Y^4 Z^2 - 18 W^2 X^9 Y^4 Z^2$$

We now call the function FACTOR on (D1) and

(C2) FACTOR(D1);
Time= 111998 msec.

$$(D2) \quad - X^6 Y^3 Z^2 (3 Z^3 + 2 W Z - 8 X^2 Y + 14 W^2 Y^2 - Y^2 + 18 X^3 Y)$$

$$(12 W^2 X Y Z^3 - W^2 Z^3 - 3 X Y^2 - 29 X^2 + W^2)$$

MACSYMA factors this massive expression in about two CPU minutes. One can also extend the field of factorization to the Gaussian integers or other algebraic fields (10).

Simplification. A very important feature of MACSYMA is its ability to simplify expressions. When I studied plane-wave metrics for a new gravitation theory (11, 12), one particular calculation produced an expression with several hundred thousand terms. From geometrical arguments I knew the expression must simplify and indeed, using MACSYMA, the expression collapsed to a small number of pages of output. The following expression occurred repeatedly in the course of the calculation and caused the collapse of the larger expression during simplification.

$$(D1) \quad \frac{(SQRT(R^2 + A^2) + A)\ (SQRT(R^2 + B^2) + B)}{R^2}$$

$$- \frac{SQRT(R^2 + B^2) + SQRT(R^2 + A^2) + B + A}{SQRT(R^2 + B^2) + SQRT(R^2 + A^2) - B - A}$$

```
(C2) RATSIMP(D1);
Time= 138 msec.
(D2)                              0
```

When the canonical simplifier RATSIMP is called on (D1) above
it returns zero. At first I did not believe that (D1) is zero, and
I spent 14 minutes verifying it by hand (almost exceeding my 15
minute limit). It is not easy to prove. Combining the expressions
over a common denominator results in a numerator that contains 20
terms when fully expanded, and one must be very careful to assure
cancellation. Try it by hand!

Indefinite Integration. MACSYMA can handle integrals involving
rational functions and combinations of rational, algebraic
functions, and the elementary transcendental functions. It also has
knowledge about error functions and some of the higher
transcendental functions.

Below is an integral that is quite difficult to do by hand. It
is not found in standard tables in its given form although it may
transform to a recognized case. It is especially difficult to do by
hand unless one notices a trick that involves performing a partial
fraction decomposition of the integrand with respect to LOG(X).
However, MACSYMA handles it readily.

$$(D1) \quad \int \frac{LOG(X) - 1}{LOG^2(X) - X^2}\ dX$$

```
(C2) INTEGRATE(D1,X);
Time= 744 msec.
```

$$(D2) \quad \frac{LOG(LOG(X) + X)}{2} - \frac{LOG(LOG(X) - X)}{2}$$

Definite Integration. Definite integration is far more difficult to
code than indefinite integration because the number of known
techniques is much larger. One has the added complication of taking
limits at the endpoints of the integral. MACSYMA has impressive
capabilities for definite integration. Here is an example of a

function whose definite integral does not appear to be tabulated:

$$(D1) \qquad X^2 \, \%E^{-U X^2} \, LOG(X)$$

(C2) INTEGRATE(D1,X,0,INF),FACTOR;
Time= 138442 msec.

$$(D2) \qquad - \frac{SQRT(\%PI) \; (LOG(U) + 2 \; LOG(2) + \%GAMMA - 2)}{8 \; U^{3/2}}$$

In (C2) above we have asked MACSYMA to integrate (D1) with
respect to X from 0 to infinity. In the answer, %GAMMA is the
MACSYMA syntax for the Euler-Mascheroni constant = 0.577215664.. .

In addition to definite integration, MACSYMA can perform
numeric integration using the Romberg numeric integration procedure.
There are a number of other numeric techniques available. And, one
has the ability to evaluate expressions numerically to arbitrary
precision.

Taylor/Laurent Series. The Taylor (Laurent) series capability is
very impressive. Below we ask for the first 15 terms of the series
of (D1) about the point X = 0. Notice that MACSYMA computes this
expression in less than 1/2 CPU second.

$$(D1) \qquad A \; SIN(X^3) + \frac{B \; LOG(X^2 - X + 1)}{X^5}$$

(C2) TAYLOR(D1,X,0,15);
Time= 365 msec.

$$(D2)/T/ \quad -\frac{B}{X^4} + \frac{B}{2\,X^3} + \frac{2\,B}{3\,X^2} + \frac{B}{4\,X} - \frac{B}{5} - \frac{B\,X}{3} - \frac{B\,X^2}{7} + \frac{(B + 8\,A)\,X^3}{8}$$

$$+ \frac{(2\,B)\,X^4}{9} + \frac{B\,X^5}{10} - \frac{B\,X^6}{11} - \frac{B\,X^7}{6} - \frac{B\,X^8}{13} + \frac{(3\,B - 7\,A)\,X^9}{42}$$

$$+ \frac{(2\,B)\,X^{10}}{15} + \frac{B\,X^{11}}{16} - \frac{B\,X^{12}}{17} - \frac{B\,X^{13}}{9} - \frac{B\,X^{14}}{19} + \frac{(6\,B + A)\,X^{15}}{120} + \ldots$$

The program can also compute Taylor (Laurent) series in several variables.

Ordinary Differential Equations. Another powerful feature is the MACSYMA program ODE. ODE is a collection of algorithms for solving ordinary differential equations. It was built over several years by E.L. Lafferty, J.P. Golden, R.A. Bogen and B. Kuipers, and its capabilities are described in the MACSYMA Reference Manual (6) in V2-4-14.

In (C1), we first declare that Y is a function of X. This assures that the derivative (2nd) of Y with respect to X will not vanish when (C2) is evaluated.

(C1) DEPENDS(Y,X)$

(C2) (1+X^2)*DIFF(Y,X,2)-2*Y=0;

$$(D2) \qquad\qquad (X^2 + 1)\, Y_{X\,X} - 2\,Y = 0$$

We now ask the system to solve (D2) for Y as a function of X using the ODE command. The general solution with the two integration constants, %K1 and %K2 is given in (D3) in about two CPU seconds. The program can also find powerseries solutions for some differential equations when it can solve the recurrence relation. It does this in (D4). MACSYMA can be used to check the answer (D3). In (C5) we tell the system to substitute (D3) into (D2), differentiate the result and simplify.

(C3) ODE(D2,Y,X);
Time= 2068 msec.

$$(D3) \qquad Y = \%K2\,(X^2 + 1)\left(\frac{ATAN(X)}{2} + \frac{X}{2X^2 + 2}\right) + \%K1\,(X^2 + 1)$$

(C4) ODE(D2,Y,X,SERIES);
Time= 8766 msec.

$$(D4) \qquad Y = \%K1\,(X^2 + 1) - \%K2\,X \sum_{I=0}^{INF} \frac{(-1)^I X^{2I}}{(I - \frac{1}{2})(I + \frac{1}{2})}$$

(C5) D2,D3,DIFF,RATSIMP;
Time= 2051 msec.
(D5) $\qquad\qquad\qquad 0 = 0$

MACSYMA is a computer system which performs many highly
sophisticated computations that will amaze people who use
mathematical tools. For many types of calculations MACSYMA offers
enormous advantages over numeric systems. In this paper I have
shown but a few of the capabilities of MACSYMA. It is difficult to
present many capabilities in a few pages. References (5, 13) provide
many more examples as well as motivating the use of MACSYMA in
several fields of research and development.

Literature Cited

1. Moses, J. MACSYMA - the fifth year. Proceedings Eurosam 74
    Conference, Aug. 1974, Stockholm.
2. Pavelle, R.; Rothstein, M.; Fitch, J.P. Computer Algebra.
    Scientific American 1981, 245.
3. van Hulzen, J.A.; Calmet, J. Computer Algebra Systems. In
    "Computer Algebra, Symbolic and Algebraic Manipulation";
    Buchberger, B.; Collins, G.E.; Loos, R., Eds.; Springer-Verlag:
    Wien - New York, 1983, p. 220.
4. Yun, D.Y.Y.; Stoutemyer, D. Symbolic Mathematical Computation.
    In "Encyclopedia of Computer Science and Technology" 15; Belzer,
    J.; Holzman, A.G., Eds.; Marcel Dekker: New York - Basel, 1980,
    p. 235.
5. Pavelle, R., Ed., "Applications of Computer Algebra"; Kluwer:
    Boston, 1985.
6. "The MACSYMA Reference Manual (Version 10)"; Massachusetts
    Institute of Technology and Symbolics, Inc.: Cambridge, MA, Dec.
    1984.
7. Proceedings of the 1977 MACSYMA Users' Conference, R.J.
    Fateman, Ed., Berkeley, CA, July 1977. NASA: CP-2012,
    Washington, D.C.
8. Proceedings of the 1979 MACSYMA Users' Conference, V.E. Lewis,
    Ed., June 1979, Washington, D.C.
9. Proceedings of the 1984 MACSYMA Users' Conference, V.E.
    Golden, Ed., July 1984, General Electric Corporate Research and
    Development, Schenectady, NY.
10. Wang, P.S. Math. Comp., 1978, 32, 1215.
11. Mansouri, F.; Chang, L.N. Phys. Rev. D 1976, 13, 3192.
12. Pavelle, R. Phys. Rev. Lett. 1978, 40, 267.
13. Pavelle, R.; Wang, P.S. J. Symbolic Computation, 1985, 1,
    69-100.

RECEIVED January 24, 1986

A Rule-Based Declarative Language for Scientific Equation Solving

Allan L. Smith

Chemistry Department, Drexel University, Philadelphia, PA 19104

Procedural languages for scientific computation are briefly reviewed and contrasted with declarative languages. The capabilities of TK!Solver are explained, and two examples of its use in chemical computations are given.

Most of the applications of artificial intelligence in chemistry so far have not involved numerical computation as a primary goal. Yet there are aspects of the AI approach to problem-solving which have relevance to computation. In scientific computation, one could view the knowledge base as the set of equations, input variable values, and unit conversions relevant to the problem, and the inference engine the numerical method used to solve the equations. This paper describes such a software system, TK!Solver.

Brief Review of Software for Scientific Computation

Since the beginning of electronic computing, one of the major incentives for developing computer languages has been to improve the ease of solving mathematical problems arising in science and engineering. Many such problems can be reduced to the solution of a set of N algebraic equations - not necessarily linear - in N unknowns. The earliest ways of doing this involved direct hand coding in hexadecimal machine language or in assembly language mnemonics, specifying in excruciating detail the procedures needed to transform input data into results. My first experience with computers (1) was on a Bendix laboratory computer, generating three-component polymer-copolymer phase diagrams in assembly language. After a summer of this I became quickly convinced that there must be a better way.

In the early 1960's the first compiled procedural programming language for scientific computation, FORTRAN, became widely used in the US, with a parallel development of the use of ALGOL in Europe. Later in the decade, the interpretive procedural language BASIC emerged, followed by the powerful algebraic notational language APL. The first structured, procedural language developed to teach the concepts of programming, Pascal, appeared in 1971, followed later in the decade by the C language.

In all of these procedural languages (also called imperative languages (2), one of the basic elements of syntax is the assignment statement, in which an algebraic

0097–6156/86/0306–0111$06.00/0

expression is evaluated and stored in a named storage location called a variable. Although both BASIC and FORTRAN use the equality symbol " = " for the assignment statement, Pascal emphasizes the procedural nature of the assignment statement by using the symbol " := " , thus distinguishing it from an algebraic equation. Another characteristic of procedural languages is that they specify in detail the procedures and flow of control needed to solve a problem, using such structures as conditionals and loops.

Parallel to, but largely independent of, this development of procedural computational languages was the evolution of non-procedural or declarative languages used for symbolic processing. Eisenbach and Sadler (2) have reviewed the evolution of declarative languages, which began with LISP in 1960 and includes such recent languages as Prolog. One of the characteristics of declarative languages is that problems are defined in terms of logical or mathematical relationships, rather than assignment statements and flow of control, and that the language itself then decides how best to solve the problem posed and in what order to use the information provided. Declarative languages have not so far been widely used in scientific computation because of their computational inefficiency.

There have also been important developments in the past decade in scientific applications software as scientists and engineers have looked for other ways of solving problems than by writing a program in FORTRAN or another procedural language. Libraries of mathematical procedures commonly used in science and engineering became available for those who wanted to write their own procedural software but needed robust numerical algorithms in an easily used form. One of the first full scientific software packages which freed the user from writing in a compiled or interpreted procedural language was RS/1, which evolved as a part of the Prophet Network established by NIH for its research grantees in the 1970's and now runs as a separate package on DEC superminis and personal computers (3). Another was the electronic spreadsheet, first embodied in its simple tabular format in Visicalc but now enhanced with plotting and sorting capabilities in Lotus 1-2-3 and several other packages. A third example is statistical software such as SPSS (4) or Minitab (5). Symbolic processing languages such as LISP led to the development of symbolic mathematics packages such as MACSYMA; their use in chemistry has been reviewed by Johnson (6). A recent ACS Symposium on symbolic algebraic manipulation contains a full description of MACSYMA among other systems, and a variety of applications in chemistry (7).

Scientific applications software packages are often characterized by close attention to the design of the user interface, sometimes at the expense of program size or execution time. By far the dominant computational idiom in these packages, however, is procedural. For example, RS/1 has an internal language called RPL, modelled after the procedural language PL/1, in which specialized procedures and functions not available in the package may be written by the user. In spreadsheets, the cell is the basic storage location for either data or formulae. Cells are provided with data by an assignment process, and formulae reference other cell locations as variables.

TK!Solver

TK!Solver (8) is a high-level computer language for solving sets of algebraic equations and tabulating or plotting their results. In TK!Solver, equations are viewed as relationships or rules, not as assignment statements, and in that sense it may be viewed as a declarative language. The basic computational approach taken by TK!Solver grew out of the research of textile engineer Milos Konopasek in the 1970's. It was realized early on by Konopasek and Papaconstadopoulos (9) that a high level computational langauge need not be procedural but could be declarative;

this point has been recently amplified by Konopasek and Jayaraman (10), who also make the case for TK!Solver's being an expert system for equation solving.

To produce TK!Solver, the problem-solving methodology implemented by Konopasek in his Question Answering System (9) was combined with the experience in designing full-screen user interfaces of Software Arts, Inc. (the originators of the electronic spreadsheet). The goal of the language was to obviate three of the time-consuming stages of procedural program development (11): (1) algebraic transformations necessary for formulating assignment statements; (2) sequencing assignment statements to secure desired flow of information through the program; and (3) setting up input and output statements. The capabilities of TK!Solver, which runs on a number of different personal computers, are as follows (10,11):

(1) It parses entered algebraic equations and generates a list of variables.
(2) It solves sets of equations using a consecutive substitution procedure (the direct solver).
(3) It solves sets of simultaneous (non-linear) algebraic equations by a modified Newton-Raphson iterative procedure when consecutive substitution fails (the iterative solver).
(4) It searches through tables of data and evaluates either unknown function values or arguments when required in solving.
(5) It performs unit conversions with definable conversion factors.
(6) It detects inconsistencies in problem formulation and domain errors.
(7) It generates series of solutions for lists of input data and displays results in tabular or graphical form.

To see how such a language can speed up the process of equation-solving, consider the steps needed to solve a set of algebraic equations when using a procedural language. First, you must identify the variable or variables for which you need to solve. Next, you must use algebraic substitution methods to express the variables to be solved for in terms of the known variables using assignment statements. Finally, you must write a program to input values for the known variables, evaluate the unknown variables, and output the results. There are several disadvantages to this method. If a different combination of variables serves as input for another similar problem based on the same set of equations, the algebra must be reworked to solve for those new variables. In many cases it may not be possible to obtain analytic expressions usable in assignment statements, so you must find some numerical approximation algorithm suitable for the problem at hand and either obtain or write the code based on that algorithm.

A Chemical Example: The van der Waals Gas

Take, for example (12), the problem of solving for the P-V-T properties of a real gas obeying the van der Waals equation of state,

$$P = n R T / (V - n b) - n^2 a / V^2 \qquad (1)$$

where **a** and **b** are coefficients characteristic of a given gas. Solving for **P**, given **n**, **V**, and **T** is a simple assignment statement, but solving for **n** given **P**, **V**, and **T** requires considerable algebraic manipulation, followed either by applying the formula for the roots of a cubic equation or by using a numerical technique for determining roots (the latter usually requires more mathematical analysis - for example, finding first derivatives using the Newton-Raphson method).

Figure 1 shows the Rule Sheet for a TK!Solver model **REALGAS.TK** (12). The first rule is the van der Waals equation of state. The second defines the gas constant, and the third rule defines the number density. The fourth defines the compressibility factor **z**, a dimensionless variable which measures the amount of departure of a real gas from ideality. The next three rules give the critical pressure, molar volume, and temperature of a van der Waals in terms of the coefficients **a** and **b**. The Van der Waals equation can be recast in a form which uses only reduced, dimensionless variables; these are defined in the next three rules. The last two rules provides values for the van der Waals coefficients **a** and **b** when the name of the gas is given (user-defined functions with symbolic domain elements and numerical range elements can be used in any model which requires reference to built-in data tables).

S Rule	
- ----	
"Equation of State of a van der Waals Gas. Chap. 4. Model name: REALGAS.TK	
* R = 0.0820568	"Value of gas constant
* nd = n / V	"Number density
* z = P * V / (n * R * T)	"Compressibility factor
* Pc = a / (27 * b^2)	"Critical Pressure
* Vc = 3 * b	"Critical Molar Volume
* Tc = 8 * a / (27 * b * R)	"Critical Temperature
* Pred = P / Pc	"Reduced pressure
* Vred = V / Vc	"Reduced volume
* Tred = T / Tc	"Reduced temperature
* a = acoeff (gas_name)	"Function for Van der Waals a coefficient
* b = bcoeff (gas_name)	"Function for Van der Waals b coefficient

Figure 1. Rule Sheet for Model REALGAS.TK

A typical use for this model would be to solve for the number of moles of a gas, given its identity, pressure, volume, and temperature. The iterative solver is used for this purpose. You must decide which variable to choose for iteration and what a reasonable initial guess is. Real gases approach ideal behavior at low pressure and moderate temperatures. Since the compressibility factor z is 1 for an ideal gas, and since knowing z along with P, V, and T allows a calculation of n, we choose z as the iteration variable and 1.0 as the initial guess.

The Variable Sheet with the solution to such a problem is shown in Figure 2. Unit conversions from psi to atmospheres, from cubic feet to liters, and from Fahrenheit to Kelvins have been built into the model via the Units Sheet. For input values of 100 cubic feet of acetylene at 300 psi and 66°F, there are 728.9 moles of acetylene and the value of z of 0.874 indicates that the deviation from ideality is 12.6%.

Another Example: Acid Rain

Problems in chemical equilibrium with many reactions involving many species often generate mathematical models containing large sets of simultaneous, nonlinear equations which must be solved by numerical means. TK!Solver is a good tool for solving such problems. For example, consider the acid-base chemistry of a raindrop. Vong and Charlson (13) have developed an equilibrium model which predicts the pH of cloud water, assuming an atmosphere with realistic levels of three soluble, hydrolyzable gases: SO_2, CO_2, and NH_3. Also included is the effect of acidic dry aerosols, particles of sub-micron diameter containing high concentrations of sulfuric and nitric acid.

St	Input	Name	Output	Unit	Comment
--	----	----	----	----	------
	300	P		psi	pressure
	100	V		ft^3	volume
		n	728.92419	mol	number of moles
		R	.0820568	l*atm/(mo	gas constant
	66.000000	T		oF	temperature
	'acetylen	gas_nam	'text	name of the gas	
		z	.87417992	decimal	compressibility factor
		nd	.97443505	mol/l	molar density
		a	4.39	atm*l^2/m	van der Waals a coefficient
		b	.05136	l/mol	van der Waals b coefficient
		Pc	61.638310	atm	critical pressure
		Vc	.15408	l	critical molar volume
		Tc	308.63925	K	critical temperature
		Pred	.33118688	decimal	reduced pressure
		Vred	4854.9325	decimal	reduced volume
		Tred	.94624676	decimal	reduced temperature

Figure 2. Variable Sheet for REALGAS.TK with Solution

There are five laws of chemical equilibrium relevant to the Charlson-Vong model: (1) the ideal gas law, relating gas species density to its temperature and partial pressure; (2) Henry's law, relating the partial pressure to the concentration of dissolved gas; (3) the law of mass action, giving equilibrium constant expressions for the hydrolysis reactions of the dissolved gases; (4) conservation of mass for species containing sulfur(IV), sulfur (VI), carbon(IV), nitrogen(V), and nitrogen(-III); and (5) conservation of charge. Applying these laws, Vong and Charlson were able to calculate the pH of a raindrop by solving a set of 17 equations in 29 variables (cloud water content, temperature, partial pressures, and species concentrations) and 9 parameters (Henry's law constants, equilibrium constants, and the gas constant). They wrote a FORTRAN program which solved all equations but one, that of charge conservation. The pH at electrical neutrality was determined by a graphical method, in which the total positive and negative charge concentrations were calculated and plotted for a series of assumed pH's and the crossing point found.

A TK!Solver model called **RAINDROP.TK** has beendeveloped to incorporate the full Charlson-Vong model of cloud water equilibrium (12), including the temperature dependence of all equilibrium constants. The iterative solver makes it possible to compute the pH at charge neutrality without having to make plots of intermediate results. The Rule Sheet is shown in Figure 3.

The Unit Sheet contains a number of conversions necessary to accommodate the variety of units used in experimental atmospheric chemistry. The Variable Sheet is arranged so that the variables at the top are the ones normally chosen as input variables. Since the usual goal of running the model is to determine the pH of the raindrop, the variable **pH** is chosen as the one on which to iterate.

The following problem, taken to match the conditions in Figure 2 of reference 13, is typical of those solved in less than one minute on an IBM PC with this model: "a cloud at 278 K contains 0.5 grams of liquid water per cubic meter of air. The atmosphere of the cloud contains 5 ppb sulfur dioxide, 340 ppm carbon dioxide, 0.29 $\mu g/m^3$ of nitrogen base, 3 $\mu g/m^3$ of sulfate aerosol, and no nitrate aerosol. What is the pH of the cloud water?" Figure 4 shows the Variable Sheet after solution.

Rule

"Equilibrium pH of a Raindrop, Charlson-Vong model. Chap. 8. Model name: RAINDROP.TK
PSO2 = NSO2 * R * T   "Ideal gas law for SO2
PNH3 = NNH3 * R * T   "Ideal gas law for NH3
PCO2 = NCO2 * R * T   "Ideal gas law for CO2

PSO2 = CSO2 * KHS   "Henry's law for SO2
PNH3 = CNH3 * KHN   "Henry's law for NH3
PCO2 = CCO2 * KHC   "Henry's law for CO2

K1S = CHSO3m * CHp / CSO2   "Mass action law for SO2 - HSO3m
K2S = CSO32m * CHp / CHSO3m   "Mass action law for HSO3m - SO32m
KB = CNH4p * COHm / CNH3   "Mass action law for NH3 - NH4p
K1C = CHCO3m * CHp / CCO2   "Mass action law for CO2 - HCO3m
K2C = CCO32m * CHp / CHCO3m   "Mass action law for HCO3m - CO32m
KW = CHp * COHm   "Mass action law for water

NTS4 = NSO2 + L * (CSO2 + CHSO3m + CSO32m) "Mass balance for sulfur(IV)
NTN3m = NNH3 + L * (CNH3 + CNH4p)   "Mass balance for nitrogen(-III)
NTC4 = NCO2 + L * (CCO2 + CHCO3m + CCO32m) "Mass balance for carbon(IV)
NTN5 = L * CNO3m   "Mass balance for nitrogen(V)
NTS6 = L * CSO42m   "Mass balance for sulfur(VI)

CNH4p+CHp=CHSO3m+2*CSO32m+COHm+CHCO3m+2*CCO32m+CNO3m+2*CSO42m
                                "Chargebalance

pH = -log(CHp)   "Definition of pH

"Temperature-dependent equilibrium constants
KHS = 0.379 * exp (-3145.99 * (278 - T) / (278 * T))
KHC = 16.6 * exp (-2367.65 * (278 - T) / (278 * T))
KHN = 7.11E-3 * exp (-3730.87 * (278 - T) / (278 * T))
K1S = 2.06E-2 * exp (2003.54 * (278 - T) / (278 * T))
K2S = 8.88E-8 * exp (1461.46 * (278 - T) / (278 * T))
K1C = 2.94E-7 * exp (-1716.92 * (278 - T) / (278 * T))
K2C = 2.74E-11 * exp (-2217.49 * (278 - T) / (278 * T))
KB = 1.5E-5 * exp (-685.59 * (278 - T) / (278 * T))
KW = 1.82E-15 * exp (-7057.27 * (278 - T) / (278 * T))
R = 0.0820565   "Ideal gas constant

Figure 3: Rule Sheet for Model RAINDROP.TK

Summary of Other Chemical Applications

In addition to the two examples above, I have developed TK!Solver models for the
ideal gas, for two-component mixture concentrations, for acid base chemistry
(including the generation of titration curves), for transition metal complex equilibria,
for general gaseous and solution equilibria, and for linear regression (12).
    Drexel undergraduate students in both the lecture and the laboratory of physical
chemistry have been using TK!Solver for such calculations as least squares fitting of
experimental data, van der Waals gas calculations, and quantum mechanical
computations (plotting particle-in-a-box wavefunctions, atomic orbital electron
densities, etc.). I use TK!Solver in lectures (on a Macintosh with video output to a
25" monitor) to solve simple equations and plot functions of chemical interest.

TK!Solver has also had heavy use in the material balance course in chemical engineering, and in a mathematical methods course in materials engineering. Graduate students in chemistry are using it in research projects in spectroscopy and kinetics.

In the teaching of quantum mechanics, TK!Solver has proved especially useful. For example, Berry, Rice, and Ross (14) give several problems on the regions of

St	Input	Name	Output	Unit	Comment
--	-----	---	-----	---	------
	278	T		K	temperature
	.5	L		g/m^3	liquid water content of the cloud
	5	PSO2		ppb	partial pressure of SO2
	340	PCO2		ppm	partial pressure of CO2
	.29	NTN3m		ug(N)/m^3	total nitrogen base concentration
	3	NTS6		ug(SO4)/m	sulfate aerosol concentration
	0	NTN5		ug(NO3)/m	nitrate aerosol concentration
		pH	4.0190385	decimal	pH of water in the cloud
		PNH3	2.9020E-4	ppb	partial pressure of NH3
		CNO3m	0	mol/lw	concentration of NO3 anion
		CSO42m	.0000625	mol/lw	concentration of SO4 anion
		NSO2	2.192E-10	mol/la	concentration of gaseous SO2
		CSO2	1.3193E-8	mol/lw	concentration of dissolved SO2
		CHSO3m	2.8395E-6	mol/lw	concentration of HSO3 anion
		CSO32m	2.6344E-9	mol/lw	concentration of SO3 anion
		NCO2	1.4905E-5	mol/la	concentration of gaseous CO2
		CCO2	2.0482E-5	mol/lw	concentration of dissolved CO2
		CHCO3m	6.2915E-8	mol/lw	concentration of HCO3 anion
		CCO32m	1.801E-14	mol/lw	concentration of CO3 anion
		NNH3	1.272E-14	mol/la	concentration of gaseous NH3
		CNH3	4.082E-11	mol/lw	concentration of dissolved NH3
		CNH4p	3.2197E-5	mol/lw	concentration of NH4 cation
		CHp	9.5711E-5	mol/lw	concentration of hydrogen ion
		COHm	1.902E-11	mol/lw	concentration of hydroxide ion
		NTS4	2.206E-10	mol/la	total concentration of sulfur (IV)
		NTC4	1.4905E-5	mol/la	total concentration of carbon (IV)
		KHS	.379	atm*la/mo	Henry's law constant for SO2
		KHC	16.6	atm*la/mo	Henry's law constant for CO2
		KHN	.00711	atm*la/mo	Henry's law constant for NH3
		K1S	.0206	decimal	equilibrium constant for SO2 - HSO3
		K2S	8.88E-8	decimal	equilibrium constant for HSO3 - SO3
		K1C	2.94E-7	decimal	equilibrium constant for CO2 - HCO3
		K2C	2.74E-11	decimal	equilibrium constant for HCO3 - CO3
		KB	.000015	decimal	equilibrium constant for NH3 - NH4
		KW	1.82E-15	decimal	ionization constant for water
		R	.0820565	la*atm/(m	ideal gas constant

Figure 4 : Variable Sheet for Solution to Sample Problem

bonding and anti-bonding in diatomics, one of which requires the calculation and plotting of contours of constant bonding force. They suggest calculation of the bonding force on a large grid of points and then connecting points of constant force, but with TK!Solver it is possible to solve directly the set of parametric equations in r and theta and to plot the resulting contours.

In summary, the rule-based, declarative approach to solving sets of algebraic equations presented by TK!Solver has proved to be a fruitful medium for chemical computations.

Literature Cited

1. S. Krause, A. L. Smith, and M. G. Duden, J. Chem. Phys. 1965, 43, 2144-45.
2. S. Eisenbach and C.Sadler, Byte 1985, 10, 181-87; see also other articles on declarative languages in the same issue of Byte.
3. "RS/1 Command Language Guide"; BBN Research Systems, Cambridge, Mass, 1982.
4. "SPSS X User's Guide"; McGraw-Hill, New York, 1983.
5. T. A. Ryan, Jr. ,B. L. Joiner, and B. F. Ryan, "Minitab Student Handbook"; Duxbury Press, Boston, 1976.
6. C.S. Johnson, J. Chem. Inf. Comp. Sci. 1983, 23, 151-7.
7. R. Pavelle, ed. "Applications of Computer Algebra"; Kluwer Academic Publishers, Hingham, Mass, 1985.
8. TK!Solver is a software product developed by Software Arts, of Wellesley, Mass., and introduced in late 1982. At the present time (November, 1985), the rights to distribute TK!Solver belong to the Lotus Development Corporation.
9. M. Konopasek and C. Papaconstadopoulos, Computer Languages 1978, 3, 145 - 155.
10. M. Konopasek and S. Jayaraman, Byte 1984, 9, 137-145 . See also M. Konopasek and S. Jayaraman, "The TK!Solver Book"; Osborne/Mc-Graw Hill, 1984.
11. M. Konopasek, "Software Arts' TK!Solver: A Message to Educators"; unpublished manuscript, October, 1984.
12. The examples in this paper are from A. L. Smith, "TK!Solver Pack in Chemical Equilibrium and Chemical Analysis"; McGraw-Hill, to be published.
13. R. J. Vong and R. J. Charlson, J. Chem. Ed. 1985, 62, 141-3.
14. R. S. Berry, S. A. Rice, and J. Ross, "Physical Chemistry"; John Wiley, 1980; Chap. 6.

RECEIVED January 16, 1986

A Chemical-Reaction Interpreter for Simulation of Complex Kinetics

David Edelson[1]

AT&T Bell Laboratories, Murray Hill, NJ 07974

Simulation of the kinetics of complex chemical systems is finding ever increasing use for analysis of reaction mechanism as well as for process prediction and control. Software for the solution of the large number of coupled mass-action differential equations is now readily available, as are reaction rate data banks for many systems of interest. However, application of the technique is discouraged by the tedious, error-prone task of manually formulating the differential equation set and coding it for the computer. Several chemical reaction interpreters which can do this have been written over the years; this report describes our most recent version which uses modern operating systems and programming techniques to implement an interactive, user-friendly program. Portability was a prime consideration in its design so that it could be interfaced with any differential equation solving program. Although it was written in C for use on machines having a UNIX operating system, the subroutines that it produces for the equation solver are in FORTRAN, so that they can be ported to other machines, and are compatible with most simulation packages in use today. Special features include free form input, batch or interactive operation, full ASCII capability, and dynamic storage allocation. The extensive use of the "structure" data type in the C source code makes it easy to modify or enhance the interpreter to suit the needs of the current application or computing environment.

Computer simulation of chemical reaction or reaction-transport systems has long been used in chemical engineering process design, and has more recently moved into the chemical research

[1]Current address: Department of Chemistry, Florida State University, Tallahassee, FL 32306

area, where it has become a tool for the elucidation of chemical mechanism(1). The technique has also found application in the prediction of the behavior of large complex chemical systems, such as atmospheric and environmental systems(2,3), especially in the study of the effect of pollutants and strategies for the minimization of their effects(4).

The mathematical problem posed is the solution of the simultaneous differential equations which arise from the mass-action treatment of the chemistry. For the homogeneous, well-mixed reactor, this becomes a set of ordinary, non-linear, first-order differential equations. For systems that are not spatially uniform and involve material and energy transport, the chemical terms are coupled with the fluid mechanics and heat transfer to give sets of partial differential equations. Numerical techniques for solving these systems have been extensively developed(5,6), but regardless of the strategy used to solve the set of ordinary differential equations, or the spatial discretization methods employed for partial differential equations(7,8), the final task is that of solving very large matrix systems of algebraic equations. This has traditionally been the forte of the large main-frame computer, but the rapidly expanding capability of minis and micros has enabled them to handle the solution of modest problems. At the other end of scale, the expanding scope of application of these simulation methods, especially to two- and three- dimensional systems, has vastly increased the number of equations to be solved, and so has entered the realm of the supercomputer.

From the chemist's point of view as a user, these simulation techniques require him to provide computer code for the time derivative of each chemical species in the mechanism. According to the principle of mass-action, the derivative of the x-th species concentration in a mechanism of M reactions involving P chemical species is given by

$$\frac{d[N_x]}{dt} = \sum_{i=1}^{M} (\pm) \nu_{xi} k_i \prod_{j=1}^{P} [N_j]^{\nu_{ji}}$$

where k_i is the rate constant of the i-th reaction, and ν_{ji} is the stoichiometric coefficient of the j-th species in the i-th reaction. Since the differential equations are usually handled by methods appropriate to stiff equations, the partial derivatives of each of the above expressions with respect to all the P species (Jacobian matrix) are needed as well. While each term in the summation above rarely has more than three species in the product (i.e. most of the ν_{ji}'s are zero), the algebra involved in collecting all the sums and products is so large and the labor (and the possibility of error) in the coding so great, that this task is unlikely to be undertaken manually for any but small chemical mechanisms. However, a mechanism for atmospheric or combustion chemistry may easily run to several hundred reactions and species. Furthermore, in research applications it is common to test several alternate models for the system under study, and the amount of code to be written escalates greatly. Clearly a machine aid is required to make the technique simple to use so that its exploitation is encouraged.

Over the years, various approaches to this problem have been taken. In one of the earliest, each chemical species was assigned an identification number, chemical equations rewritten in these terms, and the computer constructed a symbolic reaction table which would subsequently be used in a lookup procedure to guide the computation. In those years, the computation was slow and cumbersome, and the additional overhead of the table lookup at each step of the iterative solution process greatly increased the costs of the simulation. The next step was to use this table lookup just once to write a routine which would be compiled as part of the simulation program. Fortran was the major high-level language available; it was necessary to use assembly language to write the Fortran code for the simulation package(9).

The next advance was scanning and interpretation of the text of the original reaction set, written in as close an approximation to real chemical notation as the straight-line, upper-case only format of the 72-column card would permit(*10*). Since character and string manipulation through Fortran was at that time cumbersome and inefficient, these parts of the interpretation program were written in assembly language. The interpreter used Fortran output statements to generate the simulation code in assembly language, making the simulation more efficient, but unfortunately non-portable.

The passage of time has brought vastly improved facilities for string and character handling by high-level languages. Smaller machines have come into vogue, and interactive operation has taken preference over batch systems. The growth in problem size, however, has kept the mainframe machines in the picture, and has even brought in the supercomputers, which are, at this writing, geared to batch mode operation and mostly Fortran programming. Front-end machines, however, offer a variety of interactive environments and programming languages. This suggests that a better approach would be to separate the problem interpretation and solving functions of a simulation system. This paper describes a new implementation of our previous simulation package, in which each part is done on a machine and with a language which are the most effective and appropriate.

The Bell Laboratories Central Computer Service supports a Cray-1 with Fortran as the primary high-level language for compute-bound problems. This is accessed through a number of front-end machines, mostly Vaxes operating under UNIX, which support several high-level languages for interactive use. Because of the number of different tasks the chemical interpreter is required to perform in addition to elementary string manipulation, we chose C as the language in which to write it. This offers a degree of portability, as C or C-like compilers are to be found on a large number of machines and operating systems.

Input Language

Chemical notation is mostly the result of historic precedent, and was certainly never intended to be interpreted by a computer. However, in order to maintain the greatest ease of operation by a chemist, the input language should be designed to be as close to the normal notation for reaction equations. The basic input record is a chemical equation; reactants on the left are separated from products on the right by an arrow ($\rightarrow$, read 'yields'), and are in turn separated from each other by plus (+) signs. Substituting the equal sign (=) for the arrow and the ampersand (&) for the plus results in a minimal sacrifice of readability for the chemist, but eliminates ambiguities for the machine. Subsidiary fields for identification numbers are separated from the reaction expression by tab characters. Input is in free form, with embedded spaces ignored (except in text expressions, see below).

Compounds are expressed by their symbolic formulas. The use of the full ASCII character set allows the elements to be expressed by their usual one or two character names, with the upper or lower case context providing the character count. More elaborate designations for atoms (including superscripts denoting isotope number, for example) are accommodated by enclosing the expression in appropriate quotes. Upshift and downshift metacharacters can be used here to denote appropriate character placement on output devices (such as plotters and typesetters) which allow for partial line spacings; line printers would ignore them.

Unfortunately, terminal input does not allow for the subscripts and superscripts used by chemists. A rigid format is therefore enforced to distinguish subscripts (indicating number of atoms) from superscripts (indicating valence state or charge) by

preceding the latter by a + or − sign. Charge indication can be
by appropriate repetition of the sign, or by a single sign
followed by numerical indication. Parenthesized expressions are
accommodated and expanded in the usual way, and nesting to
several levels is allowed.

Where it is felt that clarity (for the chemist) is better
served by using compound names rather than formulas, text input
is accepted by surrounding it with quotation marks. This text is
not subject to lexical analysis; subsidiary tasks such as syntax
checking cannot be performed in this case. Quoted text can also
be attached to a compound expressed by formula; the formula is
interpreted and the text passed through unchanged.

Syntax Analysis

As each reaction equation is entered, several checks are
performed to catch errors in formulation or typing: the correct
number of tabs, equals, balanced quotes or parentheses, and
conformance to the syntax rules which allows the equation to be
separated into reactants and products, and these in turn to be
decomposed into atoms. Each time a new chemical species is
encountered it is reported to the user, who can determine whether
a valid name has been entered. The equation is checked for
balances in atomic elements and charges, and discrepancies listed
for corrective action. Species names may be either formulas or
text; in the latter case the balance checking feature must be
turned off to avoid false errors.

When the input list is finished, the interpreter checks the
equations against each other, to ascertain that a reaction has
not inadvertently been entered more than once, even with
permutations of the reactants or products. Finally, if all input
has been error-free, the interpreter continues by lexically
sorting the atomic elements and species names, assigning final
identification numbers, and providing a list for the user.

Should an error be encountered during input, the interpreter
will not complete its task. However, it does copy all input to a
file from which it may be retrieved, edited and resubmitted in
batch mode, making it unnecessary to retype all the equations.

Data Structures

Scanning of the reaction input leads to the generation of three
types of data structure (in the C sense(11)), one dealing with
reactions, one with chemical species, and the last with chemical
elements. Each reaction structure contains the appropriate
identification numbers, and a symbolic representation of the
reaction itself in terms of pointers to the chemical species
structures of the reactants and products. The chemical species
structures in turn contain pointers to chemical element
structures, as well as text strings to be used in printed,
plotted, or typeset output. The chemical element structures
similarly contain identifying numerical and text information.
Storage for these structures is allocated as needed, and they are
chained to each other by pointers. As each participant in a
reaction is examined, the database is searched and a structure
created for any newly encountered species. Since the number and
size of these searches may become quite large for extensive
reaction mechanisms, the species structures are ordered lexically
in a binary tree to keep the search time to a minimum. After
input is complete, final identification numbers are assigned to
the elements and species according to a lexical sort.

Program Output

The principal task of the interpreter is to provide two
subroutines for use by the simulation program, which is a solver
for ordinary or partial differential equations. Since chemical
systems are for the most part "stiff" as a result of negative
feedback(12) the interpreter expects the simulation package to

use an implicit differential equation solver requiring calculation of both the function (*i.e.* the mass-action expression of the net rate of change of the species) and its partial derivatives with respect to all species (Jacobian matrix). The former is computed stepwise: first the rate of each reaction is calculated; then these terms are combined to give individual formation and removal terms for each species, and finally these are algebraically added to give the derivatives. This strategy makes available to the user additional information that is often helpful in interpreting the mechanism. The Jacobian terms are calculated in one step. Fortran code for these subroutines is written using direct addresses for each member of the appropriate arrays, it being assumed that these are stored in the same order as that provided by the lexical sort above. The Fortran compiler is thus burdened with the task of calculating the variable addresses, relieving the simulation program of this task and so improving the run-time economy.

It is also possible to use the information which has been stored to write programs for other tasks. A useful one, for example, keeps track of the stoichiometry (*i.e.* total atom counts) of the system. For a closed system, stoichiometry should be automatically maintained by linear predictor-corrector solvers, and the stoichiometry program provides a diagnostic of numerical errors (and others) which have accumulated. In other than closed systems, it gives an independent check on the sources and sinks which are being modeled.

Various databases can also be output by the interpreter, *e.g.* lists of element and species names, text files for labeling printed and plotted output, and a symbolic reaction matrix. This information is distributed to individual ASCII files, from which they may be read by subsequent parts of the simulation package for use in the appropriate task.

Adaptability

The interpreter was designed to be independent of the simulation program for which it eventually will serve. A structured programming language such as C is therefore ideal for coding it since it is simple to add code to perform additional tasks (as for example writing the variable dimension specifications) which might be specific to the application. The use of structure data types also allows the expansion of the type of supplementary information carried along with each variable with little additional coding effort and with no danger of breaking the already existing code. Communication of information from the interpreter to the simulation program is through individual files of information, which can be input to subsequent programs and stored to be used as needed. The simulation system is thus freed from dependencies on the operating system environment.

Conclusion

The interactive interpretation of chemical equations in conjunction with the simulation of chemical reaction systems has been implemented by a C-language program which can be run on a small machine that is independent of the machine on which the simulation program resides. The flexible string and character manipulation capabilities of this environment enables the chemist to use an input language similar to the natural language of chemical kinetics, and checks syntax and consistency of the input as well. The interpreter provides verified code for any simulation program using standard differential equation solvers, and also facilitates the display of the results in chemical notation. These interpreters have been used successfully for many years and have fostered the growth of simulation techniques in many areas of chemistry and chemical engineering.

Literature Cited

1. D. Edelson, *Science,* 214, 981 (1981).
2. H. G. Booker, *et al.,* *Environmental Impact of Stratospheric Flight,* National Academy of Sciences, Washington D. C., 1975.
3. H. S. Gutowsky *et al., Halocarbons: Effects on Stratospheric Ozone,* National Academy of Sciences, Washington D. C. 1976.
4. J. H. Seinfeld, *Air Pollution; Physical and Chemical Fundamentals,* McGraw-Hill, New York, 1975.
5. D. D. Warner, *J. Phys. Chem.,* 81, 2329 (1977).
6. N. L. Schryer, *J. Phys. Chem.,* 81, 2335 (1977); and references cited therein.
7. D. Edelson and N. L. Schryer, *Computers and Chemistry,* 2, 71 (1978);
8. G. A. Nikolakopoulou, D. Edelson and N. L. Schryer, *Computers and Chemistry,* 6, 93 (1982).
9. D. McIntyre, in *A Technique for Solving the General Reaction-Rate Equations in the Atmosphere, Appendix B.* (T. J. Keneshea, author); AFCRL-67-0221, April 1967.
10. D. Edelson, *Computers and Chemistry,* 1, 29 (1976).
11. B. W. Kernighan and D. M. Ritchie, *The C Programming Language,* Chapt. 6; Prentice-Hall, Englewood Cliffs, N.J. 1978.
12. C. F. Curtiss and J. O. Hirschfelder, *Proc. Natl. Acad. Sci. U. S. A.* 38, 235 (1952).

RECEIVED December 17, 1985

Applying the Techniques of Artificial Intelligence to Chemistry Education

Richard Cornelius[1], Daniel Cabrol[2], and Claude Cachet[2]

[1]Department of Chemistry, Lebanon Valley College, Annville, PA 17003
[2]Department of Chemistry, Université de Nice, 06034 Nice, France

The computer program called GEORGE is a "problem-solving partner" for introductory chemistry students. The program has no problems to present to students; students give problems to GEORGE and he solves the problems. He explains the solution using ordinary English and then sketches a diagram to show how data are combined and relations are applied to give the solution. GEORGE operates on problems involving three fundamental quantities, mass, volume, and number of moles, and other quantities that can be expressed as ratios of these fundamental quantities.

The power of the computer holds the promise for far-reaching changes in education, but that promise remains unrealized. Most of the applications of computers in chemical education have been adaptations of teaching strategies used in other media; there are many tasks that have been done better or faster on the computer but little really new has been developed. There is a quote that summarizes the situation in which we find ourselves today: "After years of growing wildly the field of [educational] computing is finally approaching its infancy." This quote is nearly twenty years old, having been taken from the report of the 1967 President's Science Advisory Commission (1). The quote, however, is as true today as it was nearly twenty years ago. We stand on the threshold of exciting new applications for computers both within the field of education and elsewhere. The subject of this paper is a computer program which represents one totally different approach for the use of computers in chemical education. We hope that it is only one new approach out of many that we will see in the future.

One of the most important advantages of computers in education is the capacity of software to adjust the pace or nature of activities on the basis of input from the student. Tutorial or drill and practice programs available today do in fact make some adjustments based upon student responses. These programs are

0097-6156/86/0306-0125$06.00/0

limited, however, by the ingenuity of the person that wrote the
software in considering all possible student responses and in
designing appropriate action on the part of the software. A student
cannot explore areas which the author of the software failed to
consider. Thus, these programs are "instructor-driven." The author
of the software serves as a surrogate instructor, creating a
particular sequence of activites for the student. However
sophisticated the branching in the program may be, the student cannot
take the initiative; initiative is exercised only by the author of
the software.

It is useful to identify two software categories distinguished
by the identity of the person in charge of the educational acivities
that the software supports. The first category is Computer-Assisted
Instruction (CAI). In CAI the role of the software is to decide
which activities the student should pursue. Most existing software
for chemical education falls into this category. We may also,
however, consider a category of software that could be labeled
Computer-Assisted Learning (CAL). In such software, the student
makes decisions about what he or she will investigate while using the
software. Simulations fall into this category. Professor John
Gelder's ideal gas law program (2) is a classic example of using
simulations in chemical education. In using that program the student
has control over the parameters, and by exploring the model could
potentially learn aspects of the behavior of ideal gases unknown to
the author of the program. Other simulations may also fall into the
category of computer-assisted learning. Apart from simulations,
examples of software with which the student is in control and are
difficult to find.

This paper describes an example of a different style of program
which is under the control of the student. The project began in the
fall of 1983 when Dick Cornelius spent part of a sabbatical at the
Universite de Nice working with Daniel Cabrol and Claude Cachet. The
first task there was to write a chapter on microcomputers in chemical
education for a book on computers in chemistry. During the course of
writing this chapter we described programs available in the different
software styles: page turners, drill and practice, tutorial dialogs,
simulation, pre-laboratory activities, and problem-solving. In the
area of problem-solving, however, there was little that we could
discuss. Some software could be used for problem-solving, but there
were no examples of programs written for the primary purpose of
helping students learn general problem-solving techniques. It was to
this area, then, that we turned our programming attention. The
result was a program that we called GEORGE (3) that runs on the
Apple II series of computers. GEORGE differs very much from most
programs available for chemical education: GEORGE asks no questions
of students. Instead, students take problems to GEORGE. GEORGE
solves the problems that students provide and, most importantly,
explains the solutions using both text and diagrams. If insufficient
or contradictory information is available, GEORGE can provide
diagnostic comments to help the student.

The domain in which GEORGE operates is a small but important one
for introductory chemistry. He works with problems involving the
fundamental quantities mass, volume, and number of moles. He can
also work with derived quantities such as density, molar mass, molar

concentration, etc. The specific quantities with which GEORGE can
work are presented in Figure 1. That figure is taken from the
on-screen documentation and also gives the abbreviations that
students may use as shorthand to identify the quantities to GEORGE.
GEORGE works with the units g, L, mol. For derived quantities he
understands the ratios of these units such as g/L for density. He
also understands the numerical prefixes p, n, μ, m, c, d, and k.
He can work with these prefixes in ratios of units such as g/mL or
nmol/mL, and he can also accept dm^3 or cm^3 for volume.

The Logic

The basic approach that GEORGE uses to solve problems is dimensional
analysis, the same technique that many of us use in our own
classrooms to teach students how to solve problems. Instead of
having numerous formulas for different kinds of problems, GEORGE
simply contains a set of heuristic rules which he follows to search
for a solution. One result of using these heuristic rules is that he
can solve problems never worked by the authors of the program.
Another result is that GEORGE may be able to make progress toward a
solution even if incomplete information is available. In such an
instance, GEORGE may be able to respond with a statement such as "If
you could give me the density of alcohol, then I could solve the
problem." The rules are very simple in concept. First GEORGE
examines the various pieces of data available. He examines all
possible pairs of data to see whether any pair can be multiplied or
divided to give immediately the solution. If he cannot find a
solution in that way, he checks to see whether he can apply a
relation to generate a new piece of data. If GEORGE cannot apply a
relation, he searches for intermediate results that might represent a
step toward the solution. GEORGE can search for two types of
intermediates. The preferred type is the result of units cancelling
to yield a fundamental quantity. Thus dividing the mass of a
substance by its molar mass is a preferred method to form an
intermediate result. Less desirable is the formation of an
intermediate result which is not a fundamental quantity but which
represents information expressed in a manner not represented by other
data or intermediates. Each time GEORGE calculates a new quantity,
he begins again to look for an immediate solution. These are all the
rules that GEORGE needs to find solutions to millions of different
problem statements. The result is usually a solution approached in
the same manner that a teacher might use for an explanation.

The Program

The primary menu for GEORGE is shown in Figure 2. This is the way
that the menu appears when no information has been given to GEORGE;
more options are available after a problem has been defined. To
understand how GEORGE operates we will first consider an option that
is outside the primary thrust of GEORGE, namely, option C, Calculate
Molar Mass. The student sees a screen which says "Type the formula:"
There a student may type a formula as simple as NaCl or more complex
such as $Mg(ClO_4)_2 \cdot 6H_2O$. GEORGE calculates the molar
mass, explaining to the student how the calculation is done as shown

```
                    Instructions (page 1)
........................................................

George understands 11 different
quantities.  Each of these quantities
has a symbol and a name:

    Symbol              Name
    ..........          ..........
      M                 mass
      n                 no. of moles
      p                 no. of particles
      v                 volume
      d                 density
      c                 molarity
      MC                mass conc.
      M                 molar mass
      Mr                mass ratio
      nr                molar ratio
      vr                volume ratio

........................................................
       Press the space bar to continue.
```

Figure 1.   Quantities with which GEORGE can work.   (Reproduced with
permission from Ref. 3. Copyright 1985 COMPress.)

```
                    Available Options
........................................................

    D)   Enter or Modify Data

    R)   Enter a Relation

    L)   Load a Problem from Disk

    C)   Calculate Molar Mass

    ?)   See Instructions

........................................................
       Press the key of your choice.
```

Figure 2.   The primary menu screen from GEORGE.   (Reproduced with
permission from Ref. 3.   Copyright 1985 COMPress.)

in Figure 3. The ability of GEORGE to explain what he has done is the primary reason for his existence. Other programs, such as TK! Solver (4), have a much larger domain but fail to explain the logic that leads to the answer. The emphasis within GEORGE is on the solution as a process rather than upon the answer as a number.

Let us consider next the data page which is used to define a problem. The first task for the student is to identify the desired quantity. This action is a desirable first step for a student solving a problem with or wihout the aid of a computer. GEORGE understands his domain. If a student identifies "time" as the desired quantity, GEORGE will respond "Unknown quantity." Each quantity needs a label. So, for example, a student may tell GEORGE to find the mass of a hair. Here "hair" is the label and is used by GEORGE in the dimensional analysis to determine which data can be used together. Students must specify a consistent unit. If mol is given as the unit for the mass of a hair, GEORGE will reply "Unit does not agree with quantity."

The simplest kind of problem that GEORGE can work and explain is a metric unit conversion. For example, if GEORGE is asked to find the mass of a hair in milligrams, the student could supply on data line A the mass of that hair in grams. The numerical values may be entered in decimal or exponential notation with the exponential notation appearing with a superscript just as one would write it on paper. When GEORGE is asked to solve this problem he states that the answer was supplied in data line A. This statement is true, but a student working a unit conversion problem needs to have a better explanation. GEORGE displays the worked arithmetic showing the unit conversion. An example of such a display is shown in Figure 4.

As an example of a slightly more difficult problem, consider a question which asks for the density of ethanol in g/mL. A student might provide GEORGE with the mass of a particular sample of ethanol. The mass could be, for example, 25 grams. If a student tells GEORGE to solve the problem with only this piece of information, he will quickly reply that he cannot solve the problem without some information related to the volume of ethanol. If the student then supplies the volume of ethanol, GEORGE explains in plain English how to get the answer: "Solution found by dividing the mass of ethanol by the volume of ethanol to give the density of ethanol." GEORGE works internally with the units g, L, and mol. Thus, after he completes the calculation it is necessary for him to convert the answer to the units requested when the desired quantity was specified. Although GEORGE has provided a textual explanation of the solution process, it may be helpful for the student to see a diagram of how the pieces of information fit together. The diagram for the problem involving the density of ethanol is shown in Figure 5. The symbols A and B in this diagram refer to the lines on the data page and are further identified to the student when the letters A and B are pressed on the keyboard.

A student can save problems on the disk for later use and the disk is initially supplied with a set of complete problems. One example which comes on the disk involves calculating the molarity of aniline in solution. The available data are:

```
                    Calculate Molar Mass
.................................................................
The molar mass is 331.297 g/mol.

Mg(ClO4)2·6H2O

Cl:     1 x   35.453
O :     4 x   15.9994
              ..............
Subunit:      99.4506  x 2 =   198.9012

H :     2 x    1.0079
O :     1 x   15.9994
              ..............
Subunit:      18.0152  x 6 =   108.0912

Mg:     1 x   24.305        =    24.305
                                .........
Total:                          331.297
.................................................................
          Press the space bar to continue.
```

Figure 3.   Screen explaining the calculation of molar mass.
(Reproduced with permission from Ref. 3.   Copyright 1985 COMPress.)

```
                         Network
.................................................................
You supplied the result on data line A.
You need only make the proper conversion
of units:
```

$$.0034 \text{ g} \times \frac{10^3 \text{ mg}}{1 \text{ g}} = 3.40 \text{ mg}$$

```
.................................................................
          Press the space bar to continue.
```

Figure 4.   The display that GEORGE uses to show the solution to
a unit conversion problem.   (Reproduced with permission from Ref.
3.   Copyright 1985 COMPress.)

A. Volume of $C_6H_5NH_2$ 3.00 mL

B. Volume of solution   0.100 L

C. Density of $C_6H_5NH_2$ 1.022 g/mL

D. Molar Mass of $C_6H_5NH_2$

The student does not need to supply the numerical value for aniline
if the formula of aniline is used as a label. GEORGE can do the
calculation of the molar mass from the formula, but the student must
specify that the molar mass is a piece of information to be used in
the problem. When GEORGE is told to solve this problem, he first
finds an intermediate result by multiplying the density of aniline by
the volume of aniline to give the mass of aniline. He then divides
the mass of aniline by the molar mass to give the number of moles of
aniline in solution. Finally, he divides the number of moles of
aniline by the volume of solution to get the desired quantity, the
molarity of aniline. The diagram he draws to explain this solution
is shown in Figure 6. In this figure the letters represent
information found on the data page, while the numbers represent
intermediates calculated while finding the solution. Pressing one of
the letters or numbers shown brings to the top of the screen an
identification of that particular quantity.

   The data page is not the only way to provide GEORGE with
information. Consider for example a simple acid-base titration. On
the data page a student could specify the desired quantity as the
molarity of HCl in acid solution and give as available data the
molarity of NaOH in base solution, the volume of base solution and
the volume of acid solution. This information is insufficient to
support a solution to the problem. The student must also specify the
stoichiometric relation between the number of moles of HCl and the
number of moles of NaOH. An example of a relation page showing this
definition is shown in Figure 7. Once this information is available,
GEORGE can solve the problem, explaining as he works what information
is combined to find an intermediate result and at what point the
relation is used. The use of relations greatly increases the number
of different kinds of problems that GEORGE can handle.

   As an example of a more complex problem, consider one used by
Johnstone (5) in an article discussing problem-solving published
last year in the Journal of Chemical Education: "What volume of
1.0 M hydrochloric acid would react with exactly 10.0 g of chalk?"
The desired quantity is the volume of solution. The available data
are the molarity of HCl in solution, the mass of chalk, and the molar
mass of $CaCO_3$. In addition, two relations are required. One
identifies chalk as calcium carbonate by stating that the mass of
chalk equals the mass of $CaCO_3$. Another gives the stoichiometry
by saying that two times the number of moles of $CaCO_3$ equals the
number of moles of HCl. The diagram showing the solution in this
case occupies several screens; a separate screen is used to show each
application of a relation.

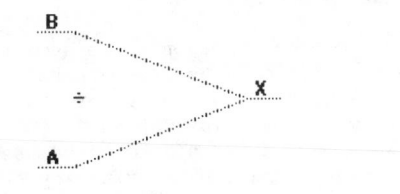

Network

Here is a diagram of how I used the
various pieces of information to reach a
solution.

For details type the relevant letter or
number. ESC displays menu.

Figure 5. Diagram showing that the density of ethanol is obtained
by dividing the mass of ethanol by the volume of ethanol. (Reproduced with permission from Ref. 3. Copyright 1985 COMPress.)

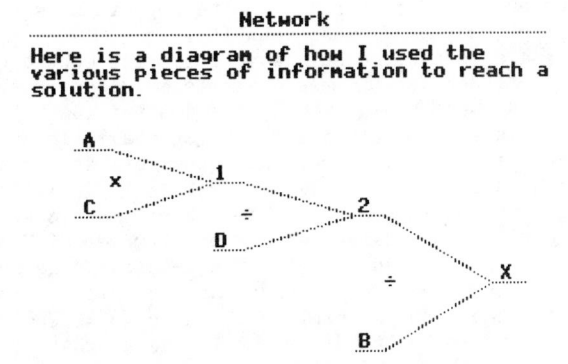

Network

Here is a diagram of how I used the
various pieces of information to reach a
solution.

For details type the relevant letter or
number. ESC displays menu.

Figure 6. Diagram showing how pieces of data are used together to
find the solution to the problem involving the molarity of
aniline. (Reproduced with permission from Ref. 3. Copyright 1985
COMPress.)

Relation 1

Coef.   Quantity

        no. of moles
        of HCl

=   no. of moles
        of NaOH

Press the space bar to continue.

Figure 7. A sample of how a relation can be defined. (Reproduced
with permission from Ref. 3. Copyright 1985 COMPress.)

Extending the Domain

The extension of the domain of application for treating a greater variety of problems requires releasing GEORGE from the constraint of using only dimensional analysis to solve problems. For this extension we are shifting to a system of logic which permits freely the definition of both quantitative and qualitative relations. In addition, the heuristic rules can be represented within the same formalism. The programming language Prolog (6) has been used in creating a prototype of a more capable problem solver. Currently the prototype handles all of the problems that GEORGE can handle, but in addition can deal with gas law problems and physical transformations.

Conclusion

We see three levels of use for GEORGE. At the first level, students benefit from the rigor required even to supply information to GEORGE. Students must identify the desired quantity, label the quantity, and supply the unit. For each piece of available data the student must be just as rigorous. This rigor should help students develop good habits for approaching problems. At the second level, GEORGE acts in much the same way that a roommate might act when helping a fellow student with a problem. We can imagine a roommate saying "Yes, I will show you how to do this problem, but you do the next one by yourself," or "You work the problem first, and then I will show you how I would have done it." In this sense GEORGE acts as a "problem-solving partner" (7). At the third level of use, a student is proficient at working the kinds of problems that GEORGE can solve. GEORGE then becomes a tool as an aid to solving even larger problems. The program frees the student from the tedium of working through the arithmetic and lets the student concentrate on the chemistry of the larger problem. As teachers, we seek ways to help students with those larger problems. A program such as GEORGE is one approach that we could use to help students expand their problem-solving capability to deal with problems that could be tedious indeed with only a calculator as a tool.

Most other software for chemical education is of the kind that an instructor would select for a class. Implicitly or explicitly the instructor says "Student, go use this program." GEORGE may lie in a very different realm in which the students rather than the instructors are the ones who choose the program. The difference could be one small step toward fulfilling the promise that computers hold for far-reaching changes in education.

Literature Cited

1. The Pierce Report, "Computers in Higher Education"; A report of the President's Science Advisory Committee, February 1967.
2. Gelder, J.; Snelling R. "Chem Lab Simulation 2"; High Technology Inc., Tulsa, OK, 1979.
3. Cornelius, R.; Cabrol D.; Cachet C. "GEORGE - A Problem-Solver for Chemistry Students"; COMPress, Wentworth, NH, 1985.
4. TK! Solver, Software Arts, Wellesley, MA, 1983
5. Johnstone, A. H. J. Chem Educ. 1984, 61, 847.

6. Colmerauer, A.; Janoui, H.; Caneghen, M. "Prolog, bases
    theoriques et developpements actuels"; <u>Technique et Science
    Informatique</u> 1983, 4, 271.
7. Cabrol D.; Cachet, C.; Cornelius R., "De nouveaux outils pour
    apprendre: les partenaires de résolution de problèmes: GEORGE et
    sa descendance"; Methodes Informatiques dans l'Enseignement de la
    Chimie, September 17, 1985, Lille, France.

RECEIVED December 17, 1985

HANDLING MOLECULAR STRUCTURES

12

Analogy and Intelligence in Model Building

W. Todd Wipke and Mathew A. Hahn

Department of Chemistry, University of California, Santa Cruz, CA 95064

This paper describes a new approach to building molecular models using methods of expert systems. We are applying symbolic reasoning to a problem previously only approached numerically. The goals of this project were to develop a rapid model builder that mimicked the manual process used by chemists. A further aim was to provide a justification for the model as a chemist would justify a particular conformation. The AIMB algorithm reported here is extremely fast and has a complexity that increases linearly with the number of atoms in the model.

This paper describes the first application of analogy and intelligence to molecular model building. It represents a departure from previous methods, a new approach aimed at rapid, automatic, accurate molecular model building.

Background

Current approaches to molecular model building involve either manual construction or energy minimization. Manual construction of models is performed using programs like COORD.(1) The user specifies internal coordinates (bond length, bond angle, and dihedral angle) and the program converts the internal coordinates into Cartesian coordinates. Programs like COORD are frequently used to generate initial Cartesian coordinates for refinement by other computational programs. The cumbersome data entry of COORD can be simplified by having the program automatically select bond distances and bond angles based on the atom types and bond types specified.(2) Using manual construction from internal coordinates, constructing chains is easy, but constructing closed rings is difficult. One must know exactly the correct set of bond lengths, angles, and dihedral angles to force the chain to close as a perfect ring.

An extension of the atom construction method allows adding groups or predefined templates rather than just atoms. Chemlab II,(3) MOLBUILD,(4) MMSX,(5) Sybyl, and Chemgraph all use this method. Preformed rings can be added as templates, thus avoiding the ring closure difficulty. These methods result directly in a

0097–6156/86/0306–0136$06.00/0

three-dimensional model, but are not automatic. Building a model
that has the right stereochemistry is an additional problem since
specification of stereochemistry in these systems is not simple.
      The first automatic model builder was PRXBLD,(6) a module of the
SECS synthesis planning program(7) and later distributed as a
stand-alone program. PRXBLD takes a two-dimensional structural
diagram with stereochemistry and minimizes the 2-D structure to a 3-D
structure. It has been incorporated in the PROPHET, DENDRAL, and
ADAPT systems and is distributed by Molecular Design Limited. PRXBLD
was the first molecular modeling program to integrate symbolic
intelligence and heuristics with numerical methods. Some of the
heuristics are: 1) Ignore hydrogen atoms, expand carbon to include
the volume that the hydrogens should occupy. 2) Ignore low energy
terms and avoid expressions with large exponents when the structure
is badly distorted. 3) Use four stages of refinement, change stages
by the strain energy per atom. 4) Include a pseudo potential to
force minimization to the stereochemistry specified. 5) Use analogy
to select parameters for force constants that are not available.
PRXBLD never balked for lack of a parameter, thus always gave an
answer. Using PRXBLD, the chemist could, for the first time, obtain
a three-dimensional model by simply drawing the two-dimensional
structural diagram. Although it was the fastest model builder of its
time, certain types of structures still required considerable
computation because PRXBLD used numerical minimization.
      More recently, the SCRIPT program by Cohen(8) also takes a
drawing as input and uses a limited library of ring conformations to
generate approximate geometry for minimization. Dolata, using PROLOG
and predicate calculus methods (analogous to those used in our QED(9)
work) developed an expert system called WIZARD(10) to select a
reasonable set of internal coordinates for an acyclic molecule. From
these internal coordinates Cartesian coordinates are derived which
are then given to MM2 for refinement. WIZARD has not yet handled
cyclic systems.
      There is a need for quick 3-D model generation. Models are
required where knowledge of molecular shape is essential to the
understanding of structure-activity and structure-reactivity
relationships. Most certainly there will be programs in the future
that hypothesize structures; these programs will need rapid model
generation in order to evaluate 3-D constraints. For these
applications, the models must be created automatically, without
interactive intervention. We also envision the vast libraries of
molecular structures stored in chemical data bases will need to be
converted to 3-D geometry libraries in order to use these data bases
in designing new 3-D structures.

Goals of AIMB

The goals of AIMB are listed in Figure 1. We wish to generate models
rapidly and symbolically. Chemists have over the years acquired a
great deal of knowledge about the structure of molecules; we plan to
give AIMB the advantage of access to that knowledge. Similarly,
chemists have used various methods to reason by analogy:
iso-electronic structures, valence electron model, the periodic
table, etc. We wish to incorporate such knowledge in AIMB. The
avoidance of molecular mechanics is a negative goal; perhaps it is

more correct say our goal is to demonstrate that it is possible to
build good models without molecular mechanics. After all, chemists
build very good models manually and mentally without minimization.

```
1. Build 3-D Model RAPIDLY, SYMBOLICALLY
2. Use Knowledge and Analogy like Chemist
3. Avoid molecular mechanics
4. Provide support for results:
      a) Literature precedent
      b) Causes of uncertainty
      c) Quality assessment
      d) Next best models
5. Extendible to conformational search
```

Figure 1. Goals of AIMB.

A significant goal is to have our model builder explain and
justify its answer. Computing methodology now enables us to show how
an answer is derived, the most notable example of explanation
capabilities is MYCIN, a medical diagnosis program.(11) Chemists
have the same need for explanation of computing results that doctors
have. Users of model building programs are frequently non-experts.
They need to know whether the program that someone else wrote is
applicable to their problem, and to what degree the program can
handle their problem. They should receive, for example, indications
of the literature precedent showing that the method can handle that
case well, or an example of an evaluated answer for a "similar"
problem. Every answer in science carries an uncertainty, but current
numerical programs do not reveal this uncertainty to the user. The
model builder should explain what the causes for the uncertainty are
as well as the probable magnitude. It is desirable to obtain an
overall quality assessment of the model and some indication which
parts of the model are most strongly supported and which parts are
most tenuous. Thus the quality assessment must also apply to the
individual components of the model when possible. Another excellent
way to justify the "answer" is by presenting for comparison the "next
best" models.

If we succeed with these objectives, the final goal is to apply
the same methods with minor modification to generate all "reasonable"
conformers of a compound, i.e., to develop a symbolic conformational
search capability.

Components of AIMB

We envisioned in our design of AIMB that we could use a library of
X-ray crystal structures as our "experience" or knowledge of three-
dimensional models of molecular structures. It could, of course, be
a library of computed structures, but we favor having an experimental
basis for our inferences. In our reasoning we need to duplicate the
learned chemical rules of analogy; in this case, those analogies that
preserve the three-dimensional structure of the molecule. A module
to analyze the problem to be solved, and a module to construct the
final three-dimensional model seemed obvious. Graphical input of the
problem and output of the result seemed obligatory. In fact, we even
rely heavily on graphics for observing the operation of the program

for debugging purposes. A model or three-dimensional inference evaluator is also important. Lastly, to construct an explanation, the program needs to extract from its inferences and knowledge base the trail of logic and supporting data for the resulting three-dimensional model. It also needs to retain runners-up models and the supporting data for them.

```
1  Create library of models
2  Enter structural diagram of target
3  Perceive target
4  Target or analogs in library?
5  No, Divide into subproblems, solve each
6  Assemble solved subproblem parts
7  Compute degree of fit of subparts
8  Prepare supporting data
9  Display completed model
```

Figure 2.   AIMB algorithm.

AIMB Procedure

The sequence of events in AIMB is summarized in Figure 2. First the library of experience (known structures) is processed into a form for rapid retrieval. In this paper we used a 2000 compound library from the Cambridge Crystal File. Each of these structures represents an experimental result with the precision of the crystal structure refinement and literature reference to the original paper. These are not "averaged" templates, although nothing in our design precludes using a library of averaged templates or theoretically calculated structures. The library is processed once. New experimental results can be entered incrementally by processing just the new structures.

Model building begins with the chemist drawing the two-dimensional structural diagram of the desired structure with stereochemistry, now a well-established practice.(12),(13) The target is perceived to identify rings,(14) chains, aromaticity, and stereochemistry,(15) but currently not functional groups. The canonical SEMA name is also generated.(16)

AIMB first determines if the target or close analogs are contained in the knowledge base. This access to the knowledge base is instantaneous because we use hash coding methods.(17) If something is found, it selects the most relative experience (known model from the knowledge base) and uses that geometry for the problem. We will discuss how AIMB evaluates "closeness" of analogies in a moment. If there is no close analog to the whole target compound, AIMB uses a "divide and conquer" strategy. The problem is divided into subproblems each of which is treated as a new problem. As in general systems analysis, the best subdivisions of a system are those that minimize connections (interactions) between subsystems. In our case we select ring assemblies(14) and chains as subdivisions. When the component is carved out, we also retain information about the context in which the component resides.

These components or subproblems are prioritized to solve the largest, most rigid ring assemblies first. This will force an early fail, if fail we must. AIMB seeks the closest analogies to this subproblem present in the knowledge base. First it looks for components in an identical environment with identical structure, next with atom type analogies, then with different environment and different atom types, relaxing the matching constraints until one or more analogs are found.

The solved subparts are then assembled by the constructor. The degree of fit is computed and retained for later evaluation. The process continues until all subproblems are solved. Then the explanation module prepares the supporting explanation for the model from the reasoning trail and evaluation results. The final model is then displayed and the explanation presented.

Our system architecture is illustrated in Figure 3 with the input screen shown in Figure 4. Graphical input is handled primarily by the PS300 system with occasional messages passing between the PS300 and the VAX. The small window is a rotatable three-dimensional view of the developing model.

We will take 7-benzyl-2-norbornanone as an example for the model building process. Figure 4 shows the compound as it has been drawn in 2-dimensions by the user. The program does not find an exact match or an analog match for the compound. Therefore, the compound is divided into subcomponents with the norbornyl group becoming the first subproblem. We can in Figure 5 observe AIMB evaluating the relevance of a bromocamphor compound as an analog for the norbornyl group. Debug switches have been set to show internal scoring and conclusions. This structure ends up scoring the best.

Now we turn to the scoring function. The function for evaluating the closeness of analogy involves a sensitivity parameter for the atom class, $\underline{s}$, times the dissimilarity parameter, $\underline{d}$:

$$D = \sum_{I}^{NA} \sum_{K}^{M} \left[s_K(A_I) * D_K(A_I, A_J') \right]$$

$$\text{WHERE} \quad A \in \text{TARGET,} \quad A' \in \text{ANALOG,} \quad J = \text{MAP}(I)$$

Summation occurs over all atoms and all attributes. There are three atom classes: normal atoms, origin atoms (where components join) and dummy atoms (partial context of neighboring component environment). Attributes include atom type, bond type, stereochemistry, etc. For atom type dissimilarity, there is a series of atom analogy classes of varying degree of closeness. For example, if atom types are equal, $\underline{d}$ = 0, if a_i and a_j belong to {Cl, Br, I} $\underline{d}$ = 3, if a_i and a_j belong to {F, Br, Cl, I} $\underline{d}$ = 5, and if a_i and a_j are not members of any analogy classes then $\underline{d}$ = 10. Aromatic bonds are considered analogous to double bonds. Since we are still exploring the analogy heuristics, we should take these only as examples of the approach. Returning to our example, Figure 6 shows the scores computed for two analogous bicyclo[2.2.1]heptanes. The bromo compound at 1500 is better than the dimer at 1540 (lower number indicates smaller dissimilarity). Note that AIMB recognizes enantiomers and is able to use a reflection of the model. In this case the carbonyl is on the wrong side of the molecule.

The acyclic component is found in two compounds (Figure 6) one having an oxygen in place of the central carbon (540), the other having a carbon (220). In both cases the compounds have an aromatic

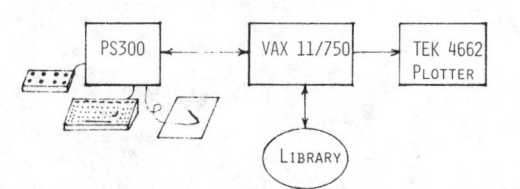

Figure 3.   Hardware configuration used with AIMB.

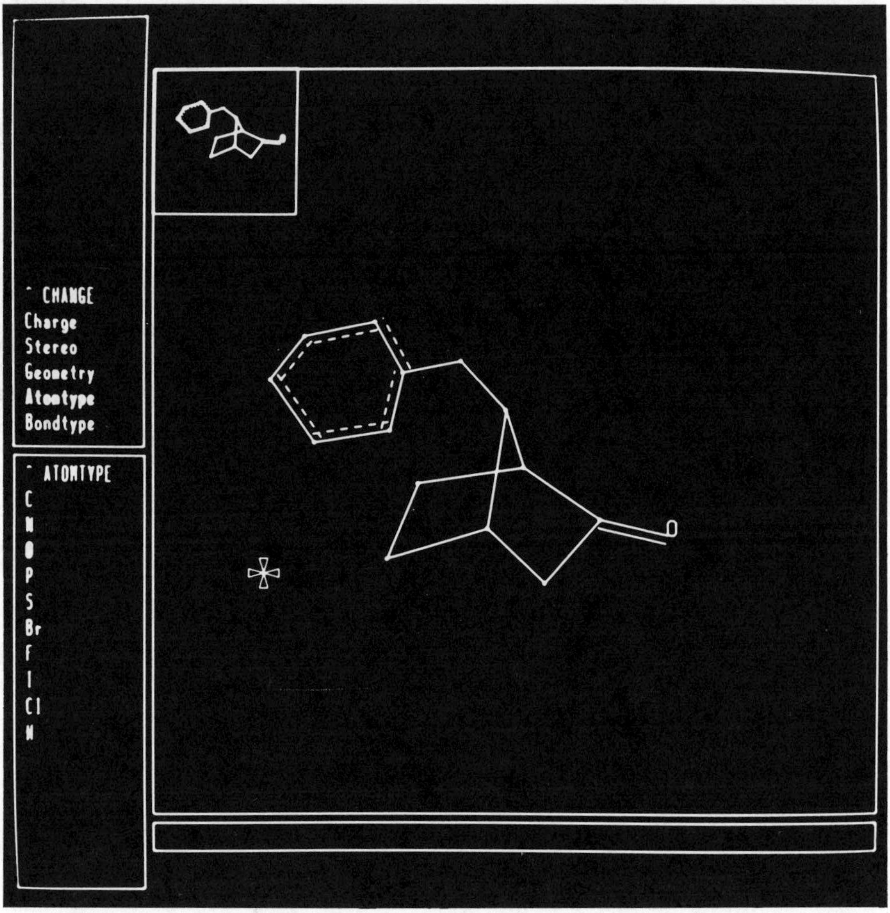

Figure 4.   Input on E&S PS330 of molecule to be modeled by AIMB.

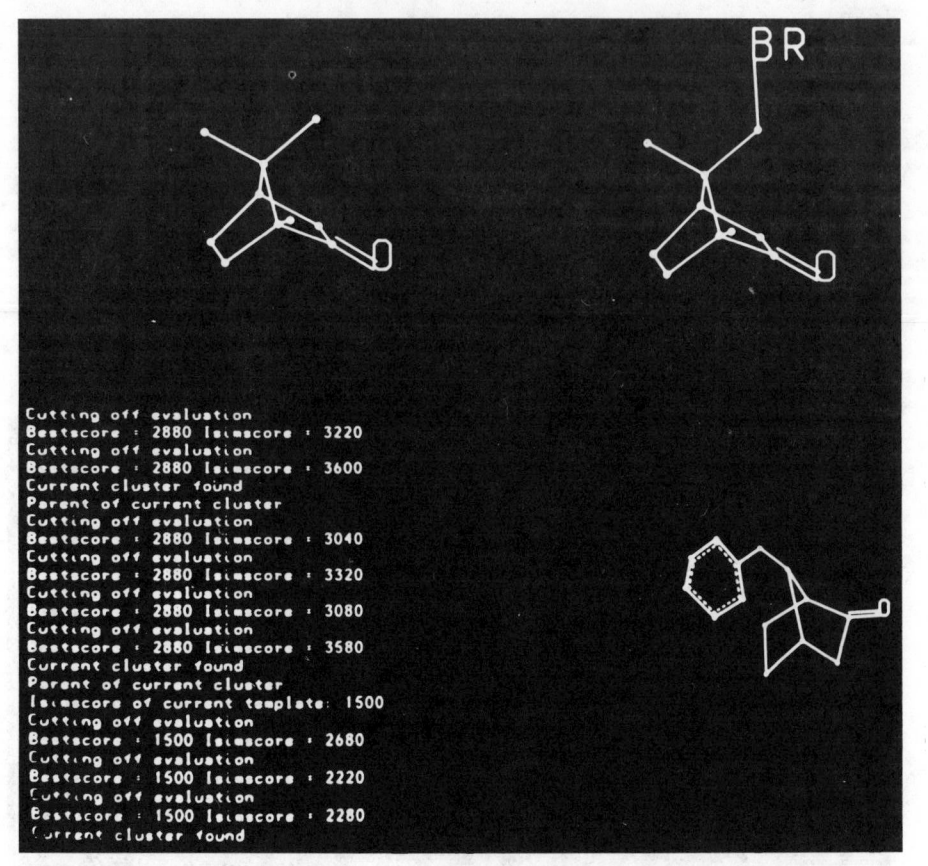

Figure 5.   AIMB evaluates relevance of bromocamphor compound to problem.

Figure 6.   Dissimilarity scores for analogies relevant to problem.   The lower score is the better analogy.

ring joined by one atom chain to an aliphatic five-membered ring system. Finally for our example, AIMB assembles the components in three-dimensions. The explanation module is under development, but the reasoning trace and quality of model evaluation is implemented. In this example the fit of the phenyl group to the chain is within 0.01 $\overset{\circ}{A}$, the fit of the norbornane skeleton to the chain is 0.11 $\overset{\circ}{A}$ (this reflects the fact that the analogy for the chain was connected to a less constrained five-membered ring). The actual accuracy of the resulting model is much better than these values indicate because the discrepancy in fit is only an indication of the similarity of the analogy and is not incorporated into the model). The chain analogy comes from Cryst. Struct. Commun. **8**, 553 (1979); norbornane skeleton from Acta. Crystallogr. Sect B **31**, 903 (1975). The reported precision of the experiment for the crystal structures is also available.

An ORTEP plot of the AIMB-built model is shown in Figure 7. Allinger's molecular mechanics program MM2(18) was then used to refine the model constructed by AIMB. Superposition of the AIMB model with that refined by MM2 is shown in Figure 8. MM2 did not change the dihedral angles of the benzyl group from those proposed by AIMB. Let's recall that AIMB has no internal knowledge of structural chemistry, but only knows how to use analogies and a knowledge base of known models. AIMB does not currently know about any kind of non-bonded interatomic interactions, yet AIMB built a correct model of the example target compound because the knowledge of interactions and how to minimize them is embedded in the knowledge base of known models. Thus AIMB built a minimum energy model (verified separately by MM2) yet AIMB did this symbolically by reasoning rather than minimization.

Table I. Speed of building model of 7-benzyl-2-norbornanone

Method	Time (seconds)
AIMB	40
Human being	118
PRXBLD	644
MM2	4436

Several significant points can now be made. First as Table I shows, on the VAX 11/750, AIMB took only 40 seconds to construct the model. A chemist took 145 seconds to assemble a Fieser model of the compound and when completed, the chemist did not know the dihedral angles of the benzyl group. PRXBLD took 644 seconds to build the model. This molecule is difficult for PRXBLD because adjustment of the chain angles requires movement of the two large groups of atoms, but PRXBLD does not recognize that the groups can be moved as a unit. Allinger's MM2 took 4436 seconds to converge to the default CHEMLAB criteria and that was when given a very good input structure (PRXBLD model).

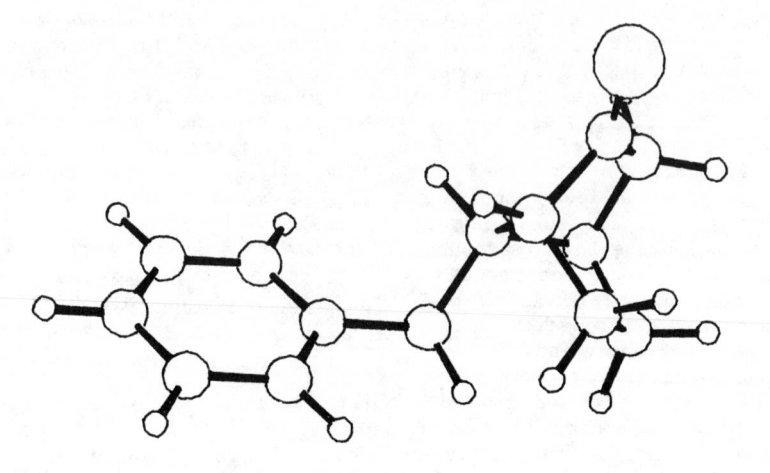

Figure 7.   ORTEP plot of final AIMB model of target molecule.

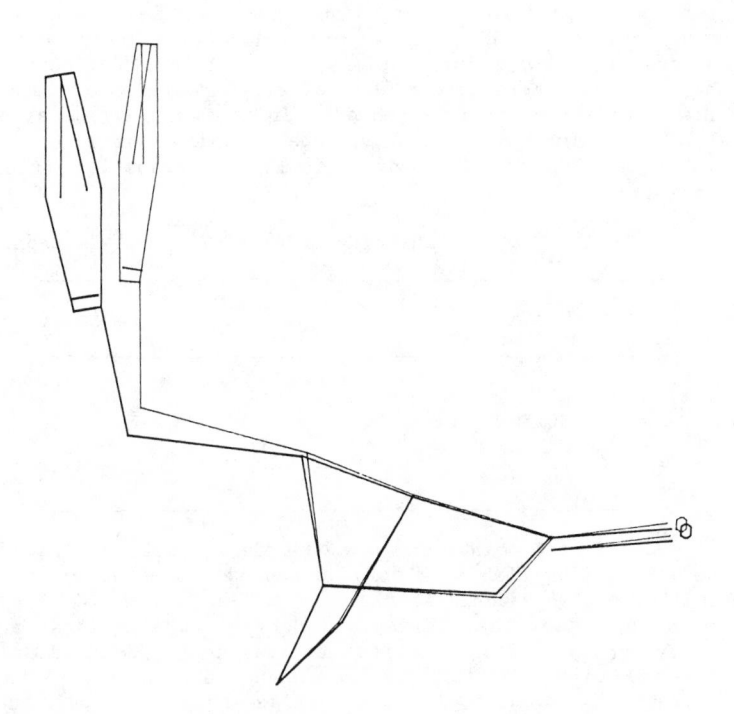

Figure 8.   Superposition of AIMB model and result of MM2
refinement.

Conclusion

We have shown that analogy and intelligence applied to model building leads to a fast, accurate algorithm. Using prior knowledge is efficient. This method is applicable to complex functionality where the forces or interactions may not be well understood, e.g., inorganics and organometallics, but where there are many known crystal structures. While we based our knowledge on crystal data, one could also use computed structures separately or in conjunction with crystal data. The process we described is easy for any chemist to understand. AIMB does not involve force fields or complicated mathematics. The models AIMB generates are supported by experimental data and highly justified. Finally, while energy minimization methods increase in time exponentially as the number of atoms in the problem increase, the AIMB algorithm increase in time is <u>linear</u> with increasing numbers of atoms.

Although AIMB is a working prototype, we have many questions remaining to be answered. We would like to explore the detailed heuristics and study the effect of changing these on the final models constructed. We are interested in seeing how the size of the knowledge base is related to the quality of results and speed of operation. Finally, we would like to explore its application in areas where conventional methods simply can not be used.

Acknowledgments

This work was supported by PRXBLD users and in part by a Faculty Research Grant from the University of California.

Literature Cited

1. Mueller, K. "COORD: Interconversion of Cartesian and Internal Coordinates (QCPE 419)". QCPE Bull. **1981**, 1, 37.
2. Program MDCORD does this. Personal Communication, Douglas Hounshell.
3. Potenzone, R., Jr.; Cauicchi, E.; Hopfinger, A. J.; Weintraub, H. J. R. "Molecular Mechanics and the CAMSEQ Processor". Computers and Chemistry **1977**, 1, 187.
4. Liljefors, T. "MOLBUILD: An Interactive Computer Graphics Interface to Molecular Mechanics". J. Mol. Graphics **1983**, 1, (4), 111.
5. Humbolt, C. "MMS-X Modeling System User's Guide"; Technical Memo 7, Washington Univ., St. Louis MO, Jan. 1980.
6. Wipke, W. T.; Verbalis, J.; Dyott, T., "Three-Dimensional Interactive Model Building", Presented at the 162nd National Meeting of the American Chemical Society, Los Angeles, August 1972.
7. Wipke, W. T. "Computer-Assisted Three-Dimensional Synthetic Analysis". In Computer Representation and Manipulation of Chemical Information; Wipke, W. T.; Heller, S. R.; Feldmann, R. J.; Hyde, E., Eds.; John Wiley and Sons, Inc.: 1974, pp 147-174.
8. Cohen, N. C.; Colin, P.; Lemoine, G. "SCRIPT: Interactive Molecular Geometrical Treatments on the Basis of Computer-Drawn Chemical Formula". Tetrahedron **1981**, 37, 1711-1721.
9. Dolata, D. P. QED: Automated Inference in Planning Organic

Synthesis, PhD dissertation, University of California, Santa Cruz 1984.

10. Dolata, D. P., "WIZARD--Artificial Intelligence in Conformational Analysis", Presented at the Drug Information Workshop, Feb. 4-6, 1985.

11. Edward Hance Shortliffe "Computer Based Medical Consultations: MYCIN"; American Elsevier, New York: 1976.

12. Corey, E. J.; Wipke, W. T. "Computer-Assisted Design of Complex Molecular Syntheses". Science 1969, 166, 178.

13. Corey, E. J.; Wipke, W. T.; Cramer, R. D.; Hower, W. J. J. Am. Chem. Soc. 1972, 94, 421.

14. Wipke, W. T.; Dyott, T. M. "Use of Ring Assemblies in a Ring Perception Algorithm". J. Chem. Inf. and Comput. Sci. 1975, 15, 140.

15. Wipke, W. T.; Dyott, T. M. "Simulation and Evaluation of Chemical Synthesis. Computer Representation of Stereochemistry". J. Am. Chem. Soc. 1974, 96, 4825, 4834.

16. Wipke, W. T.; Dyott, T. M. "Stereochemically Unique Naming Algorithm". J. Am. Chem. Soc. 1974, 96, 4834.

17. Wipke, W. T.; Krishnan, S.; Ouchi, G. "Hash Functions for Rapid Storage and Retrieval of Chemical Structures". J. Chem. Inf. and Comput. Sci. 1978, 18, 32.

18. Burkert, U.; Allinger, N. L. "Molecular Mechanics"; American Chemical Society: ACS Monograph, Vol. 177, 1982.

RECEIVED January 24, 1986

Computer-Assisted Drug Receptor Mapping Analysis

Teri E. Klein[1], Conrad Huang[1], Thomas E. Ferrin[1], Robert Langridge[1], and Corwin Hansch[2]

[1] Computer Graphics Laboratory, University of California, San Francisco, CA 94143
[2] Department of Chemistry, Pomona College, Claremont, CA 91711

KARMA is an interactive computer assisted drug design tool that incorporates quantitative structure-activity relationships (QSAR), conformational analysis, and three-dimensional graphics. It represents a novel approach to receptor mapping analysis when the x-ray structure of the receptor site is not known. KARMA utilizes real time interactive three-dimensional color computer graphics combined with numerical computations and symbolic manipulation techniques from the field of artificial intelligence.

Many problems in chemistry may benefit from developments in the field of Artificial Intelligence (AI), particularly the area now known as knowledge engineering. Knowledge can be described as that which includes both empirical material and that "which is derived by inference or interpretation". (1) It may consist of descriptions, relationships, and procedures in some domain of interest. (2) We are now incorporating methods from knowledge engineering research in computer assisted drug design.

Molecular modeling with interactive color computer graphics in real time is a powerful method for studying molecular structures and their interactions. Display and manipulation of computer generated skeletal and surface models provide efficient methods for the chemist to examine steric interactions of many ligands with the binding sites in their receptors. We have combined x-ray crystallographic results, quantitative structure-activity relationships (QSAR), and interactive three-dimensional graphics in earlier attempts to design better ligands for enzyme binding. (3,4) We are applying knowledge engineering techniques provided by the software KEE (Knowledge Engineering Environment (5)) to the development of rational drug design methods without having x-ray crystallographic results in hand.

Our integrated system, KARMA, KEE Assisted Receptor Mapping Analysis, uses knowledge sources, including QSAR and conformational analysis, in a rule-based system to create an annotated visualization of the receptor site. This is then used in an iterative manner to guide the investigator in generating rules, hypotheses, and new candidate structures for drug design. This approach to receptor mapping and drug design differs from the traditional approach used by chemists in two significant ways. Classically, in computerized drug design, one superimposes a set of structurally related molecules (congeners) so that their bioactive functional groups coincide, yielding a pharmacophore. A surface is then derived based on the composite molecule supposedly yielding a complementary shape of the receptor. (6) This approach has met with limited success because compounds that act as substrates or inhibitors of certain receptors do not necessarily bind similarly. It is our belief that the commonality of the binding mode must be established. The other shortcoming of the traditional approach is that it provides little information on the qualitative character of the enzyme surface. The classical lock and key concept of ligand-receptor emphasizes structural geometry and may neglect the importance of interactions such as hydrophobicity. Processing of the binding data using QSAR prior to receptor mapping analysis yields information not only about the hydrophobic and polar nature of the surface model, but also about the steric and electronic properties of the data. (7)

System Design

KARMA is a set of programs residing on several machines connected by a high bandwidth network (see Figures 1 and 2). The main program resides on the Lisp machine and controls all processing. The controlling program on the Lisp machine is implemented on top of KEE which embodies many knowledge engineering techniques. KEE provides a set of software tools that allows for very rapid software prototyping, evaluation, debugging, and modification. Specifically, KARMA takes advantage of KEE's capabilities that include frame based knowledge representation with inheritance, a rule-based inference system, a graphical interface for debugging and displaying knowledge bases, and a flexible interface that allows for the integration of outside methods. (5)

Input to the controlling program consists of congener sets and their related QSAR equations. A satellite program, based on the Pomona MedChem Software SMILES (Simplified Molecular Input Line Editor System) is used for input of the structures. (8) SMILES creats a unique identifying code for each chemical structure which is useful for searching for structures and physiochemical parameters, and minimizing duplication of structural information. These structures are passed to satellite programs, including distance geometry (9) and energy minimization (10) , to generate multiple conformations that are then displayed so that users may select those of interest. These structures, which constitute the basis set, are used to define the receptor model.

The receptor model is represented graphically by a set of surfaces. These surfaces are defined by a set of control points which are calculated on the compute server. Control points, which are based on minimized structures, are then manipulated by KARMA's rules system. These rules provide detail to the receptor surface model. During this process, KEE provides a graphical interface showing which rules and derivations are being accepted as true. The user can also interact with KARMA's rule system during this time. The surface model is displayed using the control points to form bicubic patches on the graphics workstation. The user can then manipulate the surface as well as modify the structure. These modifications are sent back to the controlling program for refinement by the rules. This iterative process continues until the user is satisfied with KARMA's results.

As seen in Figure 1, our hardware is connected by an Ethernet. (11) The control server is a Symbolics 3600 Lisp Machine and the compute server is a DEC VAX 8600. The three dimensional graphics workstations include the Silicon Graphics IRIS 2400 and the Evans and Sutherland PS350. Electronic communication with collaborating scientists at other institutions is available through the VAX 750 via several networks including the ARPAnet and CSnet.

System Implementation

Input to the controlling program is achieved through a series of "pop-up" menus in the Karma Window (see Figure 3a). For example, if the user is interested in entering a set of congeners, the user would select the *molecule* editor. KARMA will then display the molecule editor layout in the current window. Users can then enter the chemical structures selecting *structure* from the molecule editor menu (see Figure 3b). Structures are currently entered using the tree structure of SMILES (see Figure 3c). (The molecule editor will be expanded to allow for graphical input in the future.) KARMA then displays the two-dimensional structure for user verification (see Figure 3d). Coordinates for the three-dimensional structures are saved in a knowledge base in KEE. The three-dimensional structures are based on x-ray crystallographic data, standard bond angles, and bond lengths. All congener data, including physiochemical parameters such as log P or MR (calculated or experimental), can easily be entered and revised in the molecule editor (see Figure 4).

Three-dimensional coordinates for the congener set are passed to the distance geometry and minimization programs. These satellite programs provide efficient methods for searching conformational space. Distance geometry programs includes subroutines for controlling ring planarity of aromatic rings and orientation of the molecules based on a common group of atoms. (12)

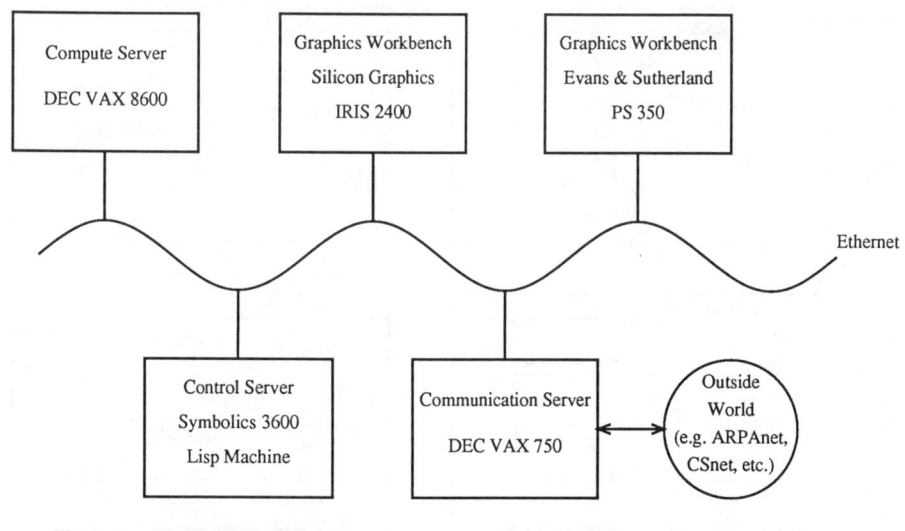

Figure 1. Hardware Configuration.
Copyright © 1985, Regents of the University of California/Computer Graphics Lab.

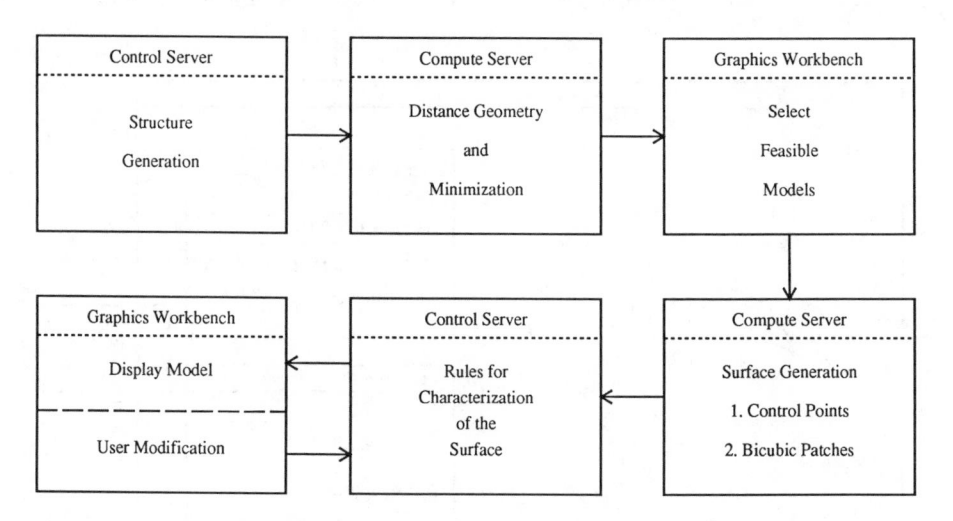

Figure 2. System Architecture.
Copyright © 1985, Regents of the University of California/Computer Graphics Lab.

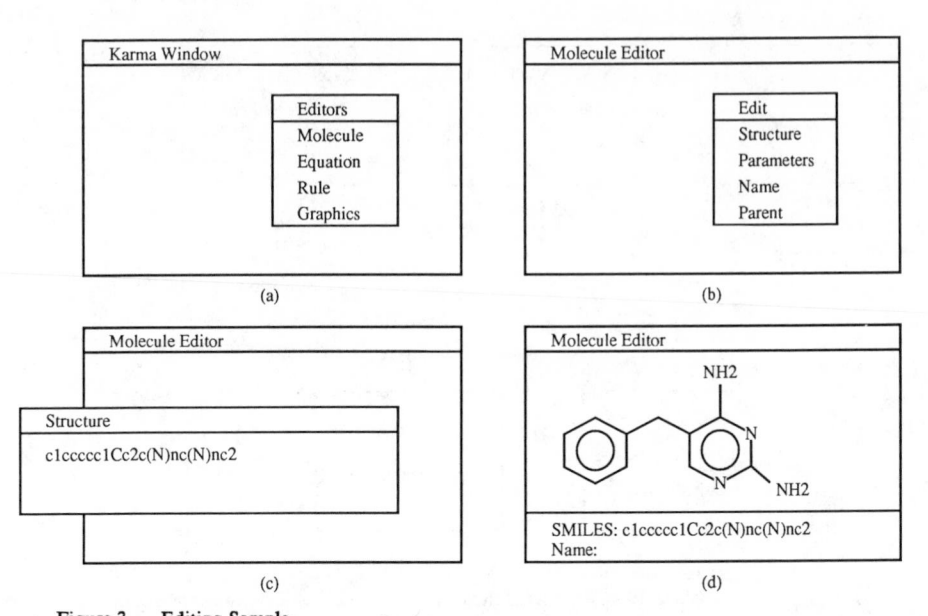

Figure 3.   Editing Sample.
Copyright © 1985, Regents of the University of California/Computer Graphics Lab.

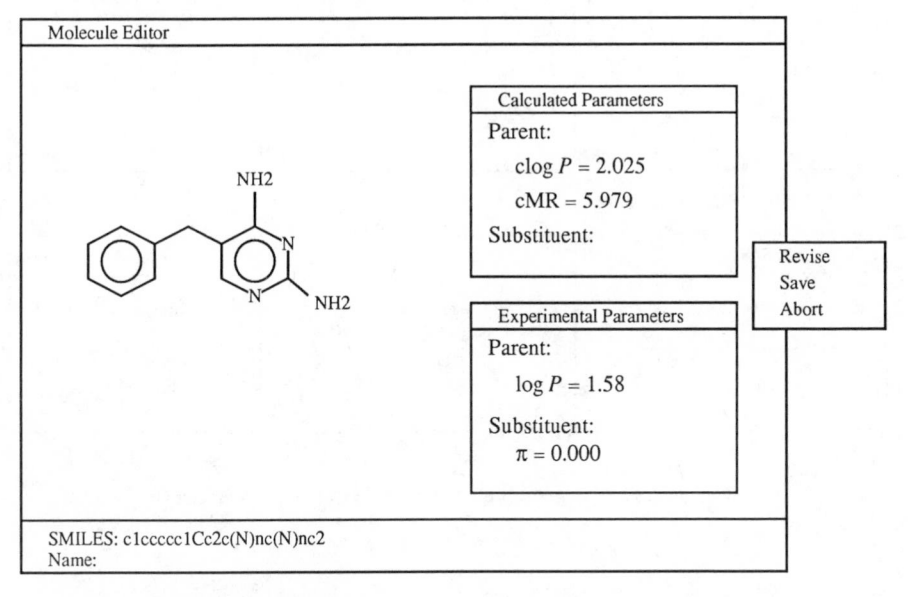

Figure 4.   Editing Sample (continued).
Copyright © 1985, Regents of the University of California/Computer Graphics Lab.

The output of the distance geometry and minimization programs is passed to the graphics program EDGE (Easy Distance Geometry Editor). The structures are displayed three-dimensionally so users may select structures to represent conformational space. Models are easily selected pointing at the desired structure (see Figure 5). X, Y, and Z rotations and translations, depth cueing, color, and labeling have been incorporated in EDGE. EDGE also provides a RMS matching routine for N arbitrary atoms designated by the user. The selected models are then used for surface generation.

Surface generation is based on a set of points derived from the outcome of distance geometry programs applied to the basis set of structures. The basis set of points, P, is defined as:

$$P = \bigcup_i P_i - \bigcup_{i,j} g\,(P_i P_j)$$

where P_i is the uniformly distributed set of points over a sphere corresponding to atom i, and, $g\,(P_i P_j)$ is the overlap of the two sets of points. The density of points/angstrom^2 can be arbitrarily set by the user. If the density is relatively high, a *large* number of bicubic patches with small area are generated; to address each bicubic patch at a high density would be time-consuming and difficult at best. If the density of points is low, the patches become too large and don't yield enough detailed information about the surface model.

The control points are defined by the basis set of points P. These control points define the parametric bicubic patches which form the surface model. Advantages of the parametric bicubic surface include continuity of position, slope, and curvature at the points where two patches meet. All the points on a bicubic surface are defined by cubic equations of two parameters s and t, where s and t vary from 0 to 1. The equation for $x(s,t)$ is:

$$x(s,t) = a_{11}s^3t^3 + a_{12}s^2t^2 + a_{13}s^3t + a_{14}s^3$$

$$+\, a_{21}s^2t^3 + a_{22}s^2t^2 + a_{23}s^2t + a_{24}s^2$$

$$+\, a_{31}st^3 + a_{32}st^2 + a_{33}st + a_{34}s$$

$$+\, a_{41}t^3 + a_{42}t^2 + a_{43}t + a_{44}$$

Equations for y and z are similar. (13) Either cardinal spline or B-spline bicubic patches can be used as they differ only by the starting coefficients. (14) Overlapping sets of control points allow for the joining of patches. Sixteen points define a bicubic patch. To determine which points define which patches, an initial triangle is formed from three nearest neighbors. The next triangle shares one side of the initial triangle and is connected to its next nearest neighbor. This process is continued iteratively until all points are accounted for. The internal edge of two triangles is then dropped to form a quadrilateral. Each internal edge is used only once. Nine quadrilaterals define a single patch. These patches are combined to form the surface model and are manipulated by both KARMA's rule system and the investigator at the graphics station.

System Core

The information contained in KARMA's knowledge bases is based upon quantitative structure-activity relationships (QSAR), kinetic data, and structural chemistry. The combination of QSAR and kinetic data allows for the study of enzyme-ligand interactions. The Hansch approach to QSAR, based on a set of congeners, states:

Biological Activity = f(physiochemical parameters)

Physiochemical parameters are used to model the effects of structural changes on the electronic, hydrophobic, and steric effects for organic molecules. (15) Examples of physiochemical parameters include, among others:

σ, an electronic constant based on the Hammett equation for the ionization of substituted benzoic acids;

π, the hydrophobic parameter for a chemical substituent based on the octanol-water partition coefficient log P;

MR, the molar refractivity, which parameterizes polarizability and steric effects; and

Verloop's parameters, which are steric substituent values calculated from bond angles and distances.

Using multivariable linear regression, a set of equations can be derived from the parameterized data. Statistical analysis yields the "best" equations to fit the empirical data. This mathematical model forms a basis to correlate the biological activity to the chemical structures.

KARMA describes the interactions for enzyme-ligand binding using QSAR equations and parameters, and the structural information of the congener data. These interactions, with illustrative examples, are shown below:

Interaction	Example
enzyme $\rightarrow$ specific enzyme	(DHFR* $\rightarrow$ chicken DHFR)
congener $\rightarrow$ specific enzyme	(benzylpyrimidines $\rightarrow$ chicken DHFR)
congener $\rightarrow$ specific congener	(inhibitors** $\rightarrow$ benzylpyrimidines)
substituents $\rightarrow$ specific congener	(3,4,5 OMe $\rightarrow$ benzylpyrimidines)
equations $\rightarrow$ congeners	(equation $\rightarrow$ benzylpyrimidines)
variable $\rightarrow$ substituents	(4-Cl $\rightarrow$ 4-Br)
specific enzyme $\rightarrow$ specific enzyme	(chicken DHFR $\rightarrow$ *L. casei* DHFR)

*DHFR - Dihydrofolate Reductase
**inhibitors - triazines, benzylpyrimidines, etc.

The data used for the above interactions is contained in KARMA's knowledge bases, ChemData and KarmaData. These knowledge bases contain information about classes of objects or about the objects themselves. Objects and their attributes are represented as individual "knowledge frames" which are linked together to form a hierarchal structure. Consistency among the objects in both knowledge bases is obtained through inheritance rules.

ChemData is one of several data bases available in KARMA. This data base contains chemical information pertaining to chemical elements and molecular substituents. Elemental data includes atom type, atomic radii, hybridization, molecular weight, etc. Substituent data consists of unique identifying codes, physiochemical parameter data, and x-ray crystallographic data. For each substituent, where known, there are values for the hydrophobic parameter, *i.e.*, π, an electronic parameter, *i.e.*, σ, and a steric parameter, *i.e.*, MR. The associated x-ray crystallographic data is used for building the small molecules in the congener set. This data is also used for specifying constraints used in the distance geometry calculations.

KarmaData contains information which the user enters, *e.g.*, QSAR equations, congener set, as well as information about previously studied enzyme-ligand binding complexes. KarmaData contains several classes and subclasses. For example, in KarmaData, there is a class called *proteins*, a subclass in *proteins* called *dehydrogenase*, a particular member of *dehydrogenase* called *DHFR*, and a specific instance of *DHFR* called *chicken* (*vide infra*). Chicken DHFR contains those attributes which are specific to itself, and inherits properties from units *DHFR*, *dehydrogenase*, and *proteins*.

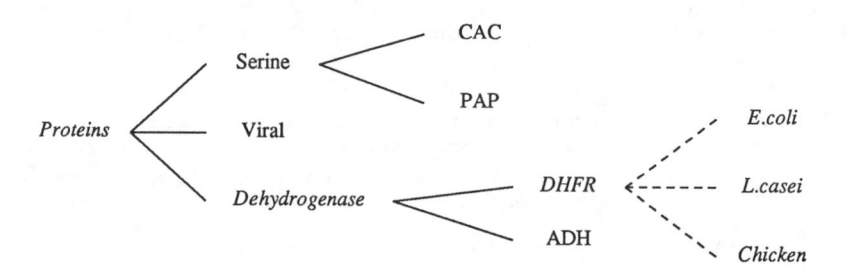

KEE provides many different mechanisms for inheritance. KEE has the ability to constrain the type and number of values assigned to attributes for consistency and description in the knowledge base. (5)

Currently, KARMA's rules are formulated in an if-then format. A rule may have multiple conditions, conclusions, and actions. KARMA takes advantage of both the forward and backward chainers for derivation of the three-dimensional receptor model. For example, two types of rules, generic and specific, can be defined empirically from the results of QSAR as well as from molecular structure.

Generic rules are based on the QSAR equations and their coefficients. Forward chaining using these rules yields basic characteristics for the receptor site model. For instance, an abstracted generic rule may take the form:

If the coefficient of the hydrophobic parameter is approximately equal to one, then expect complete desolvation about substituent X of the ligand.

This rule was derived empirically from some recent work on several species of alcohol dehydrogenase (ADH). (16) The following equations were found:

Compounds	Enzyme	Equations
	Horse ADH	$\log 1/K_i = 0.89 \log P + 3.56$ $n = 11, r = 0.960, s = 0.197$
	Horse ADH	$\log 1/K_i = 0.98 \log P - 0.83\, \sigma^* + 3.69$ $n = 14, r = 0.937, s = 0.280$
	Human ADH	$\log 1/K_i = 0.87 \log P - 2.06\sigma_{meta} + 4.60$ $n = 13, r = 0.977, s = 0.303$
	Rat ADH	$\log 1/K_i = 1.22 \log P - 1.80\, \sigma_{meta} + 4.87$ $n = 14, r = 0.985, s = 0.316$
	Horse ADH	$\log 1/K_i = 0.96 \log P + 5.70$ $n = 5, r = 0.990, s = 0.207$

where X is the substituent and $\log P$ is based on the octanol-water partition.

The average of the coefficients of the hydrophobic term is approximately equal to one (average = 0.97) suggesting complete desolvation about substituent X. Figure 6 shows complete desolvation by the enzyme ADH (hydrophobic space - red; polar space - blue) around substituent X of the pyrazole (green).

Another example of a rule dealing with hydrophobicity may take the form:

If the coefficient of the hydrophobic parameter is greater than 0.5 and less than 1.0, then expect a concave surface about substituent X of the ligand.

This type of rule is empirically based on the enzyme-ligand binding such as that of carbonic anhydrase c (CAC) and sulfonamides. (4) The following equation was found:

Compound   Equation

$$\log K = 1.55 \, \sigma + 0.64 \log P - 2.07 I_1 - 3.28 I_2 + 6.94$$
$$n = 29, \, r = 0.993, \, s = 0.190$$

Figure 7 shows how the solvent accessible surface of the enzyme CAC (hydrophobic space - red; polar space - blue) is slightly concave about the substituent X of the sulfonamide (green). Similar rules exist for the coefficients which describe other aspects of hydrophobicity, as well as polar space, which help define the basic shape, *i.e.*, cleft or hole, of the surface receptor model.

Specific rules are based on the attributes of congeners, including the physiochemical parameters used to determine the QSAR equation, the biological activity, and the molecular structure. Backward chaining, using these rules with specific instances of substituents, yields detailed shape and character for the receptor model. For instance, an abstracted specific rule may take the form:

If the biological activity of compound Y is less in enzyme A than that of related enzyme B, expect possible steric hindrance about substituent X.

One possible interpretation of this type of rule is the enzyme ligand binding of trimethoprim with bacterial DHFR and chicken liver DHFR. (17,18)

DHFR Species	Binding Affinity (log $1/K_i$)
L. casei	8.87
E. coli	6.88
chicken	3.98

This data shows a noticeable drop in binding affinity for trimethoprim and chicken liver DHFR. Figure 8 illustrates steric interaction between the *5-OMe* of trimethoprim (green) with the sidechain of *Tyr 31* of native chicken liver DHFR (red). There is no steric interaction seen between the *5-OMe* of trimethoprim (green) and the sidechain of *Phe 30* of *L. casei* DHFR (red). (Right view: chicken liver DHFR; Left View: *L. casei* DHFR) It is known from x-ray crystallographic results that the sidechain of *Tyr 31* of chicken liver DHFR rotates to accommodate trimethoprim. (18)

A specific rule can also be based upon comparisons of bond lengths and van der Waals radii, and biological activities. For instance,

If the biological activity of substituent X_1 is less than the biological activity of substituent X_2, and, X_2 is atomically larger than X_1, then expect possible steric hindrance with the receptor wall about X_2, provided that other factors are equal.

This rule can be exemplified by two compounds that differ by the type of the substituent, *i.e.*, a chlorine and a bromine atom. If the binding affinity for the bromine compound was lower (and

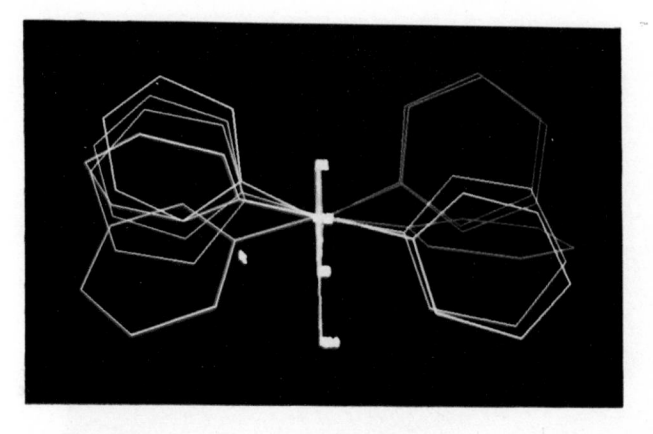

Figure 5.   Output of EDGE.
Copyright © 1985, Regents of the University of California/Computer Graphics Lab.

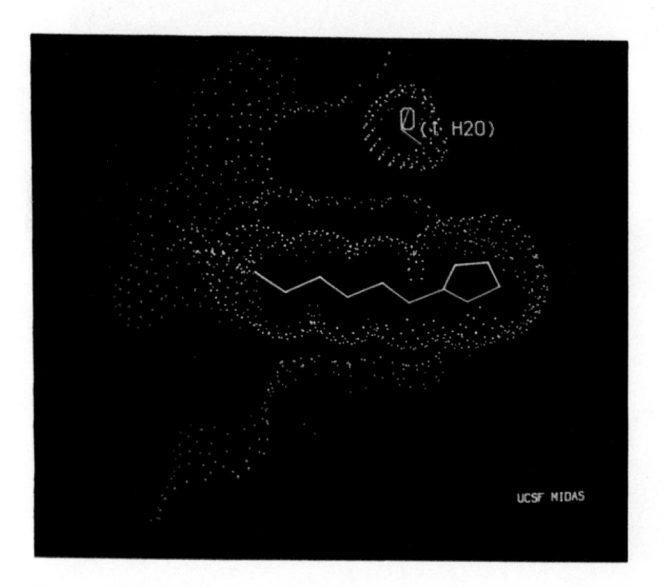

Figure 6.   Enzyme-Ligand Complex for Alcohol Dehydrogenase and a substituted pyrazole.
Copyright © 1985, Regents of the University of California/Computer Graphics Lab.

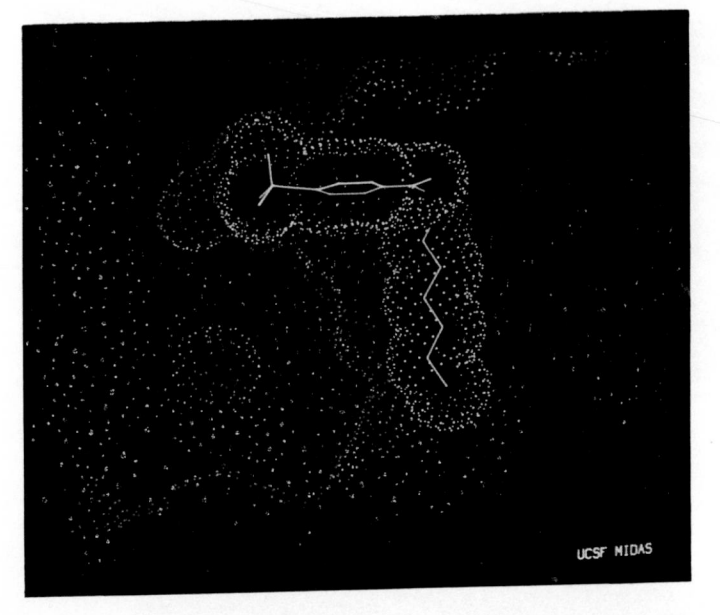

Figure 7.   Enzyme-Ligand Complex for Carbonic Anhydrase C and a substituted sulfonamide.
Copyright © 1985, Regents of the University of California/Computer Graphics Lab.

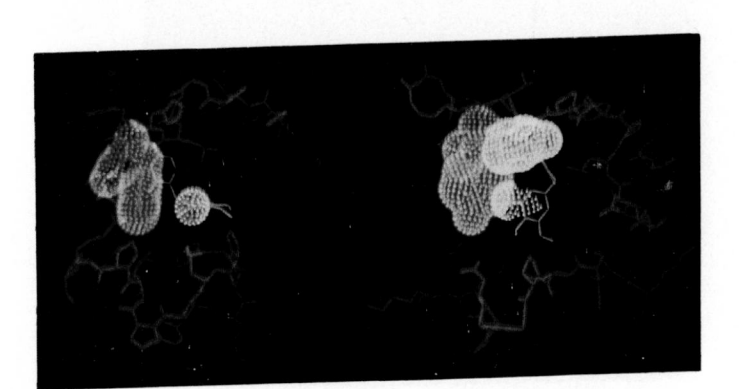

Figure 8.   Enzyme-Ligand Complex for Dihydrofolate Reductase and trimethoprim.
(*L. casei*: left, chicken liver: right).
Copyright © 1985, Regents of the University of California/Computer Graphics Lab.

possibly even lower for the iodine compound), it would suggest that the wall of the receptor model is contacted by the ligand at the bond distance of the chlorine atom and its related van der Waals radius. Therefore, one could assume that the larger bromine atom represents an intrusion into the receptor wall.

The above examples used to illustrate the specific rules for backward chaining are similar to other attempts at receptor mapping. (6) However, these other methods do not account for interactions that may be based on a combination of effects such as hydrophobicity and ligand potency. For instance, a rule that might apply to a compound with a substituted phenyl ring may take the form (19)

If a *meta* disubstituted compound is symmetrical, and the biological activities differ between hydrophobic and polar substituents, then expect possible ring rotation to maximize hydrophobic and polar interactions between the ring substituents and the hydrophobic and polar surface.

Many rules can be derived from the molecular structures and biological activities as seen from the above examples, which add both shape and character to the surface model.

Graphics Interface

KARMA presents the results from the rule system on a three-dimensional graphics workstation. The bicubic patches of the surface model are displayed graphically and may be manipulated by the user. The user may also modify the model and return to the control server for another iteration in the rule system if the results are not satisfactory.

The bicubic patches are characterized with different colors, intensities and line textures to show attributes such as hydrophobicity and steric properties. Only one attribute may be displayed at a time, with color and intensity representing the value of the attribute, and line texture representing KARMA's confidence level in the information. For example, when displaying hydrophobicity, red patches are hydrophobic space while blue patches are polar space. Patches drawn with solid lines represent areas which are well explored while patches with short dashes contain little information. Displaying information using multiple cues allows the user to examine various aspects of the surface model without having to deal with large amounts of numerical data.

The graphics interface is also the appropriate place to alter the model since it lets the user look at an overall picture of the model as it is modified. The graphics interface provides user-friendly tools for this purpose, including a pointing device for selecting the modification site and a hierarchical menu system to guide the user through the actual process of making changes. Thus, the user may select a control point on one of the bicubic patches with the pointing device; pop up a menu of permitted modifications; select an operation, *e.g.*, move the control point outwards along the surface normal. After the control point data has been modified, the graphics interface will recalculate and redraw the bicubic patches of the surface model based on the new data. After modifying the model to the desired state, the user may simply return to the control server and initiate the rule system for further refinement.

Conclusion

Currently, KARMA is in the prototyping phase. Although the hardware is connected via the high bandwidth network, it is necessary to implement the servers for data communications. Additionally, a completely new graphics package is in development for KARMA. The next two steps in terms of development are the turnkey and production versions of KARMA.

Current methods in computer-assisted drug design are most successful if the structure of the receptor is known. Our goal is to aid the investigator in those situations where the structure of the receptor may or may not be known. KARMA emphasizes two critical factors. First, three dimensional graphics presents the results from the rule-based system in a manageable format. Second, KARMA provides a means for the user to inject knowledge about the model. KARMA is designed as a tool to aid the chemist and the ability to incorporate ideas from the user is a very important aspect. It is our goal to successfully look at computer assisted drug design from a new perspective using KARMA.

Acknowledgments

This work was supported in part by NIH RR-1081, DAAG29-83-G-0080, Evans and Sutherland, Silicon Graphics and IntelliCorp. We also wish to thank Dennis Miller, I.D. Kuntz, Don Kneller, Greg Couch, Ken Arnold, and Willa Crowell for help and discussion.

Literature Cited

(1)   Morris, W., Ed. In "The American Heritage Dictionary of the English Language"; American Heritage and Houghton Mifflin: New York, 1969; p. 725.

(2)   Hayes-Roth, F.; Waterman, D.A.; Lenat, D.B., Eds. "Building Expert Systems"; Addison-Wesley: USA, 1983;

(3)   Blaney, J.M.; Jorgensen, E.C.; Connolly, M.L.; Ferrin, T.E.; Langridge, R.; Oatley, S.J.; Burridge, J.M.; Blake, C.C.F. *J. Med. Chem.* 1982, 25, 785-790.

(4)   Hansch, C.; McClarin, J.; Klein, T.; Langridge, R. *Molec. Pharm.* 1985, 27, 493-498.

(5)   KEE User's Manual. 707 Laurel Street, Menlo Park, California, 94025 KEE is a registered trademark of IntelliCorp.

(6)   Marshall, G.R. "Computer Aided Drug Design". First European Seminar and Exhibition on Computer-Aided Molecular Design. October 18-19, 1984.

(7)   Blaney, J.M.; Hansch, C.; Silipo, C.; Vittoria, A. Chem. Rev. 1984, 84, 333.

(8)   We wish to thank Dr. David Weininger and Dr. Albert Leo at the MedChem Project, Department of Chemistry, Pomona College for providing us with the SMILES software.

(9)   Crippen, G.M. "Distance Geometry and Conformational Calculations"; Research Studies: New York, 1981.

(10) Weiner, P.K.; Kollman, P.A. *J. Comp. Chem.* 1981, 2, 287-303.

(11) Metcalfe, R.M.; Boggs, D.R. *Comm of the ACM* 1976, 9. Ethernet is a registered trademark of Xerox Corporation.

(12) We wish to thank Dr. Gordon Crippen from Texas A&M University and Dr. Jeffrey M. Blaney from E.I. DuPont de Nemours & Company for providing us with the Distance Geometry software.

(13) Foley, J.D.; Van Dam, A. "Fundamentals of Interactive Computer Graphics"; Addison-Wesley: 1982; Chap. 13.

(14) Clark, J.; "Parametric Curves, Surfaces, and Volumes in Computer Graphics and Computer Aided Geometric Design," Technical Report No. 221, Computer Systems Laboratory, Stanford University, 1981.

(15) Hansch, C.; Leo. A. "Substituent Constants for Correlation Analysis in Chemistry and Biology"; Wiley-Interscience: 1979.

(16) Hansch, C.; Klein, T.; McClarin, J.; Langridge, R.; Cornell, N. *J. Med. Chem.* (in press).

(17) Hansch, C.; Li, R.; Blaney, J.; Langridge, R. *J. Med. Chem.* 1982, 25, 777-784.

(18) Selassie, C.; Fang, Z.; Li, R.; Klein, T.; Langridge, R.; Kaufman, B. *J. Med. Chem.* (in press).

(19) Smith, R.N.; Hansch, C.; Kim, K.I.; Omiya, B.; Fukumura, G; Selassie, C.D.; Jow, P.Y.C.; Blaney, J.M; Langridge, R. *Arch. Biochem. Biophys.* 1982, 215, 319-328.

RECEIVED December 17, 1985

An Intelligent Sketch Pad as Input to Molecular Structure Programs

Carl Trindle

Chemistry Department, University of Virginia, Charlottesville, VA 22901

The programming and manipulation of chemical graphs is
awkward in most familiar programming languages. LISP,
the Esperanto of artificial intelligence research,
makes possible a representation of chemical structural
formulas which is much more nearly analogous to the
chemist's view of such graphs. This is a considerable
computational advantage as well as a convenience for
the user.
We have designed a "functional fragment" representation
of structural formulas, applicable to any molecule,
which will resolve a crude sketch of a chemical structure into a list of fundamental fragments. Exploiting
the PROPERTY feature of LISP and the distance geometry
algorithms of Crippen we can recover Cartesian coordinates for each atom, suitable for input to molecular
mechanics programs, or to ab initio electronic structure packages.
Besides local geometries, the intelligent sketchpad
can contain any local properties, including bond types
and strengths, chromophore optical spectra, and nuclear
magnetic resonance and infrared spectra characteristic
of a local chemical environment.

Computational chemists have developed several remarkably powerful
and reliable computer codes, capable of describing the relative
stability of various conformations of macromolecules, and details
of the electronic structure of molecules of more modest size ($\underline{1}$).
The properties of molecules which can be obtained by use of these
programs correlate with important features of chemical reactivity
and the properties of materials. Molecular design, in pharma-
ceuticals, photochemistry, and general materials science can be
made much more efficient by the routine use of these computational
systems. However, their use is at present not widespread; it is
limited to a few large chemical companies.

0097-6156/86/0306-0159$06.00/0

One of the obstacles to wider use of the well-tested and power-ful programs such as Allinger's molecular mechanics (2) and Pople's GAUSS80 (3) is that the programs require such elaborate and awkward input. Users must ordinarily prepare a list of Cartesian coordi-nates of each atom. This is cumbersome for molecules of even moderate size. But more significantly, chemists' powerful sense of three-dimensional molecular structure is never expressed in Cartesian coordinates. Instead chemists think more naturally of "internal coordinates," that is bond lengths, primary valence angles, and local dihedral angles. Of course a full set of internal coordinates de-fines in principle the set of Cartesian coordinates (4). Unfortu-nately, the usual algorithms for generating Cartesian coordinates from internal coordinates are sensitive to small errors. These errors accumulate and can perpetrate enormities such as leaving rings unclosed, or forcing unrealistically short separations be-tween nonbonded atoms. In the chemist's conceptual picture, realistic bond distances for rings are maintained, even if distor-tions in normal valence angles are required.

The problem is to transform the pictorial view of molecules which is the daily companion of the chemist, to the numerical form required by programs, WITHOUT FORCING THE USER TO EFFECT THE TRANS-LATION. We must not ask the chemist to do much more than identify the atoms, their connectivity, and some gross features of the stereochemistry. The structural formula is the medium by which such simple yet richly evocative information is conveyed. The structural formula does after all suffice for the chemist's work day to day. It should be adequate to convey the essential information to useful computer programs.

There will be two major stages to the translation of information from the chemist's pictorial image to the rigidly formatted input file required by molecular mechanics or molecular orbital programs. First the sketch is impressed on a digitizing tablet (perhaps as simple as a Koala Pad (R), or a more accurate digitizing tablet). Then the graph must be interpreted and a trial geometry generated.

Accepting the Sketch. The (computationally) most convenient way to enter a structural diagram is to use a digitizing tablet with a mouse or stylus. Our experience has been with the Houston Instruments HIPAD (R). The software accompanying this (and most ordinary) digitizing tablet accepts and stores local coordinates of particular points, and a set of pointers designating which vertices are to be connected (5). In this way the molecular topology can be specified with no novel analysis or programming.

It would be more interesting from the point of view of Arti-ficial Intelligence research to interpret a sketch already on paper, by the analysis of dark and light elements (6). We have made only small progress in this task, but some preliminary remarks can make the difficulties clear. The field of view is resolved into picture elements, and an optical scanner would assign a numerical value corresponding to the darkness of the sketch at that location. Heavy lines would be easy to recognize, by the sequence of adjacent dark spots detected by the scanner. Intersections might be harder to recognize if the grid is coarse, but knowledge of the existence of lines could guide the search, by estimates of the intersections by extrapolation. A planar graph (with no crossing lines) would seem to present few difficulties. Vertices representing generalized atoms (that is, "Me" in place of a fully detailed methyl group) would have to be more carefully specified. The chemist would use Berzelius-notation capital letters for labels, which would have to be interpreted. This is a hard task, as the post office has learned. It would be essential for the system to realize when identification of a vertex is impossible or ambiguous, and request guidance from the user. Figure 1 shows how vertices are specified.

It will be necessary to distinguish the strokes which identify single or multiple bonds from the strokes denoting lone pairs, and it will be required to supply missing hydrogens and lone pairs which are often omitted from casual sketches. This latter problem will also be encountered if the sketch is input directly by the digitizing tablet. We return to that line of approach.

Preliminary Processing of the Sketch. Even at this early stage, before different atoms are distinguished and hydrogens are fully expressed, we have much of the information needed for some kinds of analysis. All of the graph-theoretic analysis of pi systems (7), which may be considered to be based on the Huckel model, uses no more than the connectivity between equivalent centers. However powerful the graph theory has been, it cannot be denied that it suppresses much of the detail expressed in the structural diagram. Therefore we will not be content to stop at this stage.

It will be necessary at minimum to define the type of atom present at each vertex. We reduce the labor necessary for this specification by (a) suppressing hydrogens in the preliminary sketch; and (b) assuming as a default that each vertex represents a carbon atom, requiring an amendment only for heavy atoms. Our software is responsible for filling in hydrogens. This process is frequently ambiguous, given only the skeleton of heavy atoms. Therefore the computer system will sometimes interrogate the user for the number of hydrogen atoms at each vertex. With this information the task of completing a Lewis structure is left to the software, which is at least as capable of this task as the average first-year student. This is the first task that requires anything resembling Artificial Intelligence, so a few remarks on the design may not be out of place.

A Routine to Assign Lewis Structures. The procedure for assigning Lewis structures is familiar (8). Given the set of atoms, one must sum the valence electrons. In our LISP system, each ATOM can be assigned PROPERTIES which may include the number of valence electrons it contributes to the molecule, and equally important, its set of NEIGHBORS by which the skeleton of the molecule is specified. Each such link is assigned a pair of valence electrons, and a census is kept of electron pairs in the vicinity of each atom. Among the PROPERTIES of each atom is an estimate of its electronegativity, and the program assigns electron pairs to fill octets using the electronegativity to set priority. The last step is most "difficult." For each of those atoms which lack a full octet, the system must look among the NEIGHBORS for atom(s) possessing a lone pair which it might share. Of all those potential donors, one chooses the atom with the most negative formal charge. The multiple bond

A Dialogue Accompanying the Entry
of a Molecule of Moderate Complexity

SPECIFY NON-CARBON VERTICES:

NUMBER: 1   TYPE: nitrogen
NUMBER: 8   TYPE: oxygen
NUMBER: 10   TYPE: oxygen
NUMBER: 12   TYPE: oxygen
NUMBER: 0

SPECIFY NET CHARGE: +1

HYDROGENS AT VERTEX 2: 1
HYDROGENS AT VERTEX 3: 0
HYDROGENS AT VERTEX 4: 1
HYDROGENS AT VERTEX 5: 1
HYDROGENS AT VERTEX 6: 1
HYDROGENS AT VERTEX 7: 0
HYDROGENS AT VERTEX 8: 0
HYDROGENS AT VERTEX 9: 3
HYDROGENS AT VERTEX 11: 0
HYDROGENS AT VERTEX 12: 0
HYDROGENS AT VERTEX 13: 3

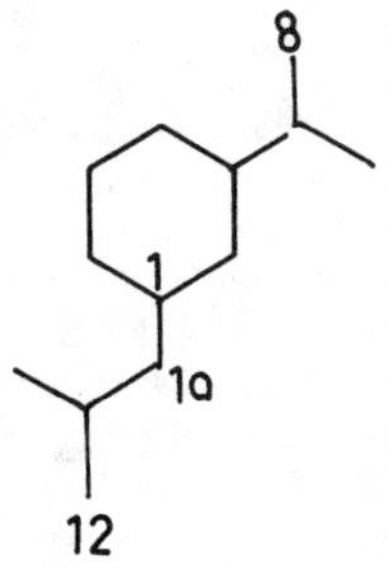

Figure 1. All vertices are first assumed to be CARBON. The
system requests that the user specify non-CARBON vertices; it
will build a set of user's abbreviations.

is represented by the appearance of the donor several times in the (revised) NEIGHBOR list. When the Lewis structure routine finds an ambiguity which we would represent by a set of resonance structures, it reports that fact and chooses the first legal structure for further processing. Figure 2 shows the procedure in practice.

Representation of the Molecule in LISP. We have used the chemist's sketch, or its Lewis structure equivalent, as the model of a data structure in LISP (9). This language has the flexibility needed to express an essentially non-numerical object, in terms of lists. LISP will permit us to organize molecular structure information in a way that mimics the human expert's knowledge. To accomplish this representation, we must develop a clear idea how the chemist assimilates the information provided directly and explicitly by the sketch, and how the properties of the molecule are recalled to the chemist's awareness.

The structural formula at minimum identifies the atoms and their connectivity. This hardly seems to be adequate in complexity to express much molecular information. This apparent paradox is resolved when we recognize that the chemist brings much of his experience to the task of interpreting the sketch, and much of the information is evoked rather than transmitted by means of the structural formula. The atoms' names--carbon, nitrogen--call up a flood of associations which (although they are almost never written explicitly in the chemist's sketch) are nonetheless part of the information it can summon. Among this data are the atomic mass, typical valencies, local geometry, perhaps a van der Waals radius, and a guide to chemical behavior, its "electronegativity."

The connectivity can define some aspects of the geometry in a useful semiquantitative way. The chemist has a very reliable idea of the range of bond lengths; CC(single), 1.54 A; CC(double), 1.33 A, etc. By counting connections and recognizing the atoms being connected, one can assign good estimates of the distances between directly bonded atoms.

The chemist's knowledge of molecular geometry extends beyond typical values of bond distances. He will also be able to predict many bond angles fairly accurately. This is equivalent to specifying a 1-3 nonbonded interatomic distance. The chemist's sketch portrays cis and trans isomerization, syn and anti, and gauche conformations which specify either torsion angles, or indirectly, a 1-4 nonbonded distance.

Besides primary bond distances and angles, and some special cases of torsional and dihedral angles, the chemist knows more global features of molecular geometry. However, such knowledge becomes more and more fragmentary; the longest distances in a molecule are most poorly defined.

A LISP Structural Recognizer. A molecule is represented in our LISP program first as a list of atoms. A numbering scheme assigns an unique label to each atom. Each atom has a collection of PROPERTIES; foremost among them is its generic NAME. The name CARBON carries with it a van der Waals RADIUS and a VALENCE. Other properties can be added as desired.

The major feature of a molecular sketch is the topology or

connectivity of the molecule. This is expressed as the property
NEIGHBOR for each atom. This is just the set of labels of other
atoms connected to a particular atom. The NEIGHBOR property is a
compact way to store the adjacency matrix used in graph theory.

The chemist's sketch, processed into the list representation
just described, is not yet very valuable; the system at the moment
is very ignorant of the structure of the molecule in question. But
the chemist knows much of the molecule from little more than the
diagram. How does the chemist "see" a complex molecular diagram?
In our judgement a chemist knows so much about a molecule because he
recognizes recurrent fragments of moderate size. Rings of varying
atomic composition, structure, and size ranging from carbonyl groups
to steroid systems, are recognized at a glance. Many stereochemically well-defined fragments, such as spiro and norbornyl systems,
are part of the chemist's conceptual toolkit. Our programming task
is to assure that our system recognizes such fragments, with all the
associated information on their structure and properties, with ease.

Somehow we must discern the presence of meaningful, familiar
fragments in the molecular list. We mimic this stock of informative
portions of molecules in our LISP system by lists called FRAGMENTS.
The FRAGMENTS, permanent members of a growing data base, each contain a set of ATOMS and a NEIGHBOR list for each atom identifying the
connectivity. Besides this topological information, the fragments
contain as PROPERTIES a stock of attributes of the fragments. The
first collection of PROPERTIES we gathered were interatomic distances gleaned from crystal structures. All interatomic distances
are defined within a fragment. The system can now assign many
(though not all) interatomic distances in an arbitrary molecule if
fragments could be discerned within the sketch.

We have developed a search technique which will scan the
MOLECULE and locate all fragments. Design of this recognition algorithm is difficult. The search routine shares some of the features
of the "knapsack problem," a classic difficulty in computer science.
We expect that we will be able to speed this step considerably. At
present we scan all stored fragments, though that is not the way an
expert would proceed. We screen out many fragments by a superficial
test that the atoms in the fragment must be a subset of the atoms in
the molecule. The fragments are subjected to more and more thorough
tests, until recognition is complete. These tests are essentially
recursive applications of the requirement that if a fragment is to
be identified in a molecule, the environment of each atom in the
fragment must be found in the molecule for the corresponding atom.
More detail on the search condition may be found in a previous
article (10). Figure 3 shows a typical fragment representation.

In this first formulation we have already established that it is
most effective to scan the largest candidate fragments first. It is
desirable to recognize overlapping fragments; more distances are
determined. However, it is inevitably the case that a substantial
number of distances will be left undefined, particularly the longest
distances which would not be incorporated into a fragment.

Distance Geometry Changes Distances to Cartesian Coordinates. Most
esperimental measures of molecular geometry provide quantities which
may be most directly interpreted as defining interatomic distances.

```
             Assignment of a Lewis Structure

FORMULA: C9 H10 N O3 (+)

COMPUTED VALENCE ELECTRONS: 63

23 PAIRS ASSIGNED TO LINKS

VERTEX 8 ASSIGNED 3 PAIR(S)
VERTEX 10 ASSIGNED 2 PAIR(S)
VERTEX 12 ASSIGNED 3 PAIR(S)
VERTEX 1 ASSIGNED 1 PAIR(S)
VERTEX 2 ASSIGNED 1 PAIR(S)
VERTEX 3 ASSIGNED 1 PAIR(S)

VERTEX 4, 5, 6, 7, 11 UNSATISFIED

SHARING BETWEEN VERTICES 4 AND 3
VERTEX 5 UNSATISFIED
SHARING BETWEEN VERTICES 5 AND 4
VERTEX 3 UNSATISFIED
SHARING BETWEEN VERTICES 3 AND 2
SHARING BETWEEN VERTICES 6 AND 1
SHARING BETWEEN VERTICES 8 AND 7
SHARING BETWEEN VERTICES 12 AND 11
NONZERO FORMAL CHARGES:
     VERTEX 1: +1
```

Figure 2. The Lewis structure routine will draw on the
PROPERTY VALENCY, which is the number of electrons each vertex
contributes to the Lewis structure. It assigns a pair of
electrons to each LINK, and satisfies the octet requirement.
In case of resonance, it will choose one of the set of
equivalent structures arbitrarily.

```
      Recognition of A Set of Known Fragments in a Molecule

SIX MEMBERED RING RECOGNIZED (BENZENOID)
BENZENE DISTANCES ASSUMED
REVISED DISTANCE 1-2
REVISED DISTANCE 1-6
REVISED DISTANCE 1-10

ACETYL GROUP RECOGNIZED

ACETYL GROUP RECOGNIZED

N-O-C SATURATED LINK RECOGNIZED

C-C SATURATED LINK RECOGNIZED
```

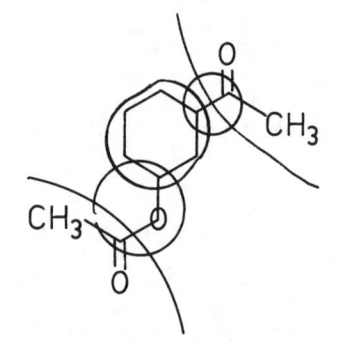

Figure 3. A typical fragment decomposition for a molecule
of moderate complexity. Roughly half of the interatomic
distances can be specified in this case by the fragment data.
The remaining distances are estimated by the distance geometry
algorithm of Crippen (11).

(In fact, usually the interpretation requires the presumption of a
rigid framework so that interatomic distances are persistent.)
There are $N(N-1)/2$ distinct distances in a cluster of N atoms, dis-
regarding symmetry-dictated equivalencies. This set of distances is
of course redundant; 3N-6 Cartesian coordinates are sufficient to
determine molecular geometry, apart from the position of the center
of mass and the orientation of the principle moments of inertia.
The larger the system, the more redundant is the full set of
distances.

Of course it is almost never the case that we have anything
resembling a full set of interatomic distances from experimental
data. Crippen has shown how one may pass not only from Cartesian
coordinates to interatomic distances, but from distances to Cartesian coordinates (11). More significant, he has shown that an
incomplete set of interatomic distances, together with even very
crude estimates of unmeasured distances, can produce helpful estimates of Cartesian coordinates. The estimates of missing distances
can be provided by "triangle conditions" which express that a 1-3
distance must be in the range from the (absolute value of the)
difference of the 1-2 and 2-3 distances to the sum of the 1-2 and
2-3 distances. By a factor analysis of the matrix of vector dot
products one obtains the best three-dimensional "imbedding" of the
geometry.

By the methods of Crippen we can use our well-known distances
within identified fragments, with crude estimates of distances
between atoms in disjoint fragments, to estimate the geometry of the
entire molecule. The factor analysis produces "statistically best"
estimates of every distance. Of course we cannot evaluate the
quality of the estimates of the missing longer distances. But the
statistically best estimates of the shortest distances (influenced
indirectly as they are by the poorly known longer distances) depart
substantially from the known fragment distances. One can improve
the overall estimates by replacing the estimates of the well-known
distances by accurate values and iterating the factor analysis.

The structure produced by distance geometry is not necessarily
the optimum energy form. It is merely a legal three-dimensional
structure reproducing the short-range structure. Of course if some
of the longer distances are known, further constraints are possible.
In our experience, however, short-range fragment properties determine much of the global form even of rather large systems. This is
particularly striking in (say, carborane) clusters and (even very
large) rings, both of which are inconvenient to describe by other
methods.

Extensions of the Functional Fragment Data Structure. In principle,
any molecular property which may be represented as a sum of contributions from fragments is natural to incorporate into the functional
fragment representation. Maksic has recently reviewed such group
additivity relations, concentrating on atoms as the fundamental
fragment (12). Magnetic and electric properties are remarkably
well represented by such methods, if a suitable hybridized electronic state is chosen for the atom in the molecular environment.
Such atomic additivity relations are the simplest form of a fragment
additivity scheme for representation of molecular properties. If

we choose slightly larger fragments, molecular properties can be better represented. Ascending the scale, we can adapt the functional fragment data structure to help us perform bond energy calculations (13). Benson shows us how to estimate thermodynamic properties given values for fragments (14). Often it is possible to estimate spectra by summing chromophore properties, so long as the absorbing centers are only weakly coupled (15). The same statement applies to chemical reactivity, so long as the functional groups interact weakly (16).

Interacting-Fragments Modeling Schemes may be Incorporated. It is not required that fragments be nearly independent parts of a molecule, and the molecular property be considered a simple sum of fragment properties. Consider for example the possibility of incorporating the quantitative perturbation - molecular - orbital method of describing the electronic distribution in molecules, which begins with MOs for fragments. The perturbation theory provides a systematic way to account for fragment interactions, and reproduces a wide variety of interesting electronic behavior at little computational cost (17). This effort still lies before us.

Summary. The everyday reasoning of the chemist is primarily pictorial and qualitative; it is analogic. The chemist can make astounding predictions of the chemical, thermodynamic, and spectroscopic properties of a substance given only an image, the structural formula. This process rests heavily on knowledge of the behavior of similar systems. We have devised a strategem whereby important molecular structure programs can be supplied the Cartesian coordinates they require, without forcing the chemist to provide much more than the structural diagram, which is a more natural language. The system interprets a sketch impressed on a digitizing tablet, and scans the structure for familiar fragments. Stored properties of each known fragment include intra-fragment interatomic distances. From these known distances, a legal three-dimensional structure can be constructed by the methods of Crippen, and supplied in the form of Cartesian coordinates to molecular structure programs.

Literature Cited

1. Quantum Chemistry Program Exchange Catalog, Chemistry Department, Indiana University, Bloomington, IN 47001.
2. Berkert, U.; Allinger, N. L. "Molecular Mechanics", ACS Monographs: Washington, D. C., 1982.
3. Clark, T. "A Handbook of Computational Chemistry"; Wiley: New York, 1985.
4. Fitts. D. D. "Vector Analysis in Chemistry"; McGraw-Hill Book Co.: New York, 1974.
5. HIPAD software copyright by Houston Instruments, Inc.
6. Cohen, P. R.; Feigenbaum, E. A. "The Handbook of Artificial Intelligence"; W. Kaufmann, Inc.: Los Altos, CA, 1982; Vol. III, p. 125. Raphael, B. "The Thinking Computer: Mind Inside Matter"; W. H. Freeman: San Francisco, 1976.

7. Graovac, A.; Gutman, I.; Trinajstic, N. "Topological
    Approach to the Chemistry of Conjugated Molecules"; Springer:
    New York, 1977.
8. Lewis, G. N. J. Am. Chem. Soc. 1916, 38, 762.
9. Winston, P.; Horn, B. "LISP"; Addison-Wesley Publ. Co.:
    Reading, MA, 1981. Johnson, C. S. J. Chem. Inf. Comp. Sci.
    1983, 23, 151.
10. Trindle, C.; Givan, R. In "Chemical Applications of Graph
    Theory and Topology"; King, R. B., Ed.; Elsevier: New York,
    1983. Trindle, C. Croatica Chem. Acta 1984, 57, 1231.
11. Crippin, G. M. "Distance Geometry and Conformational
    Calculations"; Chemometric Research Studies Press of Wiley
    Publ. Co.: New York, 1981.
12. Maksic, Z. B.; Eckert-Maksic, M.; Rupnik, K. Croatica
    Chem. Acta 1984, 57, 1295.
13. Benson, S. W., et al. Chem. Rev. 1969, 69, 279; Int. J. Chem
    Kinet. 1974, 6, 813.
14. Benson, S. W. "The Foundations of Chemical Kinetics";
    McGraw-Hill: New York, 1960; p. 665.
15. NMR and vibrational spectra of organic molecules are well
    described by group-additivity ideas; optical spectra require
    corrections to the spectra of chromophores. Cf. discussion
    of spectra by Gordon, A. J. and Ford, R. A., "The Chemist's
    Companion: A Handbook of Practical Data, Techniques and
    References"; Wiley-Interscience: New York, 1972.
16. Almost every elementary textbook of organic chemistry provides
    a systematic description of properties of functional groups
    and their characteristic reactivity; for example,
    Fessendon, R. J. and Fessendon, J. S. "Organic Chemistry";
    Willard Grant Press: Boston, 1979.
17. Dewar, M. J. S. "The Molecular Orbital Theory of Organic
    Chemistry"; McGraw-Hill: New York, 1969. Albright, T. A.;
    Burdett, J. K.; Whangbo, M. H. "Orbital Interactions in
    Chemistry"; Wiley-Interscience: New York, 1985.

RECEIVED December 17, 1985

The Similarity of Graphs and Molecules

Steven H. Bertz[1] and William C. Herndon[2]

[1] AT&T Bell Laboratories, Murray Hill, NJ 07974
[2] University of Texas at El Paso, El Paso, TX 79968–0509

A new definition of molecular similarity is presented, based upon the similarity of the corresponding molecular graphs. First, all of the subgraphs of the molecular graph are listed, and then various similarity indices are derived from the numbers of subgraphs. One of these compares favorably with the standard distance measures of sequence comparison. Measurement of similarity provides a new way to measure molecular complexity, as long as the most (or least) complex member of a set of molecules can be identified.

The concept of the similarity of molecules has important ramifications for physical, chemical, and biological systems. Grunwald (1) has recently pointed out the constraints of molecular similarity on linear free energy relations and observed that "Their accuracy depends upon the quality of the molecular similarity." The use of quantitative structure-activity relationships (2-6) is based on the assumption that similar molecules have similar properties. Herein we present a general and rigorous definition of molecular structural similarity. Previous research in this field has usually been concerned with sequence comparisons of macromolecules, primarily proteins and nucleic acids (7-9). In addition, there have appeared a number of ad hoc definitions of molecular similarity (10-15), many of which are subsumed in the present work. Difficulties associated with attempting to obtain precise numerical indices for qualitative molecular structural concepts have already been extensively discussed in the literature and will not be reviewed here.

Results and Discussion

We begin with the way chemists perceive similarity between two molecules. This process involves, consciously or unconsciously, comparing several types of structural features present in the molecules. For example, considering the five aliphatic alcohols (represented by their H-suppressed molecular graphs) in Figure 1, we note both similarities and differences: they are all four-carbon alcohols; **a**, **b**, **c** and **d** are acyclic, whereas **e** has a ring; **a** and **b** are primary alcohols, **c** and **e** are secondary alcohols and **d** is a tertiary alcohol; **b** and **c** have the same skeleton, but for the labeling of points (atoms), while the other skeletons are distinct; *etc.*

0097–6156/86/0306–0169$06.00/0

The first step in quantifying the concept of similarity is to list all subgraphs of the given molecular graphs, *e.g.* **a-e**, which has been done in the first column of Table I. The subgraphs include the vertices (atoms), all connected subgraphs, and the full molecular graphs themselves, since it can be seen that the molecular graphs for **a** and **c** are both subgraphs of **e**. Next, the number of each subgraph contained in the molecular graphs must be counted. Row 1 lists the number of C atoms, row 2 the number of O atoms, row 3 the number of C-C bonds, row 4 the number of C-O bonds, *etc.* Gordon and Kennedy (*16*) defined N_{ij} as the number of subgraphs of graph j isomorphic with graph i, and more colloquially as "the number of distinct ways in which skeleton i can be cut out of skeleton j." The entries in Table 1 are the number of ways the subgraphs can be cut out of the molecular graphs (the number of subgraphs of the molecular graphs isomorphic with the subgraphs in the first column).

In terms of the numbers of C or O atoms, **a-e** are equally complex. In terms of C-C bonds (ethane subgraphs) **a-d** are 3/4 as complex as **e**; however, in terms of propane subgraphs (row 5) **a** and **c** are 1/2 as complex as **e**. A simple algorithm that takes account of all the subgraphs involves comparison of two columns at a time, examining them row by row and dividing the smaller of the numbers by the larger. A similarity index (*SI*) can then be calculated by taking the average of the quotients. Of course, for two identical molecular graphs, $SI=1$. Inclusion of the molecular graphs in the list of subgraphs ensures that two different molecules which have the same number of each *proper* subgraph will not have $SI=1$. The values of $SI(1)$ for **a-e** are summarized in the form of a similarity matrix $SM(1)$ in Figure 2.

A simpler similarity index can be calculated by dividing the sum of the lesser of the two numbers in each row by the sum of the greater. (Only two columns of Table I are considered at a time, of course.) The values of $SI(2)$ for **a-e** are summarized in $SM(2)$, also in Figure 2. According to both $SI(1)$ and $SI(2)$, 1-butanol (**a**) and 2-butanol (**c**) are the most similar, whereas *t*-butanol (**d**) and cyclobutanol (**e**) are the least similar pair. In between these extremes there are a significant number of disagreements between these indices. For example based on $SI(1)$, **c** and **e** are more similar than **c** and **d**; however, **c** and **d** are more similar than **c** and **e** based on $SI(2)$. There are seven such pairs (out of 45 possible pairs), and each index has one "degeneracy". By considering standard measures of "distance," $SI(2)$ would appear to be the superior index (*vide infra*).

The calculations of similarity indices can also be done with labeled subgraphs of a labeled molecular graph. The points can be labeled according to the valency of the corresponding atoms (*i.e.* whether they are primary, secondary, tertiary, *etc.*), labeled with stereochemical descriptors, or labeled to reflect isotopic composition to cite but a few examples. Furthermore, the number of similarity indices can be doubled by relaxing the stricture that only connected subgraphs be considered. We have concentrated on connected subgraphs, as they are more intuitively meaningful to the average chemist; nevertheless, for some applications the inclusion of disconnected subgraphs may be desirable or even necessary.

Similarity and Distance. Two sequences of subgraphs **m** and **n** such as those in Table 1 have the property that there is a built-in one-to-one correspondence between the elements of one sequence (m_i) and those of the other (n_i). Accordingly, it is straightforward to calculate various well-known (*17*) measures of the distance d between the sequences, *e.g.* Euclidean distance $[\sum_i (m_i - n_i)^2]^{1/2}$, "city block" distance

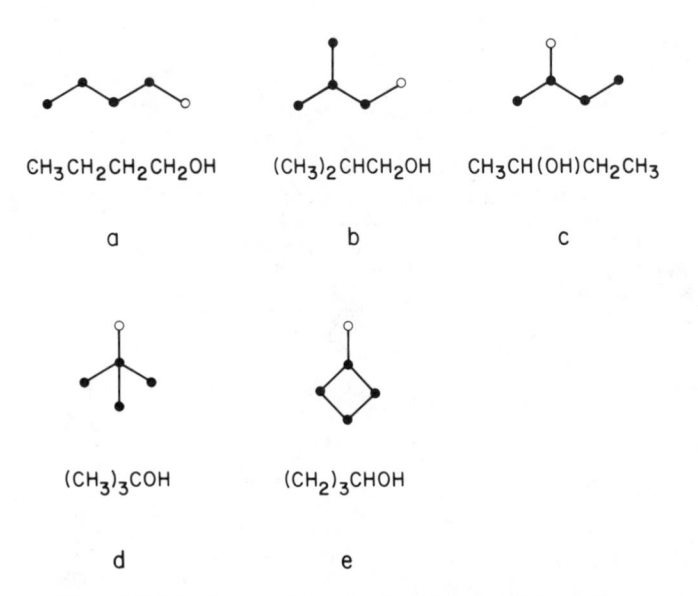

CH₃CH₂CH₂CH₂OH   (CH₃)₂CHCH₂OH   CH₃CH(OH)CH₂CH₃

a   b   c

(CH₃)₃COH   (CH₂)₃CHOH

d   e

Figure 1.   Selected four-carbon alcohols, abstracted as their H-suppressed molecular graphs: **a** 1-butanol, **b** isobutanol, **c** 2-butanol, **d** *t*-butanol, **e** cyclobutanol.

	a	b	c	d	e	
	1.000	0.561	0.682	0.417	0.462	a
		1.000	0.472	0.576	0.400	b
$SM(1) =$			1.000	0.472	0.577	c
				1.000	0.367	d
					1.000	e

	a	b	c	d	e	
	1.000	0.684	0.778	0.522	0.517	a
		1.000	0.619	0.609	0.484	b
$SM(2) =$			1.000	0.609	0.586	c
				1.000	0.441	d
					1.000	e

Figure 2.   Similarity matrices $SM(1)$ and $SM(2)$ for the graphs in Figure 1.

Table I.   Subgraph Enumeration for Some Four-carbon Alcohols.

SUBGRAPH	NUMBER IN GRAPH				
	a	b	c	d	e
●	4	4	4	4	4
○	1	1	1	1	1
●—●	3	3	3	3	4
●—○	1	1	1	1	1
●–●–●	2	3	2	3	4
●–●–○	1	1	2	3	2
●–●–●–●	1	0	1	0	4
●–●–●–○	1	2	1	0	2
(3-branch ●)	0	1	0	1	0
(3-branch with ○)	0	0	1	3	1
(square)	0	0	0	0	1
a ●–●–●–●–○	1	0	0	0	2
b (branch with ○)	0	1	0	0	0
c (branch with ○)	0	0	1	0	2
d (branch with ○)	0	0	0	1	0
e (square with ○)	0	0	0	0	1

$\sum_i |m_i - n_i|$, or Hamming distance, which counts the number of positions in which the corresponding elements are unequal. It may be noted that these are measures of dissimilarity; of course, it is easy to draw conclusions about similarity from them (*e.g.* by taking their inverse). Table II contains the distances calculated according to each of the definitions discussed above as applied to molecular graphs **a-e**. The three distance functions parallel each other quite closely: there are only two disagreements between Hamming distance and Euclidean distance, and there are no disagreements between city-block distance and Euclidean distance. There is a two-fold degeneracy within city-block distance and Euclidean distance (the same as $SI(1)$ and $SI(2)$) and a four-fold one within Hamming distance, which is the crudest measure. Both city-block and Euclidean distance have only a single disagreement with $SI(2)$, but many with $SI(1)$; therefore, it is recommended that $SI(2)$ or one of the distance measures that parallel it be used to index similarity.

Table II. Distance Measures

	Hamming	City-bl.	Euclid.	1/City-bl.	1/Euclid.	$SI(1)$	$SI(2)$
$d(a,b)$ =	6	6	2.449	0.167	0.408	0.561	0.684
$d(a,c)$ =	4	4	2.000	0.250	0.500	0.682	0.778
$d(a,d)$ =	8	11	4.359	0.091	0.229	0.417	0.522
$d(a,e)$ =	10	14	4.899	0.071	0.204	0.462	0.517
$d(b,c)$ =	8	8	2.828	0.125	0.354	0.472	0.619
$d(b,d)$ =	5	9	4.359	0.111	0.229	0.576	0.609
$d(b,e)$ =	11	16	5.657	0.062	0.177	0.400	0.484
$d(c,d)$ =	8	9	3.317	0.111	0.301	0.472	0.609
$d(c,e)$ =	8	12	4.690	0.083	0.213	0.577	0.586
$d(d,e)$ =	12	19	6.245	0.053	0.160	0.367	0.441

Similarity and Complexity. On account of the variety of features that contribute to the complexity of a molecule (*e.g.* rings, double bonds, branching, heteroatoms, *etc.*), two molecules can have the same complexity and yet be quite dissimilar, depending on the weights given to the features (*18*). In contrast two molecules which are very similar must have nearly equal complexities. Therefore, once the most complex member of a family of molecules has been identified somehow, the others can be ranked in order of complexity by calculating their similarity to it. For example, taking tetrahedrane as the most complex member of the family butane (P_4), cyclobutane (C_4), bicyclobutane (K_4-x), tetrahedrane (K_4), $SI(2)$ confirms that this is the order of increasing complexity. The same order is obtained by considering the total number of subgraphs or by counting only the number of propane subgraphs (*19*), η (Table III).
Subgraph Enumeration. The total number of subgraphs increases rapidly with the number of atoms, making hand calculations of SI impractical for large molecules. Therefore a computer program was written. Our program is based on the fact that the entries in the nth power of the adjacency matrix of a graph count paths of length n, which includes retraced pathways and, therefore, branched chains and cycles. A molecular graph is represented by the string adjacency matrix $A\$(I,J)$, where the I,J-entry is a string of characters describing a bond ($I \neq J$) or an atom ($I = J$). Matrix multiplication is defined as string concatenation. The concatenated strings are alphabetized, processed to eliminate duplicates, sorted by number of bonds, and stored for future use. (A copy of this program can be obtained by writing to WCH.)

Table III. Complexity Measures			
G	$SI(G,K_4)$	Subgraphs	η
P_4	0.156	10	2
C_4	0.266	17	4
K_4-x	0.516	33	8
K_4	1.000	64	12

$$
SM(2) = \begin{array}{cccc}
P_4 & C_4 & K_4-x & K_4 \\
1.000 & 0.588 & 0.303 & 0.156 \\
 & 1.000 & 0.515 & 0.266 \\
 & & 1.000 & 0.516 \\
 & & & 1.000
\end{array}
\begin{array}{l}
P_4 \\
C_4 \\
K_4-x \\
K_4
\end{array}
$$

Potential Applications. Quantitative structure-activity relations have been formulated on the basis of common substructures (*2,14*) and similarity indexing (*5,10*). For example, Carbó *et al.* (*10*) related phermone activity to "an electron density measure of similarity between two molecular structures." Randić *et al.* (*14*) have related pharmacological activity to the numbers of paths in the molecular graph. The extension from this one kind of subgraph to all possible subgraphs should improve the statistical correlation of properties with substructures; but, even more importantly, it will make the results easier to visualize in a way that is meaningful to a chemist. Gordon and Kennedy (*16*) observe that a physical measurable can be expressed as a linear combination of graph-theoretical invariants (N_{ij}, see above). By using all possible subgraphs in such an analysis and optimizing the coefficients the most important ones might be found.

Another important subject for similarity considerations is the planning of organic syntheses. Wipke and Rogers (*20*) point out that "chemists do not always work systematically backward but sometimes make an 'intuitive leap' to a specific starting material from a target without consideration of reactions needed for interconversion. This intuitive leap probably involves a *Gestalt* pattern recognition based on the chemist's knowledge of available starting materials and similarity between the starting material structure and the target structure." Our method should allow not only the overall similarity of target and potential starting material to be assessed, but also the similarity of portions (substructures) of the target and all or part of a starting material.

Acknowledgment. WCH is grateful to the Robert A. Welch Foundation of Houston, Texas for financial support.

Literature Cited

1. Grunwald, E. *Chemtech* **1984**, *14*, 698.
2. Crandell, C. W.; Smith, D. H. *J. Chem. Inf. Comput. Sci.* **1983**, *23*, 186.
3. Bawden, D. *Ibid.* **1983**, *23*, 14.
4. Hansch, C.; Leo, A. J. "Substituent Constants for Correlation Analysis in Chemistry and Biology"; Wiley: New York, 1979.
5. Kier, L. B.; Hall, L. C. "Molecular Connectivity in Chemistry and Drug Research"; Academic Press: New York, 1976.
6. Kubinyi, H.; Kehrhahn, O. *J. Med. Chem.* **1976**, *19*, 1040.
7. Waterman, M. S. in "Mathematical and Computational Problems in the Analysis of Molecular Sequences" (*Bull. Math. Biol.* **1984**, *46*); Pergamon: Oxford, 1984; p. 473. *Cf.* other articles in this volume.
8. Lipman, D. J.; Pearson, W. R. *Science* **1985**, *227*, 1435.

9. Sellers, P. H. *SIAM J. Appl. Math.* **1974**, *26*, 787.
10. Carbó, R.; Leyda, L.; Arnau, M. *Int. J. Quantum Chem.* **1980**, *17*, 1185.
11. Cone, M. M.; Venkataraghavan, R.; McLafferty, F. W. *J. Am. Chem. Soc.* **1977**, *99*, 7668.
12. Bersohn, M. *J. C. S. Perkin I* **1982**, 631.
13. Armitage, J. E.; Lynch, M. F. *J. Chem. Soc. (C)* **1967**, 521.
14. Randić, M.; Kraus, G. A.; Džonova-Jerman-Blažić, B. in "Chemical Applications of Topology and Graph Theory"; King, R. B., Ed.; Elsevier: Amsterdam, 1983; p. 192.
15. Seybold, P. G. *Int. J. Quantum Chem., Quantum Biol. Symp.* **1983**, *10*, 95, 103.
16. Gordon, M.; Kennedy, J. W. *J. C. S. Faraday Trans. II* **1973**, *69*, 484.
17. Kruskal, J. in "Time Warps, String Edits, and Macromolecules"; Sankoff, D.; Kruskal, J., Eds.; Addison-Wesley: Reading, MA, 1983, p. 1. *Cf.* other articles in this volume.
18. Bertz, S. H. in "Chemical Applications of Topology and Graph Theory"; King, R. B., Ed; Elsevier: Amsterdam, 1983; p. 206.
19. Bertz, S. H. *J. C. S. Chem. Commun.* **1981**, 818.
20. Wipke, W. T.; Rogers, D. *J. Chem. Inf. Comput. Sci.* **1984**, *24*, 71.

RECEIVED December 17, 1985

16

Symbolic Computer Programs Applied to Group Theory

Gordon D. Renkes

Chemistry Department, Ohio Northern University, Ada, OH 45810

Applications of symbolic computer programming to group theory will be discussed. These programs, which are written in Common Lisp, perform the symbolic manipulations involved in the generation of multiplication tables, finding the classes, taking products of groups, establishing the correlations between subgroups and supergroups, etc. This software should prove very useful for applications of group theory to the spectroscopy of non-rigid molecules, for which the molecular symmetry groups are often large, not standard point groups, and very tedious to manipulate by hand.

The symposium one year ago on symbolic computing in chemistry, and this symposium on uses of artificial intelligence in chemistry demonstrate that symbolic computation is now becoming recognized as a useful tool for chemists. Just as computer "number crunching" is now fully accepted and implemented to assist the solving of many chemical questions, it appears that eventually computer "symbol crunching" will fulfill an equally important role to assist the chemist with his thinking.

Why Symbolic Computing for Group Theory?

This paper addresses the application of symbolic programming to the symbolic manipulations of group theory. Chemists are already familiar with the standard applications of group theory as explained in the standard texts. For many applications, the useful information such as character tables and correlation tables are in their appendices. However, in certain areas of current research, such as the interpretation of the spectra of non-rigid molecules, unfamiliar and sometimes large groups which are not included in the standard tables are employed (1-4). A variety of formulations have been developed to approach the analysis of the symmetries of such species, (e.g. molecular symmetry group and the isometric group, etc.). They all share the common hazzard of many elements and

0097-6156/86/0306-0176$06.00/0

tedious manipulations for many molecules of interest. When confronted with this situation, the investigator must generate the required tables himself, if he is not lucky enough to find them published somewhere. A common clause in many papers reads "The character table for this group has already been published....". The reader can hear the author's sigh of relief that he didn't have to work it out himself. Many labor saving techniques have been devised to speed up this process, e.g. (4), but these have to be learned and executed with care. And, the amount of paper work involved can still be considerable, especially when complicated situations are considered. For example, to evaluate the classes and character table for the molecule boron trimethyl (of order 324) required 18.5 pages and 15 intermediate tables even when efficient algorithms were used (4). Upon surveying this situation, one can appreciate the convenience of computer programs which would handle the tedious details. (This would be analogous to the application of computers to numerical computations. Before computers, tedious arithmetic was minimized by use of log tables, perturbation theory, algebraic approximations, etc. With computers, computations can be executed with far fewer approximations and applied to more extensive and realistic situations.) Such programs would be useful tools, because they would free one to spend more time thinking about the problem at hand, and to quickly test out ideas without having to decide, "Is it worth the effort?"

Lisp as a Language for Implementation

Given that such programs would be useful, we must next decide which language would be most appropriate for implementation. At least three reasons justify the symbolic language Lisp.

First, Lisp is designed to be used interactively at a computer terminal. This would be very convenient for the investigator in the midst of thinking about a particular problem. Suppose a question arises which requires the use of group theory tables. Rather than digging through appendices or searching in the library, the computer programs would be employed to supply results on the spot, even if no one has ever done it before.

Second, the fundamental data structure of the Lisp language is a list of symbols. Two examples of legitimate lists are,

$$(1\ 2\ 3) \qquad \text{and} \qquad (\ (1\ 2\ 3)\ (4\ 5\ 6)\ (7\ 8\ 9)\ )$$

The first is a simple list of three integers, and the second is a list of lists, each of which is a simple list of three integers. The parentheses serve to enclose the lists. Such lists ideally match the permutation notation for group operations which are employed in these programs. For example, to represent the permutation of the integers 1 and 2 in a list of integers (1 2 3) one writes,

$$\text{operator} * \text{operand} = \text{result}$$

$$(1\ 2) \quad * \quad (1\ 2\ 3) = (2\ 1\ 3)$$

And, the product of two operators, which is equal to another single permutation operator is written, for example,

$$\text{operator} * \text{operator} = \text{operator}$$

$$(1\ 2\ 3) * (1\ 2\ 3) = (3\ 2\ 1)$$

(This matches the notation used by many spectroscopists who study non-rigid molecules (1-3). At present, a user of this software is confined to this notation, although it could be possible to expand the capability of reading and displaying standard point group notation at the terminal.)

Finally, the language provides built in devices for conveniently manipulating, categorizing, storing and recalling all the information which pertains to a group, such as the multiplication table, classes, characters of the irreducible representation, correlations between subgroups and product groups, etc. Much of the rest of this paper will summarize the details of how this is done.

Implementation Discussion

Basic Functions. The fundamental symbolic operation which is used performs the permutations on lists of numbers. The Common Lisp supplied function ROTATEF is designed to do just this. An arbitrary number of arguments can be supplied to it, and it returns a list in which the first argument is at the end, and the others have been shifted one space to the left. An example of the application of this function, and the result displayed on the screen is,

(ROTATEF '1 '2 '3)

(2 3 1)

A user function CYCLOPERATE was written to employ this Lisp function, using the operator list as a recipe for how ROTATEF should rearrange the numbers in the operand list. In this example, the list (1 2 3) is the operator, and (1 2 3 4 5) the operand. The permuted list is returned as the result of the function.

(CYCLOPERATE '(1 2 3) '(1 2 3 4 5))

(2 3 1 4 5)

Another user function, PERMUTE, applies CYCLOPERATE repeatedly when a list of permutation operators is applied successively to an operand list, to return a list of permuted numbers,

(PERMUTE '((1 2 3) (2 3 4) (3 4 5)) '(1 2 3 4 5))

(2 1 3 5 4)

These two functions, and one more which can reconstruct a

permutation operator from a permuted list, serve as the workhorses for the procedure of setting up a group multiplication table.

Some of the information pertaining to a group is stored in property lists. Table I exemplifies how this looks for the simple case of the cyclic group of order three. (This would be isomorphic to the rotational subgroup of a molecule such as methyl fluoride. The operators (1 2 3) and (1 3 2) would correspond to the permutations of the three hydrogen nucleii numbered 1, 2 and 3. NIL, the language's symbol for the empty list, serves as the identity.)

Table I.   Property lists for cyclic group, order 3.

(#:GRP-1		#:GRP-2		#:GRP-3)	
PERMOP	(NIL)	PERMOP	((1 2 3))	PERMOP	((1 3 2))
RESULTLIST	(1 2 3)	RESULTLIST	(2 3 1)	RESULTLIST	(3 1 2)
INVPERMOP	(NIL)	INVPERMOP	((1 3 2))	INVPERMOP	((1 2 3))
INVERSE	#:GRP-1	INVERSE	#:GRP-3	INVERSE	#:GRP-2
CLASS	#:CLS-1	CLASS	#:CLS-2	CLASS	#:CLS-3
#:GRP-1	#:GRP-1	#:GRP-1	#:GRP-2	#:GRP-1	#:GRP-3
#:GRP-2	#:GRP-2	#:GRP-2	#:GRP-3	#:GRP-2	#:GRP-1
#:GRP-3	#:GRP-3	#:GRP-3	#:GRP-1	#:GRP-3	#:GRP-2

The three operators are represented by the three Gensym symbols in the list (#:GRP-1 #:GRP-2 #:GRP-3) which is stretched out across the top of the table to make room for the property lists underneath. These symbols by themselves mean nothing. The useful information is contained in the property lists, which are displayed underneath in vertical tabular format for readability. The property list is a list of pairs of symbols. The first symbol of each pair is the property indicator, which allows access to the second symbol, the property value, by execution of an access function. For example, the PERMOP property is the permutation operator for each group element. If we want the permutation operator for a particular group element, we use the access function GET, to get from the appropriate GENSYM symbol the PERMOP property.

        (GET '#:GRP-2 'PERMOP)

        ((1 2 3))

The RESULTLIST property is the result of operating with that operator on an initially ordered operand list. INVPERMOP and

INVERSE are the inverse operator list, and the Gensym symbol for it, respectively. The CLASS property value is another Gensym atom which has as its value a list of all of the operators in that class. (In this simple case, the value of #:CLS-1 is the list (#:GRP-1), etc.) The remaining pairs in each property list represent the group multiplication table. For any particular group multiplication, an element of the group list at the top of Table I pertains to the right operator, the property indicator pertains to the left operator, and the property value pertains to the product. For example, for the product of the permutation operator (1 2 3) with itself,

(GET '#:GRP-2 '#:GRP-2)

#:GRP-3

And to obtain the operator itself,

(GET (GET '#:GRP-2 '#:GRP-2) 'PERMOP)

((1 3 2))

Much of the computational labor expended is used to set up these property lists, but once that is accomplished, other manipulations which need the information stored in them only have to GET the results which are stored in these property lists.

The strategy used to set up the multiplication table is handled by a function which accepts a list of operators which are a set of generators for that group. All possible products between the generators fill in a portion of the table, and usually produce new operators. Further multiplication using these new operators fills in more of the table, and may produce more new operators. This process is repeated exhaustively until no new operators are produced, at which point the closure property of groups is satisfied, and the table is complete. Following this, another function uses this table to find the conjugacy classes by application of the definition.

Terminal Display and Practical Usage. Once calculated, other user functions can extract the desired information from the internally stored representation and display it on the terminal or print it on a lineprinter. Table II shows the list of operators by classes for S3, the permutation group of degree three, as displayed on the terminal. (This is isomorphic to the point group C-3V.) The multiplication table and character table can also be displayed in appropriate formats, although the multiplication table is readable for only the smallest groups, and probably would not normally be displayed anyway. (At the present stage of development, the character table must be entered from the terminal. A function which sets it up from scratch will be written in the near future. Also at present, the classes are simply numbered.)

Other typical group manipulations can be performed once all the aforementioned information has been found. For example, direct

Table II.   Terminal Display of Classes

For the group S3

Operators by Classes

1 is (NIL).

2 is ((1 3 2)), ((1 2 3)).

3 is ((2 3)), ((1 3)), ((1 2)).

---

products can be taken between two groups, and the correlations established between the representations of the subgroups and the product group. Consider the direct product of the permutation groups of degree three and degree two, represented by the names S3 and S2, to produce the product named S3-DP-S2. The character table of the product group is exhibited in Table III. The default labels for the representations in the product group are formed by concatenating the labels of the representations in the subgroups (A and B for S2, and A1, A2 and E for S3).

Table III.   Terminal display of character table of
              direct product of S3 with S2

For the group S3-DP-S2

Group character table.

CLASS	1	2	3	4	5	6
A1A	1	1	1	1	1	1
A1B	1	1	1	-1	-1	-1
A2A	1	1	-1	1	1	-1
A2B	1	1	-1	-1	-1	1
EA	2	-1	0	2	-1	0
EB	2	-1	0	-2	1	0

A record of the correlations between the representations is constructed with association lists while taking the product, and these can be used to display character correlation tables in both

the forward sense, from a subgroup to the product group, and in the
reverse sense, from product group to a subgroup. Table IV shows the
terminal display of both forward correlations.

Table IV. Terminal Display of Character Correlation Tables

Character correlation table.

SUBGROUP S3	PRODUCT-GROUP S3-DP-S2	
A1	A1B	A1A
A2	A2B	A2A
E	EB	EA

Character correlation table.

SUBGROUP S2	PRODUCT-GROUP S3-DP-S2		
A	EA	A2A	A1A
B	EB	A2B	A1B

Products between the irreducible representation characters
within a group will produce representations which are often
reducible. A simple calculation can decompose this product to a sum
of the irreducible representation characters, as is demonstrated in
Table V for two representations from the S3-DP-S2 group.

Table V. Terminal display of the decomposition of the product
of two representations of S3-DP-S2

Within the group S3-DP-S2

EA	2	-1	0	2	-1	0
EB	2	-1	0	-2	1	0
EAxEB	4	1	0	-4	-1	0

The decomposition is

EAxEB	1 EB	1 A2B	1 A1B

Other Implementation Details. All of the information resulting from the computations described above is stored in a named record structure which is defined using the Common Lisp DEFSTRUCT facility. An example of what this looks like is shown in Table VI for the group S2, which we used earlier.

Table VI. Record structure which stores all the
           information pertaining to the group S2

ORDER	2
OPERAND-LIST	(4 5)
GROUP-LIST	(#:GRP-10 #:GRP-11)
CLASS-LIST	(#:CLS-7 #:CLS-8)
CHARACTER-LIST	(#:CHR-8 #:CHR-9)
SUBGROUPS	NIL
SUPERGROUPS	(S3-DP-S2)

The left element on each line is a name for the field (or record) which is shown in the right element. The ORDER of this group is 2. The OPERAND list used for this example was the numbers 4 and 5. The Gensym symbols for the two group elements are stored in the GROUP-LIST field. As was explained earlier, property lists were attached to each of these which contained the multiplication table and other information. The elements of the CLASS-LIST and CHARACTER-LIST fields contain the information indicated by their names. In the above examples, we did not work with the subgroup of S2, so NIL is stored there; but the direct product name S3-DP-S2 is stored in the field SUPERGROUPS. Attached to this name (not shown) is the association list for the correlation between the representations, which was used for the construction of Table IV.

The examples used above to illustrate the features of the software were kept deliberately simple. The utility of the symbolic software becomes appreciated when larger problems are attacked. For example, the direct product of S3 (order 6) and S4 (isomorphic to the tetrahedral point group) is of order 144, and has 15 classes and representations. The list of classes and the character table each require nearly a full page of lineprinter printout. When asked for, the correlation tables and decomposition of products of representations are evaluated and displayed on the screen within one or two seconds. Table VII shows the results of decomposing the products of two pairs of representations in this product group.

These programs have been coded in Common Lisp (5) which is

Table VII. Display of the decomposition of products of representations within the direct product group S3-DP-S4.

Within the group S3-DP-S4

EA1	2	-1	0	2	2	2	2	-1	-1	-1	-1	0	0
EA1	2	-1	0	2	2	2	2	-1	-1	-1	-1	0	0
EA1xEA1	4	1	0	4	4	4	4	1	1	1	1	0	0

The decomposition is

EA1xEA1   1 EA1   1 A2A1   1 A1A1

Within the group S3-DP-S4

ET2	6	-3	0	-2	2	2	-2	1	0	-1	1	0	0
ET2	6	-3	0	-2	2	2	-2	1	0	-1	1	0	0
ET2xET2	36	9	0	4	0	4	4	1	0	1	1	0	0

The decomposition is

ET2xET2   1 ET2   1 ET1   1 EE   1 EA1   1 A2T2   1 A2T1   1 A2E   1 A2A1   1 A1T2
          1 A1T1   1 A1E   1 A1A1

being promoted as a standardized dialect which should be easily transportable between different computers. (It has not been determined if it could operate on any microcomputer implementations of Common Lisp.)

Other Group Theory Software

Other published reports of computer programs applied to group theory include the following. J.J. Cannon (University of Sydney) is a mathematician who has led the writing of a large set of Fortran programs to generate and study groups from a mathematician's point of view (6). C. Trindle (Univ. of Virginia) has written programs in Basic, which execute on an Apple microcomputer (7). These programs are also intended to be used for academic instruction in group theory as well as for research work. K. Balasubramanian (Arizona State) has written programs which use the wreath product formalism to generate the permutation operators for non-rigid molecules (8), and compute nuclear spin statistical weights (9).

Future Plans

Some features which will be included in future developments of these programs include; the evaluation of semi-direct products between groups, the direct evaluation of the character tables from scratch, and storing in files the record structures which contain the information about the larger groups. The first two are necessary for useful work to be accomplished for non-rigid molecule groups, since their construction usually includes the semi-direct product combination of subgroups. The third feature is intended to avoid repeating long computations which occur for large groups, even on a computer.

Literature Cited

1. Bunker, P.R., "Molecular Symmetry and Spectroscopy"; Academic Press: New York, 1979.
2. "Symmetries and Properties of Non-rigid Molecules"; Maruani, J., Serre, J., Ed.; Elsevier: Amsterdam, 1983.
3. Ezra, G.S. "Symmetry Properties of Molecules"; Springer-Verlag: Berlin, 1982.
4. Altmann, S.L. "Induced Representations in Crystals and Molecules"; Academic Press: N.Y., 1977.
5. Steele, G.L., Jr. "Common Lisp"; Digital Press: Burlington, MA, 1984.
6. Cannon, J.J. in "Computational Group Theory"; Academic Press: London, 1984: pp 145-83.
7. Trindle, C. J. Computat. Chem. 1984, 5, 162-9.
8. Balasubramanian, K. J. Computat. Chem. 1983, 3, 302-7.
9. Balasubramanian, K. J. Computat. Chem. 1982, 1, 69-74, 75-88.

RECEIVED December 17, 1985

ORGANIC SYNTHESIS

17

A Multivalued Logic Predicate Calculus Approach to Synthesis Planning

W. Todd Wipke and Daniel P. Dolata

Department of Chemistry, University of California, Santa Cruz, CA 95064

Stereochemical principles of synthesis planning have been axiomatized using first-order predicate calculus with a multi-valued logic as implemented in the QED system. Given the definition of a synthetic target molecule as a set of axioms, QED is able to infer a synthesis plan in high-level terms without reference to reactions. Key benefits of this approach are clarity of expression and transparency of the system: all chemical knowledge used is explicit in the axioms.

The purpose of this research was to explore the representation, manipulation, and utilization of strategic knowledge in organic synthesis planning. The method we decided to explore was to create an axiomatic theory to replace our intuitive theory about chemical synthesis. This formal method of reasoning is very powerful in that it completely eliminates any questions about the method used to reach a conclusion. Since any conclusions reached would be theorems of the axiomatic theory, the acceptability of the conclusions rests completely on the acceptability of the postulates and not upon the method of reasoning. We are then free to focus on the chemical principles which are provided as postulates.

In this paper we describe the need for planning, and then develop the predicate calculus we used and the choice of multi-valued logic. Finally we briefly describe the QED program, a few rules, and an example analysis. Other papers in the QED series will cover the program and chemical results in detail.

Need For Planning

Synthesis planning programs such as SECS,(1)(2)(3) LHASA,(4)(5) or SYNCHEM(6)(7) work backward from the target molecule to be synthesized toward starting materials. By applying applicable chemical transforms (inverse chemical reactions) to the target, the first set of chemical precursors is generated. Each of these in turn can be considered a new target and processed in like manner

0097-6156/86/0306-0188$06.00/0
© 1986 American Chemical Society

recursively. This process develops a "synthesis tree", where nodes in the tree correspond to chemical structures and edges to chemical transforms. The fundamental problem is that there are many possible chemical transforms that can be applied and typical syntheses require several steps. If each molecule in the tree has ten precursors, by the time we reach the sixth level, one million precursors must be evaluated! If our program can process 10 precursors per second, this will require a day. This example actually underestimates the size of the problem, because the typical branching factor is between 100 to 200 rather than ten, and many syntheses require more than six or seven steps.

Approaches to Large Search Spaces

- Heuristics
- Macro Operators
- Abstraction
- Planning

A heuristic is a rule of thumb may lead to a shortcut in the solution of a problem. If such a heuristic removes 90% of the routes at each level, then it will eliminate 99% of the possible routes by the second level, 99.9% by third, etc.. Even a simple heuristic can make the problem far more tractable.. Gelernter's SYNCHEM II(7) is an example of a program that has focused on heuristic evaluation functions for reducing the search space that must be explored. Wang used macro operators to establish "planning islands" that can then serve as near-term objectives.(8) The macro operators make bigger jumps in the search space, thus eliminating much branching and combinatorics. Abstraction can be used to simplify chemistry so there are fewer kinds of functional groups, and fewer chemical transforms, thus a reduced search space.(9) Planning provides direction in the search space, thus permits pruning of pathways which are headed in the wrong direction and permits focusing resources in a particular direction.

Providing a Sense of Purpose. Planning is more than just a heuristic evaluation function that measures complexity. The ultimate goal of synthesis is to prepare the complex from things simple. Thus in Figure 1 our goal is to find syntheses that lead downward, but some excellent syntheses may require increasing the complexity of a precursor in order to ultimately lead to a very simple one as the path $T \rightarrow P_1^{-1} \rightarrow P_6^{-2}$ in Figure 1 illustrates. Going uphill in complexity is acceptable if there is some purpose and a plan provides that sense of purpose.

Plan Representation in SECS. The SECS Simulation and Evaluation of Chemical Synthesis program explicitly represents its plans(2) as a list structure of goal instructions with logical connectives. A goal instruction can specify one of the following:

- Introduce a functional group at a position
- Change a functional group at a position
- Make or Break a bond
- Use an atom, bond, or group

The logic instruction can include AND, OR, NOT, or XOR, and also includes an action to take if the goals beneath it are not achieved. The actions generally modify the evaluation score for a synthesis pathway or completely terminate consideration of a synthesis pathway.

SECS uses the goal list to select transforms that appear to have the potential for satisfying the goals, based on the character of the transform. The character of a transform specifies the types of architectural changes the transform may effect. The goals specify desired architectural changes in the molecule. It is the responsibility of the SECS program to find transforms that can achieve the goals. The strategy module of SECS creates a plan and writes it on the goal list. The chemist may modify those goals or add new ones.

The Strategic Basis. The strategic basis for designing a synthesis plan rests on general principles of molecular architecture construction, and is independent of reaction knowledge. Examples include symmetry of the target molecule, potential symmetry of the target molecule, the relative reactivity of functional groups in the target, consideration of potential starting materials, the connectivity of the structure, and the control of stereochemistry. A symmetry-based strategy for β-carotene is shown in Figure 2. The reaction-independent principle is to construct the molecule from identical pieces to take advantage of the symmetry of the structure. The resulting goal structure is a set of three alternative goals, each of which specifies two bonds that should be broken in the analytical direction.

Since we were interested in studying these strategies, we wanted a means for explicitly representing the principles that enabled us to easily modify them and to be able to easily understand exactly what principles the program was using. For this reason the QED project was initiated.(10) QED was to use statements of the principles together with a definition of the molecule and then infer a reasonable set of strategies for the synthesis of the molecule and write these to the SECS goal list.

First Order Predicate Calculus

We chose the first order predicate calculus (PC) as our language for representing synthetic principles. The first order predicate calculus (PC) is a "formal" system of logic.(11)(12)(13) In this context, formal means that it is the form of the arguments that is important, not the actual content. The term "calculus" comes from the meaning "a method of calculation", and does not refer to Newton's differential calculus.

To give an example, the following form of argument is known as modus ponens: "If statement **A** implies conclusion **B**, and we know **A** to be true, then we may conclude B" and may be represented as follows:

    A => B
    B
 :. B

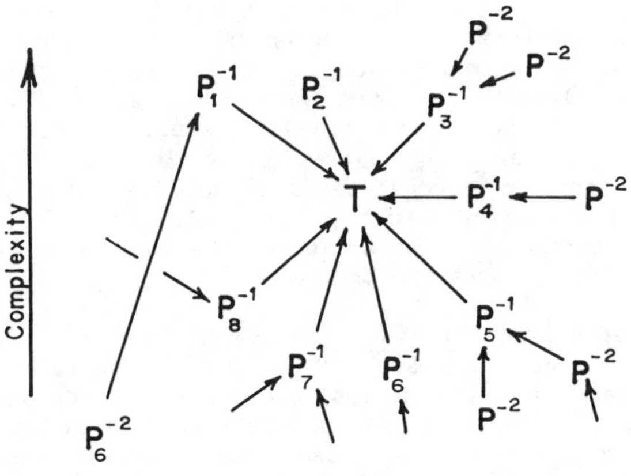

Figure 1. Plot of the complexity of synthesis tree.

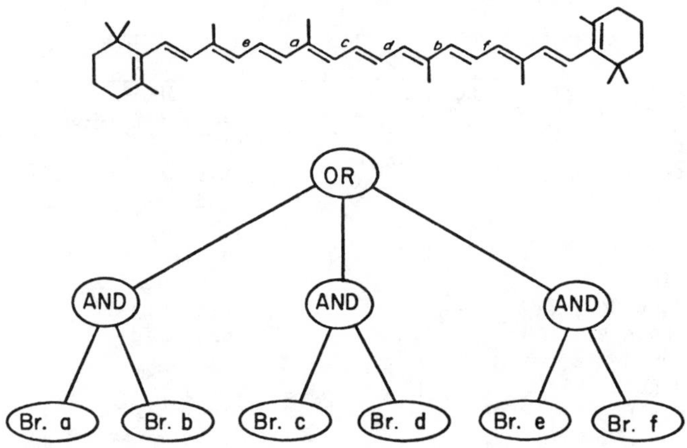

Figure 2. Symmetry-based strategies for β-carotene.

It is possible to substitute various meanings to the statements **A** and **B**, and as long as this form is followed, the conclusion is said to follow from the premises. **A** could be the statement "A ketone is present" and **B** could be "A carbonyl is present". If **A** is true, then **B** follows from the premises.

If the conclusion seems to be in error, and the chain of reasoning is valid, we have then demonstrated that one (or more) of our initial assumptions must be erroneous. For example, if one of the premises was the implication "If a ketone is present then the compound is an alkane", then the conclusion that would follow from the observation "a ketone is present" would be obviously false. Since the PC is logically correct (in fact defines "logically correct") any erroneous conclusions must arise from an erroneous axiom, and cannot be the fault of the calculating procedure used. This allows us to focus our attention on the assumptions and frees us from having to worry about the procedure.

A Working Definition of the Predicate Calculus. In a formal theory(13) the statements are written in a specially constructed symbolic language, and are manipulated in accordance with specified rules which make no appeal to any possible meaning of the symbols. A string is a finite sequence of these formal symbols. There exists a grammar for deciding if a string is a statement. There is a method for determining if a statement is an axiom. This method involves pattern matching, and perhaps rearrangement and rewriting. Given a finite sequence $S_1, \ldots , S_k$ of statements, there is a procedure for deciding if S_k follows from one or more $S_1, \ldots , S_k$ by the rule of inference. The formal symbols defined are shown in Table I.

Table I. The Formal Symbols of the PC used by QED.

Name	Normal Representation	QED Representation
For All	$\forall$	$All
There Exists	$\exists$	$Exists
And	$\land$	.and.
Or	$\lor$	.or.
Not	$\sim$	.not.
If/Then	=>	if then
If and Only If	<->	if only-if
Therefore	:.	(not used)
Parenthesis	( )	( )
Brackets	[ ]	[ ]
Curly bracket	{ }	{ }
Comma	,	,
Predicate	P,Q,R...	Initial capital letter
Identifier	i,x,z...	All small letters

To decide if a formula is a well formed formula (wff) of the PC, it must conform to the following definition:

1. There exists a set of variable symbols, a ... z, which can hold places. We can also define a set of objects, $o_1 \ldots o_n$, also known as constants. This set of objects is known as the domain.

2. For each of n = 0, 1, ... there is a possibly empty set of n-place predicates. These are denoted by $P(x_1, ... , x_n)$. This predicate may represent some property of an object, or a relationship between objects.

3. An "atomic formula" of the PC is formed by taking a n-place predicate symbol $P(x_1, ... , x_n)$, and substituting variables or constants for any of the x_j.

4. A formula of the PC is formed by combining atomic formulae using the connective symbols ∨, ∧, ~, => or <->.

5. A formula of the PC may be preceded by a quantifier, ∀ or ∃.

6. Parentheses are used to avoid confusion when the order of evaluation or the scope of the arguments is important. Parentheses, square brackets and curly brackets are all equivalent, and may be used interchangeably, as long the same type is used to open and close the same term.

If A, B and C represent atomic formulae, then the following are examples of formulae:

~ A   ∀ A
A ∨ B   ∃ x B
A ∧ B   A ∧ (B ∨ C)
A => B   ((A => B) => C)
A ∨ (B => A)

An object may be any "thing". This can include tangible items such as atoms, or bonds, or can include non-tangible things such as goals, or plans. In the first order PC, an object may not be a predicate.

Predicates are properties of objects, or relationships between objects. A predicate is derived from the predicate clause of a sentence. In the sentence "atom 5 is a carbon", the predicate clause is "is a carbon". This would be represented as Is-carbon(atom5). To avoid confusion, names of predicates and objects are written without embedded blanks or spaces.

Translating Chemical Statements into Predicate Logic.   In the following examples, we use the QED representation for connectives and quantifiers.

an atom which is a ring atom
Ring-atom (x)   (1)

atom5 is a ring atom
Ring-atom (atom5)   (2)

atom4 and atom5 are ring atoms
Ring-atom (atom4) .and. Ring-atom (atom5)   (3)

atom4 and atom5 are alpha
Alpha (atom4, atom5)   (4)

Note the difference between equations 3 and 4, where an ellipsis makes the two phrases look similar. However, if 3 is rewritten as shown in 5 the difference is obvious. In equation 5, the English word "and" is used in the logical sense, whereas in equation 4, "and" indicates items which will be related by a predicate phrase.

atom4 is a ring atom   and   atom5 is a ring atom   (5)

atom4 and atom5 or atom6 are beta
Beta (atom4, atom5) .or. Beta (atom4, atom6)   (6)

bond12 is not an appendage bond
.NOT. Appendage-bond (bond12)   (7)

$All (x) $All (y) $All (z)
   [If Atom (x) .and. Atom (y) .and. Atom (z) .and.
     Alpha (x,y) .and. Alpha (y,z) .and.
     .not. Identity (x,z) then Beta (x,z)]   (8)

Postulate 8 defines the beta relationship from the alpha predicate (alpha(x, y) is true if bond(x, y) is true). The term ".not. Identity (x,z)" must be included in axiom 8 to prevent an atom from being beta to itself (x=z).

Definition of Axiomatic Theories. An axiomatic theory is an attempt to formalize an intuitive theory. Geometry was intuitive before Euclid wrote "The Elements". An intuitive theory is defined as a body of knowledge which attempts to express relationships and causality between objects, but is not formal. Most modern science is still intuitive, even though it may represent many of it's findings in exact mathematical formulae. As long as the entire corpus of knowledge is not expressed in a single formal system, it will remain intuitive.

The following list of the steps is necessary to create an Axiomatic Theory:

1. Provide a set of symbols, and associated definitions. This set of symbols will include both objects, and predicate relations. Together with the set of symbols defined by the associated logic, these will be the only allowed symbols.
2. Provide a set of axioms (postulates). If one is attempting to create an axiomatic theory which mirrors experimental reality, then these axioms should express some fundamental properties of the system you are trying to model.
3. Choose a type of logic to associate with the symbols and definitions. This will be used to deduce all further statements of the theory. This logic will provide at least one rule of inference, and rules of combination of truth values.
4. Any new statement created by the rule of inference upon the postulates is known as a theorem of the theory and may be referred to as "valid" or "proved" within the axiomatic theory.

An axiom is defined as a principle which holds across all domains of knowledge such as, "if two objects are equal to a third object, then they are equal to each other". Postulates are statements which are given without proof, based only on the definitions provided before. Euclid called postulates "self-evident truths", and he felt that they mirrored some fundamental principle of the universe. The idea of postulating an obviously false statement such as "two parallel lines may intersect at some point" seemed useless. However, Bolyai-Lobachevsky geometry does just this and has met with success in space-time physics, where large bodies can create bends in space such that parallel lines can intersect.

In modern axiomatic theory, postulates and axioms are defined simply as given statements. By the definition of an axiomatic theory the concept of truth is not considered relevant to its construction. If we can derive a theory which seems to mirror reality as reported by our current experimental knowledge, then we consider the postulates to be "successful" in some sense of the word. If the theory derived from the postulates clash drastically with our observations, the postulates can be thrown away as "non-relevant". If the differences are slight, or if the theory predicts new experiments which should show differences from what the intuitive theory would predict, we can even call the axiomatic theory "interesting".

Why More Systems Haven't Been Axiomitized. Geometry is unique in that it can be expressed in a simple logic, the results are either true or false, and that the actual "experiments" were capable of being done with thought alone. In chemistry there was not sufficient knowledge to enumerate the basic definitions and postulates. The recent explosion of knowledge in chemistry has made it feasible to begin the process of axiomatization of chemical theories.

Geometry is also special in that most examples we wish to reason about consist of a few objects, with a limited number of relationships. Thus it is feasible to use a simpler sentential calculus form of logic which cannot reason with variables. Sentential calculus can only be used when the total space of all statements is easily numerable, since all properties about objects and relationships must be stated in separate explicit sentences.

But in chemistry, where a typical molecule will have 20 - 30 atoms, as many bonds, several rings, stereocenters, hetereoatoms, etc., a theory expressed in the sentential calculus would require thousands of statements. Thus chemistry had to await development of the predicate calculus,(14) to axiomatize the theory.

Finally, while geometry was axiomatized with bimodal logic, chemistry required creation of a new (within last few decades) branch of logic rich enough in expressive power to manipulate uncertain statements without over or understating their value. The next section introduces these new logics.

Bimodal Logic

There is a principle of logic called the principle of the "Excluded Middle", dating back to Plato. The principle states that "a statement is either True or False". The Heisenberg uncertainty principle however forces us to recognize uncertainty as a reality. Logics capable of expressing statements containing uncertainty are now available.(15) These new logics can be viewed as extensions of bimodal logic.

Redefining True, False, and the Value of Connectives. We begin by
replacing T with 1, and F with 0, mapping atomic formula onto the
range {0,1}. Then the truth value of a formula A is either 0 or 1:

Definition 1: $v(A) = 1$ "true"

Definition 2: $v(A) = 0$ "false"

For any wff F, there exists a function $v(F)$, which will map F
onto the range {0, 1}. We determine $v(F)$ by the following technique:

1. If F is an atomic formula, $P(x1, \ldots xn)$, then we can say that
    P maps the tuple $(x1, \ldots xn)$ onto the range {0, 1}. This
    mapping is generally done by explicit statement, i.e.

        Socrates is Mortal.
        Mortal(socrates) maps to 1
        v(Mortal(socrates)) = 1

2. If F is a composite formula, we define functions which extend
    the values of the terms to the formula.

Definition 3: $v(A \wedge B) = \min (v(A), v(B))$

Definition 4: $v(A \vee B) = \max (v(A), v(B))$

Definition 5: $v(A \Rightarrow B) = \min (1, (1-v(A))+v(B))$

Definition 6: $v(A \Leftrightarrow B) = \min (v(A), v(B))$

Definition 7: $v(\sim A) = 1 - v(A)$

We will examine implication (Definition 5) in detail because it
is less obvious and quite important to QED. We start by enumerating
the four possible cases for implication (Table II):

Table II. Truth table from Definition 5 for A => B.

A	$v(A)$	B	$v(B)$	$\min (1, (1-v(a))+v(B))$	
True	1	True	1	1	case1
True	1	False	0	0	case2
False	0	True	1	1	case3
False	0	False	0	1	case4

Only case 2 has a zero value for the result of the combination
formula. It is only when the antecedent A is true, but the
conclusion is false that the rule is considered to be bad. For
example, given the rule shown in equation 10, then only if the
molecule is a protein but it contains no amide is the rule shown to
be faulty. If the molecule is not a protein, the respective truth or
falsity of the assertion that it contains an amide would not affect
the truth value of the implication. An implication rule sets no
limitation upon the value of B if A is not true, so case3 and case4
are valid.

IF Is-protein(mol) then Contains-amide(mol)   (9)

In summary, the value of an implication rule is an inverse measure of how often the antecedent has a higher value than the consequent. This is very important in multi-valued logics where the truth value ranges over many numbers rather than just 0 and 1. The more "valuable" the rule, the more often it implies the correct consequence.

Lukasiewicz-Tarski Multi-Valued Logic

In chemistry, uncertainty may arise because we are ignorant of some of the underlying principles, or just have not done enough experiments yet. The normal two valued logic is not satisfactory for many complex problems, e.g., principles of chemical synthesis. The solution is to add additional values to our logic so that we can represent "I don't know." One of the more popular multi-valued logics (MVL's) was created by the Polish logicians, Lukasiewicz and Tarski (LT) about 50 years ago.($\underline{15}$) This logic, called LT, is further identified as LT_n, where n represents the number of discrete values that are covered. LT_2 is the same as the bimodal PC.

Only small changes are necessary to convert our definition of the bimodal logic into the LT logic. For a given atomic formula, $P(x_1, \ldots x_n)$, we say that P maps the tuple $(x_1, \ldots x_n)$ onto a range $\{0, 1, \ldots, n-1\}$, where n is the order of the LT. Every appearance of the number "1" in our previous definitions (1-7) is replaced by "n-1" where n is the value of the LT. For example, Definition 5 is changed from min $(1, (1-v(A))+v(B))$ to min $((n-1), ((n-1)-v(A)) + v(B))$. When n=2, the formula is the same.

We will illustrate how the LT logic works using the simplest logic, LT_3. The range of LT_3 is 0, 1, 2. These numbers can be thought of as expressing the English terms False/0, Maybe/1, and True/2. The rule for explicit evaluation would be used to assign values to the following examples:

There is an atom alpha to atom5.
v(\$Exists x Alpha(x, atom5)) = 2   (10)

There are no atoms which are alpha to themselves.
v(.NOT. \$Exist x Alpha(x,x)) = 2   (11)

Having assigned values to the atomic formulae, the modified formulae of definitions 3 through 7 are used to assign values to wffs.

Atom5 is on a ring
Atom6 might be on a ring
v(On-ring(atom5)) = 2
v(On-ring(atom6)) = 1   (12)

It might be true that both atom5 and atom6 are on a ring.
v(On-ring(atom5) .AND. On-ring(atom6)) =
min (v(A) , v(B)) = min (1, 2)   = 1   (13)

It is certainly true that at least one of atom5
or atom6 are on a ring.
v(On-ring(atom5) .OR. On-ring(atom6)) =
max (v(A), v(B)) = max (1, 2)   = 2   (14)

The value of the rule "If atom5 is on a ring, then
atom6 is also on a ring" cannot be established from
the data at hand.
v($All x $All y (If On-ring(x) then On-ring(y))) =
         max (n-1 - v(A), v(B)) = max (2-2, 1) = 1 (15)

Lets focus on the implication rule in equation 15. If the rule
is considered to be "good" in LT_2, it will have a value of 2. Thus it
has to be the case that the second term in the valuation formula
(Definition 5), (n-1)-v(A)+v(B) must be greater than or equal to 2.
Table III shows the values of B allowed for given values of A as a
function of the "goodness" of the rule. If the rule has value 0,
then any value of B is acceptable no matter what the value of A is.

Table III.   Allowed values of B for A => B in LT_3.

v(A)	v(A => B) = 2	v(A => B) = 1	v(A => B) = 0
2	2	1, 2	0, 1, 2
1	1, 2	0, 1, 2	0, 1, 2
0	0, 1, 2	0, 1, 2	0, 1, 2

In assigning a value to B from the possible range of values for
B, we must select the lowest value, since only this value is
supported by the implication, the other values are only possibilities
that cannot be ruled out. Other implications may infer the
consequent B with a higher value.

Cumulative Evidence. LT does not mirror human intuition about
accumulation of evidence. Generally if one has more pieces of
orthogonal evidence which support a deduction, then there is more
reason to believe that the deduction is true. However, this is not
the case with LT logic. For example, consider the following.

       (A1 => B) v=1   (imp1)
       (A2 => B) v=2   (imp2)
       .
       .
       (A9 => B) v=9   (imp9)

Assume that we are using LT_{11}, and that A1 ... A9 all are True
with v=10. In this case, B will be inferred from each of these rules
with v=1 to v=9 respectively. Using an LT combining function for two
values, v1 and v2, we will choose v(B) as max(v1, v2). Thus, only 1
piece of evidence (imp9) will actually matter, and A1 through A8
could assume any value from 0 to 10 and not affect the value of B.
The reader is recommended to the book by Ackermann(15) for the
complete expostulation why LT logic uses this combining function.
Basically LT logic was designed to avoid some of the many
difficulties that arise when applying MVL to mathematical domains.

Unlike chemistry, mathematics often deals with infinite domains, and infinite axiom sets. If we allow the fact that two axioms infer the same conclusion to increase the truth value of that conclusion, we must choose some increment that reflects the importance of each individual axiom. If there are an infinite number of such axioms, then each axiom becomes infinitesimally important. Thus LT logic chooses to err on the side of conservatism, assuring that the conclusions will be valid, though perhaps less strong than they could actually be.

Obviously we should not allow multiple iterations of the <u>same</u> rule to increase the value of the consequent. If this were to be allowed then one could obtain any final value by simply re-iterating the same rule sufficient times. But redundancies in rules arise in subtle ways, e.g, **B => A** and **C => A** where B **<->** C, i.e, B is another name for C. Finally, it can be shown that even if the chain of relation between B and C contains logical connectives other than **<->**, then allowing two successive inferences to increase the value of the consequence above that inferred by the strongest alone can lead to problems.

In chemistry, where the axioms are generated by formalizing finite human experience, it is reasonable to allow evidence to accumulate and we did in QED. It is probable that all our axioms will not be orthogonal. Simple perception concepts are a part of the antecedents in many different rules, hence there is some commonalty. Assuring that our rules are orthogonal in all of the many possible combinations is a difficult task.

Incremental Multi-Valued Logic (IMVL)

We utilized in QED a type of MVL which is similar to a very popular form of MVL known as Bayesian Logic. There are several unfortunate problems with Bayesian logic, including the fact that Russell showed that this logic incorporates several fatal paradoxes. Fortunately, these paradoxes only manifest themselves in infinite systems. There are still problems with finite systems, such as the ability to assign unwarranted values to conclusions if the data base is aapoorly constructed. But there are significant advantages to this logic.

Once again the logic maps a set of statements onto a range. In this case the range will be the rational numbers from $-m$ to $+m$, where $-m$ is equivalent to False, and $+m$ is equivalent to True, with complete ignorance at 0.

However, the value of an atomic formula is comprised of three parts, the confirmation value, disconfirmation value, and the combined truth value:

$$\text{confirmation value} \quad CV: \quad 0 =< \quad CV \quad =< +m$$
$$\text{disconfirmation value} \quad DV: \quad 0 =< \quad DV \quad =< +m$$
$$\text{truth value}, \quad TV = CV - DV: \quad -m =< TV \quad =< +m$$

An advantage of carrying CV and DV is that one can recognize from the magnitude of CV and DV the amount of concurrence or conflict in support of a given inference.

For any explicit assignment of an atomic formula, only the CV or DV is assigned. The TV is then calculated from that. For the value of wffs comprised of assigned terms, the following formulas are used:

Definition 8: CV $(A \wedge B)$ = min (CV (A), CV (B))

Definition 9: DV $(A \wedge B)$ = max (DV (A), DV (B))

Definition 10: TV $(A \wedge B)$ = CV $(A \wedge B)$ - DV $(A \wedge B)$

Definition 11: CV $(A \vee B)$ = max (CV (A), CV(B))

Definition 12: DV $(A \vee B)$ = min (DV (A), DV(B))

Definition 13: TV $(A \vee B)$ = CV $(A \vee B)$ - DV $(A \vee B)$

Definition 14: CV (~ A) = DV (A)

Definition 15: DV (~ A) = CV (A)

Definition 16: TV (~ A) = CV (~ A) - DV (~ A)

Definition 17: CV (A => B) = max (DV (A), CV (B))

Definition 18: DV (A => B) = min (DV (A), CV (B))

Definition 19: TV (A => B) = CV (A => B) - DV (A => B)

While these formulae are more complex than those of the LT, they give the same results as do those of LT logic. Two features of the IMVL are however quite different from the LT: 1) the rules for incrementally acquiring evidence; and 2) the rules for computing the value of a consequent of an implication when the antecedent is not fully True.

Incrementally Acquiring Evidence. Unlike LT logic, the IMVL allows successive inferences about a fact to increase the truth value of that fact. One way of viewing the way that the IMVL deals with inferences is to say that an inference in support of a theorem decreases our ignorance about that theorem. Thus, when the theorem is first proposed, the ignorance is maximal, the values for CV, DV, and TV are all 0. The amount of ignorance about the CV (or DV) could be said to be m.

The first inference of a theorem with value v, (v =< m), then reduces our ignorance by v. If the value v was in confirmation of the theorem, then the values become DV = 0, CV = v, and TV = v. We have reduced our ignorance about the CV to m-v. Further confirmatory evidence for the theorem is applied to the remaining measure of ignorance. The truth value is calculated from CV and DV as normal. The formulae for this are:

Definition 20: CV(A, given A1, A2) = CV(A1) + CV(A2)*[m - CV(A1)]/m

Definition 21: DV(A, given A1, A2) = DV(A1) + DV(A2)*[m - DV(A1)]/m

Definition 22: TV(A, given A1, A2) = CV(A, given A1, A2) – DV(A, given A1, A2)

It can be shown that CV and DV approach the maximal value m asymptotically and that the order of acquiring truth values does not matter.

Implication in IMVL. The second major difference is in how the value of the consequent of an implication is calculated. The method is multiplicative as compared to the minimum function of LT MVL. The formulae involved are shown below where TV'(A) = TV(A) iff TV(A) > 0; otherwise 0.

Definition 23: CV (B, when A => B) = TV'(A) * TV(A => B)/m

Definition 24: DV (B, when A => B) = unchanged

Definition 25: CV (B, when A => ~ B) = unchanged

Definition 26: DV (B, when A => ~ B) = TV'(A) * TV(A => ~B)/m

With this background on the theoretical representation issues, we will now briefly describe the QED program that implemented this IMVL PC logic.

The QED Program

QED was implemented on the SUMEX-AIM DEC 2060 TOPS-20 system. The source code consists of about 18000 lines for FORTRAN code and 1500 lines of macro code. A block diagram of the program modules is shown in Figure 3. QED itself contains no chemical information. The chemical knowledge is stored as postulates in a formal first order predicate calculus language. The grammar for this language is also explicitly described in the BNF notation. The PARSER interprets the postulates and interaction with the user, both for entering questions and also for entering new rules interactively. The QED EXEC handles opening of files, entry of a molecule, and debugging aides. The AGENDA EXEC creates, prioritizes, selects, performs, and deletes tasks. The INFER EXEC selects rules, examines the data base, instantiates predicates and interprets the logic. All information, including postulates, rules, dictionary, instantiations, tasks, etc., is stored in an associative relational data base. The ANSWER EXTRACTER and FORMATTER communicates the answer to a question in a form the chemist can understand and that SECS can understand. The design of the system is very much like the Japanese 5th Generation Computer System design which is also based on logic.

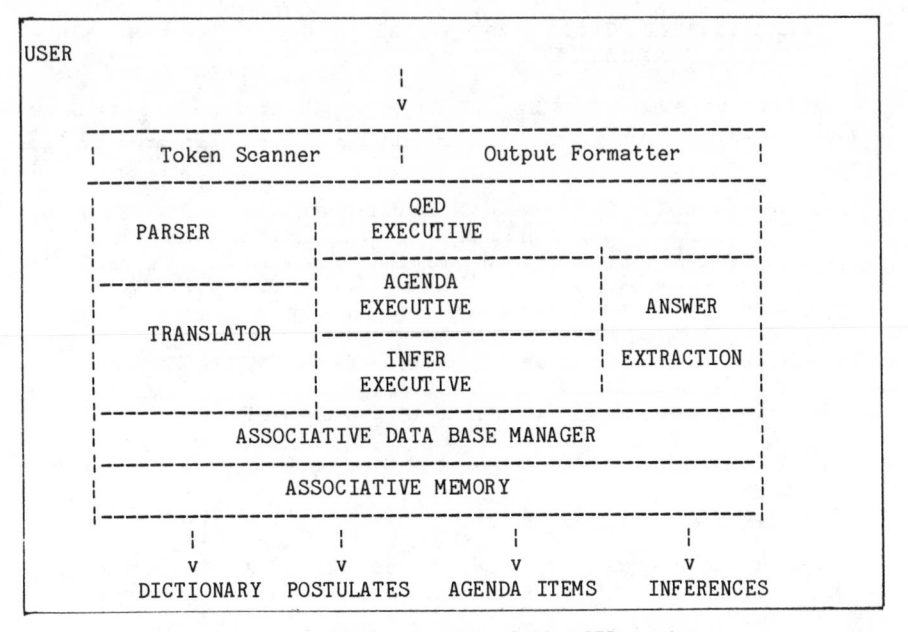

Figure 3. Block diagram of the QED system.

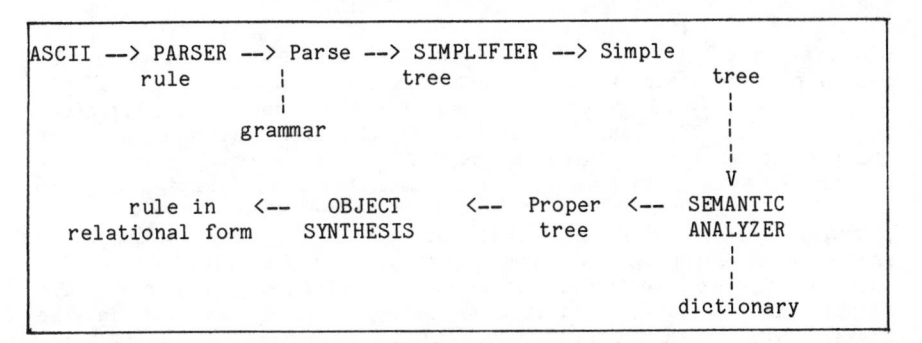

Figure 4. Compilation process for rules.

QED Rule Parsing

Since FORTRAN (unlike LISP) cannot easily accept ASCII representations of rules and use them directly, they must be read, parsed, analyzed and translated to the form QED can interpret. The general flow of the compiler is shown in Figure 4. As an example, lets follow the processing of the rule "ALPHA-TO-SC" that defines sites where stereochemical induction may occur:

```
Rule ALPHA-TO-SC
        $All Atom(x) $All Atom(y)
        [IF   Stereocenter(x) .AND. Alpha (x,y)
        THEN Alpha-anisotropic (x)] CF 0.7
```

The rule is parsed in a top-down fashion(16) using a BNF driven parser according to the explicit grammar, a portion of which is shown in Figure 5. This is a fairly simple context free grammar, written in the BNF (Backus Normal Form) style.(17) A condensed version of the parse tree for the sample rule is shown in Figure 6. Semantic analysis checks for:

- Recursive rules with identical bindings
- Unbound variables
- Variables that are improperly scoped
- Predicates and functions having the incorrect number of arguments
- Predicates and functions having improper types of arguments
- Quantifiers incorrectly scoped
- Predicates and functions incorrectly defined

Once this has been done, the translation to internal form can be performed. Finally the rule is added to the axiom data base.

```
<z> ::= <rule> ';' ;
<rule> ::= <ruleid> <quants> <implication> <certfact> ;
<ruleid> ::= (  'Rule' ! 'rule' ! 'RULE' )  <identifier> ;
<quants> ::= <quanpair> [ <quants> ] ;
<implication> ::= '[' <antecedent> <impsymbol> <consequent> ']' ;
<impsymbol> ::=  'then' ! 'Then' ! 'THEN' ;
<antecedent> ::= [ 'If' ! 'IF' ] <formula> ;
<formula> ::= <and-node> ! <or-node> ! <atomicform> !
              <quanpair> '[' <formula> ']' ;
<and-node> ::=  ( <atomicform> ! '(' <or-node> ')' )  <and-op>
              ( <and-node> ! <atomicform> !  '(' <or-node> ')'  ) ;
```

Figure 5. A Portion of the BNF Grammar for QED/s language.

```
          <rule>
           /
          /
     <quants><quants><implication><CF value>
       /      /           \           \
$All Atom x    /           \         CF 0.7
             /              \
   $All Atom y       <antecedent> <impsymbol> <consequent>
                           /           !           !
                          /            !           !
                  <conjunction>      THEN      Alpha-anisotropic x
                     /   !   \
                    /    !    \
   Stereocenter y    .AND.    Alpha x y
```

Figure 6. Simplified parse tree of the ALPHA-TO-SC rule.

Associative Relational Data Base

Everything in QED is stored in "triples" like in the LEAP language.(18) Each triple consists of an index, an attribute, and a value. QED maintains pointer lists to entries that have the same index, attribute, or value so that it can quickly retrieve relations given any combination of I, A, or V. The triples are stored in QED's software implemented virtual memory that is mapped to disk. The internal form of the ALPHA-TO-SC rule is shown in Table IV.

Table IV. The internal form of the ALPHA-TO-SC rule.

Index	Attribute	Value
1	isa	rule
1	Rule-id	"Alpha-to-SC"
1	son	index # 2
2	isa	quant-description
2	quantifier	$All
2	variable	"x"
2	parent	index # 1
2	son	index # 3
3	isa	quant-description
3	quantifier	$All
3	variable	"y"
3	parent	index # 2
3	son	index # 4
4	isa	inference
4	antecedent-son	index # 5
4	consequent-son	index # 6
4	CF	value 0.7
4	parent	index # 3
5	isa	conjunction
5	formula-son	index # 7
5	formula-son	index # 8
5	parent	index # 4
7	isa	atomic-formula
7	Predicate	"Stereocenter"
7	variable-1	"x"
7	parent	index # 5
8	isa	atomic-formula
8	Predicate	"Alpha"
8	variable-1	"x"
8	variable-2	"y"
8	parent	index # 5
6	isa	atomic-formula
6	Predicate	"alpha-anisotropic"
6	variable-1	"x"
6	parent	index # 4

Agenda List Control

QED puts problems to be solved on the Agenda List. Initially the top goal is the task to find a plan for a molecule. Vital information such as task type, task pointer, parent, priority, clock tick, item depth, and the current unification set with truth values is stored as part of the task, again as associative relations. QED uses heuristics to prioritize the tasks by examining the number of terms in a rule and the types of connectives and quantifiers and estimating the amount of work required to complete the task. Easy tasks and tasks which may fail early are chosen first in order to truncate extensive search. We use the agenda list to provide "best first" control.

Example Rules

Before we can consider an example application of QED, we need to present some of the rules we have developed for stereochemical control in chemical synthesis and then see how they are used in developing a plan.

Rule Suggest-control-sc
  $All Atom (x)
  [IF Stereocenter (x) THEN Control-sc (x)] CF 0.8 ;

Rule Connect-to-control
  $All Atom (x) $All Atom (y)
  [IF   Control-sc (x) .and. Anisotropic (y)
  THEN Connect (x,y)] CF 0.8 ;

Rule Connect-apps-for-control
  $All Atom(z) $All Appendage (y) $All Ring (r)
  [IF Root-of-appendage (z,y)   .and.
     Control-sc (z) .and. Atom-of-ring (z,r)
  THEN Reconnect-app (y,r)] CF 0.8 ;

Rule **Suggest-control-sc** says simply if there is a stereocenter at an atom, then it is important to control stereochemistry there. **Connect-to-control** tells one way for controlling stereochemistry at a center, namely to connect that center to another center that is sterically differentiated. **Connect-app-for-control** states that if the center to be controlled is on a ring and is the root of an appendage, then it might be a good idea to reconnect the appendage to the ring to form a new ring. Currently, QED has rules for reconnection of appendages, removal of stereocenters, making transannular bonds, breaking appendage bonds, increasing steric hindrance, and using functional groups.

Example of Analysis

A very simple dialog with QED will be presented for the target molecule shown below:

Appendage1 consists of atoms 1-4, Appendage2 consists of atoms 9-12, and Appendage3 = atoms 5, 14-16. The molecule is initially represented as a standard SECS MOLFILE, i.e., a connection table. In the QED dialog on this problem in Figure 7, the user typing is shown underlined. The **molin** command reads the connection table 3app.mol then converts the molecule to QED logical predicates, e.g., Atom(atom3) TV 100; Atom(atom4) TV 100; Bond(atom3, atom4) TV 100; Stereocenter(atom4) TV 100; Root-of-appendage(atom4, appendage1) TV 100; appendage(appendage1) TV 100; etc. Thus the molecule is represented as a set of premises within QED which are known to be completely true. The user then asks QED to infer a plan for all atoms x. Connecting two atoms Connect(x,y) is one possible part of a plan.

```
@QED
    - QED -
    For commands type HELP
QED: molin 3app.mol
QED: infer 3app-example $All Atom (x) Plan (x)
    5 User Infer Request 3app-example Plan
    To Be Instantiated by rule "Plan1"

    .
    .
  599 Rule "Plan1" - Plan
  END OF AGENDA LIST
QED: Lookup Reconnect-app-app
    appendage1, appendage2, 75
    appendage2, appendage1, 75
QED: Lookup Reconnect-app
    appendage1, atom10, 67
    appendage1, atom5,  56
QED: Lookup Connect
    atom2, atom11, 60
    atom1, atom5,  50
    .
    .
QED: What Rule Infers Connect
    Rule "where-to-reconn"
QED: Show rule where-to-reconn
    (rule is printed out)
QED: Lookup Stereo-center
    I don't know the word "Stereo-center"
    Please choose one of the following
    0) None of the following
    1) Stereocenter
    :: 1
    atom9, 100
    atom4, 100
QED: Quit
    Thus it is shown.
```

Figure 7.   Sample QED dialog.

To summarize the example, in all cases, the non-stereo appendage, appendage3, was not utilized for reconnection. The truth values for reconnecting appendage1 to appendage2 was 75, and because of symmetry of reconnection, appendage2 to appendage1 is also 75. Reconnection of the appendage to ring atoms: for connection to stereocenter, TV = 72; for connection alpha to stereocenter, TV = 56; for connection beta to stereocenter, TV = 50. We will present the chemical significance of some QED analyses separately elsewhere.

Conclusion

The multi-valued predicate calculus logic as implemented in QED has been demonstrated to be suitable for cleanly representing strategic axioms of chemical synthesis. QED is a powerful tool for exploring inference in the planning of synthesis strategies. QED helped us elucidate key strategic concepts and their interdependence and enabled us to create a consistent rule base. The clarity of the QED PC language allows anyone to easily read and understand the strategic principles and may encourage further axiomatization work by others.

Acknowledgment

This research was supported in part by NIH Grants RR01059 and ES02845, with computational support from SUMEX-AIM, NIH Grant RR-00785 and support from the users of the SECS synthesis program.

Literature Cited

1. Wipke, W. T. "Computer-Assisted Three-Dimensional Synthetic Analysis". In Computer Representation and Manipulation of Chemical Information; Wipke, W. T.; Heller, S. R.; Feldmann, R. J.; Hyde, E., Eds.; John Wiley and Sons, Inc.: 1974, pp 147-174.
2. Wipke, W. T.; Braun, H.; Smith, G.; Choplin, F.; Sieber, W. "SECS — Simulation and Evaluation of Chemical Synthesis: Strategy and Planning"; American Chemical Society: Vol. 61, 1977.
3. Wipke, W. T.; Ouchi, G. I.; Krishnan, S. "SECS: An Application of Artificial Intelligence Techniques". Artificial Intelligence 1978, 9, 173-193.
4. Corey, E. J.; Wipke, W. T. "Computer-Assisted Design of Complex Molecular Syntheses". Science 1969, 166, 178.
5. Corey, E. J.; Wipke, W. T.; Cramer, R. D.; Howe, W. J. J. Am. Chem. Soc. 1972, 94, 421, and adjacent papers
6. Gelernter, H.; Sridharan, N. S.; Hart, H. J.; Yen, S. C.; Fowler, F. W.; Shue, H. J. "The Discovery of Organic Synthetic Routes by Computer". Topics Curr. Chem. 1973, 41, 113.
7. Gelernter, H. L.; Sanders, A. F.; Larsen, D. L.; Agarwal, K. K.; Bovie, R. H.; Spritzer, G. A.; Searlemen, J. E. Science 1977, 197, 1041.
8. Wang, T.; Burnstein T.; Ehrlich S.; Evens M.; Gough, A.; Johnson, P. Y. "Using a Theorem Prover in the Design of Organic Synthesis". In Applications of Artificial Intelligence in Chemistry; Hohne, B.; Pierce, T., Eds.; American Chemical Society: Washington D. C., 1986.
9. Wipke, W. T.; Rogers, D. "Artificial Intelligence in Organic Synthesis. SST: Starting Material Selection Strategies. An Application of Superstructure Search". J. Chem. Inf. Comput. Sci. 1984, 24, 71-81.

10. Dolata, D. P. QED: Automated Inference in Planning Organic Synthesis, PhD dissertation, University of California, Santa Cruz 1984.

11. Robert R. Stoll "Set Theory and Logic"; Dover Publications: 1963.

12. A. Margaris "First Order Mathematical Logic"; Blaisdell Publishing Company: 1967.

13. Haskell B. Curry "Foundations of Mathematical Logic"; Dover Publications: 1977.

14. Frege, G. "Die Grundlagen der Arithmatik, Eine Logischmathematisce Untersuchung uber der Begriff der Zahl"; Marcus, Breslau: 1884.

15. Ackermann, R. "Introduction to Many Valued Logics"; Dover, New York: 1967.

16. Gries, D. "Compiler Construction for Digital Computers"; John Wiley and Sons, New York: 1971.

17. Cleaveland, J. C.; Uzgalis, R. C. "Grammars for Programming Languages"; Elsevier North Holland, New York: 1977.

18. Rovner, P. D.; Feldman, J. A. Massachusetts Institute of Technology, Lincoln Laboratory, The LEAP Language and Data Structure, 1968.

RECEIVED December 17, 1985

A Self-Organized Knowledge Base
for Recall, Design, and Discovery in Organic Chemistry

Craig S. Wilcox[1] and Robert A. Levinson[2,3]

[1]Department of Chemistry, University of Texas at Austin, Austin, TX 78712
[2]Department of Computer Science, University of Texas at Austin, Austin, TX 78712

The design and operation of a system which forms generalizations about organic chemical reactions and structures and uses these generalizations to organize the reactions and structures for efficient retrieval and to generate precursors to a given target molecule is presented. Approaches to computer based classificatory concept formation and organization are discussed. A new linear notation for organic reactions is described.

The complex professional tasks accomplished by organic chemists are an intriguing example of intelligent human activity. Organic chemists organize and recall a vast amount of information. In ascending order of complexity, the knowledge created and used by the organic chemist consists of individual observations, conceptual schemes and generalizations which organize this factual knowledge base, and, most importantly, procedures which describe how to use these facts and conceptual schemes to solve a given problem. We are interested in the ways in which information is organized and used for problem solving.

Our objective is to design machines which will encode reactions and structures, will automatically create generalizations based on these data, and will use these generalizations to organize the data and to solve the problem of precursor generation. Organic chemists often use structural features to classify reactions. The capacity to conceptualize is an indispensable aspect of intelligence. We wished to determine whether a computer, given a large number of structures or reactions and a small set of rules, can create useful generalizations. In designing such a program, we have faced a number of interesting issues concerning conceptualization in organic chemistry.

Organic chemistry is a unique theater for AI research because over the past 150 years organic chemists have created a powerful graphical knowledge representation scheme. This representation

[3]Current address: Board of Studies in Computer Science, University of California, Santa Cruz, CA 95064

0097-6156/86/0306-0209$06.25/0

method is the second language of all organic chemists and supports tasks ranging from mundane recall of a specific datum to the generation of highly complex, creative proposals for multi-step syntheses of previously unknown molecules. Computer science and organic chemistry have been in comfortable collaboration for the past 25 years.(1-7) A number of important programs have been developed in that time. The DENDRAL project influenced AI research in far-reaching ways.(8) Organic chemistry is an enticing arena for AI research because to a limited but important extent, in the microworld of the organic chemist, the problem of how to represent knowledge has been solved. The graphical language shared by all organic chemists for over a century is a remarkably sophisticated knowledge representation scheme which is easily adapted to contemporary techniques in computer science. The organic chemist does use many concepts (electronegativity, insights from quantum theory, and spatial relationships between molecular components) which are absent or are indirectly encoded in his graphical notation. Nevertheless, a substantial amount of knowledge at the factual level, and a useful number of higher level concepts, can be expressed as connected graphs.

Consider the following list: functional groups, the aldol reaction, the Paterno-Buchi reaction, carbon-carbon bond formation, ene reaction, esters, alkenes, elimination, enamines, Claisen rearrangement, allylic alcohols, halogenation. These words describe just a few general categories used by chemists to classify reactions or structures. These categories, some in use for over 100 years, can be described using organic stuctural formulae and find daily use in classifying chemical facts. Computer systems have used such generalizations (provided by chemists) to guide data organization, recall, and planning.

The benefits of original machine calculated generalizations will be realized when capable conceptualizing programs are available. It will be shown here that, given structures and reactions and a simple set of instructions, a computer can indeed discover generalizations, some of which are equal to the categorizations used by chemists. While the fact that some discoveries are very similar to known categories is interesting, it is more important that the computer can also discover patterns previously unknown to chemists.

In this program the generalizations about reactions and structures which are discovered by the system are used very much as man-made generalizations have been used. They organize the data, they are used during the recall procedure, and they are used to generate precursors to target structures. We hypothesize that because only a few chemistry specific heuristics are used in the generalization algorithm, this system will have more creative potential than systems which are more rigidly constructed from many special rules based on detailed chemical knowledge. In current system the answers provided to the precursor generation problem are naive because we have not yet incorporated a heuristic based module to guide precursor selection. Here, as in a child, however, this naivety is accompanied by the potential to suggest fresh approaches to solving a problem. The answers are not directed to conform to a concensus view of correctness. We seek a system of answering questions, but not a system which provides only expected answers.

The first part of this paper provides an overview of what the program does and how it works. We present an approach to representing both structures and reactions as single connected graphs. We will refer to all such labelled graphs as structural concepts or simply as concepts. Structural concepts range from the very general (carbon-carbon single bonds, carbon-oxygen double bonds) through intermediate size and generality (the aldol reaction, the pyran ring) up to the most complex real-world instances of molecules or reactions. By virtue of the graph representation scheme, reactions and structures, both real and abstract, may be stored in a single data base.

This system differs in several ways from other approaches to organic chemistry data base organization. The data organization of this system is based on machine generated structural concepts rather than pre-determined screens. The rules which guide the generalization process will be detailed. The data is heirarchically self-organized, in a partial ordering proceeding from the smallest, most general structural concepts (primitives) to the largest and most specific structures or reactions. The generalizations that are created aid retrieval and are used in precursor generation.

The idea of a heirarchical organization of knowledge has history far predating computer science.(9) (Consider for example the arbor porphyriana, a "tree of concepts" proposed by Porphyry in the 3rd century A.D.) We recognize that the heirachical organization and manipulation of graphs is a general approach to knowledge processing and should find application outside of organic chemistry.

In the second part of the paper examples of the system in action will be given. We feel that because our system uses clearly defined rules for creating generalizations, it may offer fresh insights and solutions to problems. Rules for generalization can be systematically modified. The question of how such modifications affect the problem solving capabilities of the system is unanswered.

An appendix is provided and details the new techniques used in this program. An efficient algorithm based on a partial ordering allows the recall of subgraphs, supergraphs and close-matches for any query graph. Some comparisons will be made of this algorithm with previously used screen approaches for graph retrieval.

Overview of the System

Reaction Representation. From the outset, this project was shaped by the graphical form of traditional organic reaction representations:

Reactions are invariably written this way, and obviously have a left hand side and a right hand side. To the beginning organic student, this format naturally suggests a "before and after" or "cause and effect" perception of reactions. "If the starting material is treated in this way, **then** the product will result." This perception has influenced the design of some computer programs. Reactions have been represented either as two related structures or as one structure and a set of changes required to produce the other structure.

To simplify comparisons between reactions, we sought to describe entire reactions as a single labeled graph. Just as cause and effect can be considered either as two separate events or as a unified process, changing with time, so a reaction can be perceived as two structures, as shown above, or as a single assembly of nuclei connected by bonds which change with time. The aldol-type reaction just illustrated can be rewritten as follows:

Note that bonds which are invariant with time are represented in the usual way. The dotted lines represent bonds which change over the time course of the reaction event. Each changed-bond is labeled to indicate its bond order before and after the reaction. Obviously, the unchanging bonds can also be labeled in an identical fashion. (For example, "1:1" would represent an unchanged single bond.) A second example of this representation is illustrated in Figure 1.

These formulae are unorthodox only because they contain unusual types of bonds, bonds which change with time. It is this same feature which makes the formulae very useful. The single formulae represent entire reactions and can be stored or manipulated using any of the methods already devised for the storage of static structures. We have chosen to use a bond-centered approach to encoding these graphs. The smallest structural unit is the atom-bond-atom fragment, and will be referred to as a primitive. Connected networks of these atom-bond-atom fragments define a molecule or a reaction. These networks of primitives are node labeled connected graphs and can be represented as adjacency tables wherein the nodes are labeled with numbers corresponding to primitives. Finally, these adjacency tables are stored in files as LISP lists and reside in core as arrays. Steps followed in the translation of a reaction into a LISP list structure are illustrated in Figure 1.

Reaction Generalizations Based on Specific Observations.   Organic chemists have long sought to organize their observations. Reactions represented as connected graphs can be formed into groups on the basis of common substructures (subgraphs) shared by all the members of the group. These substructures (subgraphs) are structural concepts which are more general than the specific reactions from which they are derived. These structural concepts help to organize the large numbers of known reactions.

Structural concepts derived from examples of real-world reactions may have the form of a normal reaction but are not necessarily good reactions as formulated. For example, most organic chemists would recognize the following as the generic form of the Diels-Alder reaction but few chemists expect this exact reaction to afford a high yield.

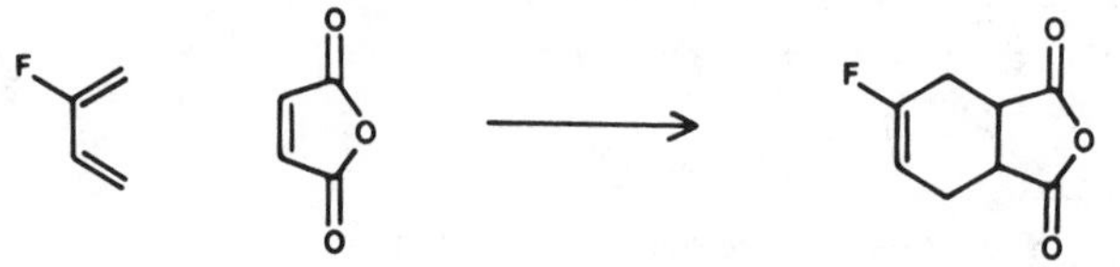

(a)

(c=:-c1-:=c:-c2=:-c3:-),c2-c4-o-c5-c3,c4=o,c5=o,c1-f.   (b)

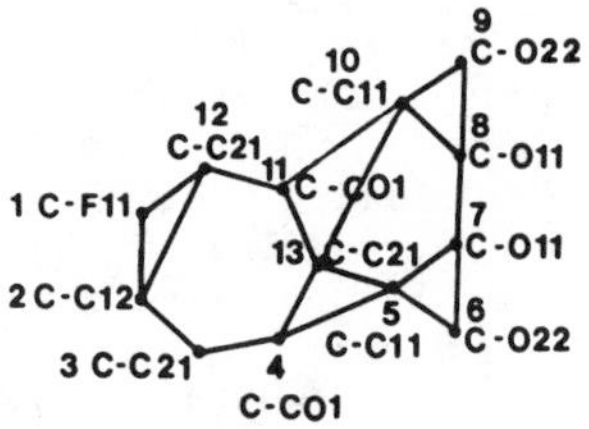

(c)

NODE #	CONCEPT #	ADJACENCIES	
1	11(C-F11)	2,12	
2	6(C-C12)	1,3,12	
3	5(C-C21)	2,4	
4	2(C-CO1)	5,13	(d)
5	1(C-C11)	4,6,7,13	
6	9(C-O22)	5,7	
7	16(C-O11)	5,6,8	
8	16	7,9,10	
9	9	8,10	
10	1	8,9,11,13	
11	2	10,12,13	
12	5	1,2,11	
13	5	4,5,10,11	

Lisp list structure:

(13 (11 2 12) (6 1 3 12) (5 2 4) (2 5 13) (1 4 6 7 13) (9 5 7)
(16 5 6 8) (16 7 9 10) (9 8 10) (1 8 9 11 13) (2 10 12 13)
(5 1 2 11) (5 4 5 10 11))   (e)

Figure 1.   Five representations of the same chemical information.
The canonical chemical reaction graph (a) can be represented in
linear notation (b, see Appendix) or as a bond-centered labeled
graph (c) by using time-variant bonds. The labeled graph affords
an adjacency table (d) and a LISP list representation (e).

To imitate this traditional and human process of generalization, we use substructure discovery to create general concepts which organize our data base. This mechanical generalization occurs whenever a new reaction is entered into the data base and is accomplished in two stages.

First, for each new reaction (R) added, two generalizations, one very general and one very specific, are calculated. These generalizations (subgraphs) of R will be referred to as the minimum reaction concept (MXC(R)) and the complete reaction concept (CXC(R)), respectively, and are defined as follows:

> MXC(R): A graph which is equal to the smallest connected-subgraph of reaction R which contains all the changed bonds in that reaction.

> CXC(R): A graph made by initializing a set C equal to all bonds in the MXC(R), adding to C all bonds adjacent to C, and then continuing to add to C all bonds which are not carbon-carbon single bonds and which are adjacent to C until this is no longer possible.

The relationship between a reaction and its MXC and CXC is more clearly illustrated in Figure 2. The MXC is a very general statement about a specific reaction. The CXC is a very specific "generalization" of that reaction. The value of the MXC is that it helps to organize the data base and it will be used later during retrieval and comparisons of reactions, and it is used in the precursor generation algorithm. The MXC will not contain everything that is necessary for the reaction to proceed. The value of the CXC is that it will very likely contain everything required for a successful reaction. The expected yield of the reaction represented by the CXC is likely to approach or even exceed the yield of the original reaction. Obviously the CXC contains much more than is necessary for the reaction. An organic chemist, if asked to define what was essential to the success of the original reaction, would probably define a subgraph larger than the MXC and smaller than the CXC.

The first stage of generalization begins, then, with calculating the MXC and CXC of a reaction and adding those graphs to the data base. A very simple heuristic used here is that generalizations about reactions will be subgraphs of reactions and will contain all the changed bonds of the reactions. The reaction itself is next added, and during that process previously known reactions which are similar to the new reaction are identified.

The second stage of generalization begins with this list of similar reactions. If a reaction (RR) on this list contains MXC(R), (that is, it has the same MXC as the original reaction, R), we calculate the largest common subgraph of R and RR which contains MXC(R). This new graph is a specific plausible generalization formed by comparing R and RR. This process results in identifying interesting reaction subgraphs of a size larger than an MXC and smaller than a CXC. An example of the effects of these algorithms for creating generalizations is illustrated in Figure 3.

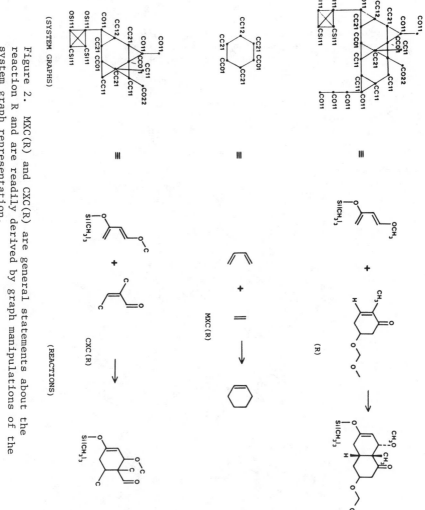

Figure 2. MXC(R) and CXC(R) are general statements about the reaction R and are readily derived by graph manipulations of the system graph representation.

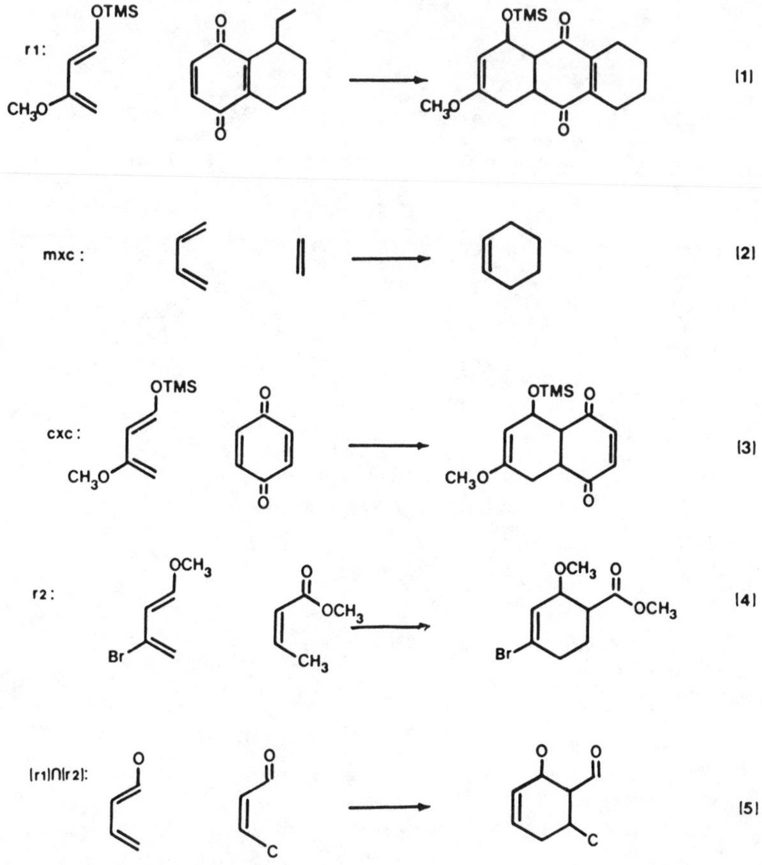

Figure 3. When a reaction (1) is added to the data base its MXC
(2) and CXC (3) are also added. The reaction is then compared
with other reactions (4) and a maximum common subgraph (5) is
added.

Calculating Generalization Validities.   The knowledge base consists of a large number of structures and reactions from the real world and an even larger number of structures or transforms which are calculated based on these original data. These calculated graphs are generalizations based on the known reactions.

If these generalizations are to be used for problem solving, then their validity is an important issue. By validity we mean the probability that the exact reaction represented by the generalization would work. If the generalization was considered as a reaction, what would be the yield of that reaction?

Measuring the validity of these generalizations is important because they are machine generated.   In systems which use human generalizations about reactions to generate precursors estimated yields or reliability factors are provided for each generalization. Our system seeks to automate this approach to machine intelligence and a calculational approach to the reliability of generalizations is required.

The estimation of one type of validity is a task faced by organic chemists every day.   In the process of reviewing research grants, experts must predict whether proposed reactions, hitherto unknown, will succeed. To make this judgement, the expert relies in part on precedent.   Previously observed reactions similar to the proposed reaction lend credence to the proposal. If many reactions (very similar to be proposed reaction) are known to proceed in high yield, the validity or likely yield of the new reaction is high.   If similar reactions are known to give low yields, then the proposed transform is of low validity.

Before precedent can be used to estimate validity, the meaning of "similar" (as it is applied to reactions) must be defined. It is not surprising that problems of conceptualization and similarity arise in the same project. Philosophers have long recognized the complexity and interdependence of comparison and concept formation. What makes one reaction a better precedent than another? Can similarity be quantified and if so can the similarity of a reaction and a proposed transform be quantified? The ways in which reactions are similar or dissimilar and the prediction of yields based on precedent are important questions which deserve further study.

At present, we calculate transform validities (estimated yields) for a generalization or an unknown reaction as follows.

Let TV(i) = transform validity of i.
Let A(i) = chemical reactivity of i.

Currently chemical reactivity is equal to the number of bonds which are not carbon–carbon single bonds. This is a crude approach to estimating the potential reactivity of i. We wish to calculate TV(r) for a newly discovered transform r based on reactions of precedent.   Let IS(r) = the set of known transforms upon which the validity of r is to be based, then:

$$CW(i) = \frac{TV(i)}{2} \cdot \left[\cos\left[1 - \left[\frac{A(r)}{A(i)} \right]^{\frac{1}{a}} \right] + 1 \right] \qquad (1)$$

$$TV(r) = \frac{\displaystyle\sum_{i \in IS(r)} CW(i)^2}{\displaystyle\sum_{i \in IS(r)} CW(i)} \tag{2}$$

$CW(i)$ is the closeness (similarity) weighted validity of transform i with respect to the new transform r. If the denominator in Equation 2 is 0, $TV(r)$ is 0. The constant a determines the magnitude of the effect of closeness $A(r)/A(i)$ on the calculated transform validity.

Equations 1 and 2 were intended to produce the following results. If there are a large number of reactive bonds in the precedent not in the proposed transform, the closeness weighted validity of the precedent is small. If there be the same number of reactive bonds in both reactions, the closeness weighted validity of the precedent is equal to its yield or known validity. This is an attempt to encompass the idea that if the precedent has reactivity similar to the proposed transform, the proposed transform is likely to work as well as the precedent. If the precedent and the proposed transform are very different, the precedent is not helpful. A weighted average of the closeness weighted validities of the precedents provides the estimated yield for the new transform. The weighting procedure favors close precedents of high yield. This follows from the chemist's usual optimism: if there are several equivalent good precedents, some of high yield and some of low yield, the proposed transform is judged to have a good chance.

The comparison of "numbers of reactive bonds" crudely measures similarity. A more appropriate but complex approach would evaluate similarity in terms of known functional groups or discovered reactive substructures shared or not shared by two reactions. Both these approaches to validity estimation are limited because they are entirely based on structure. The expert will use other factors, including theoretical considerations, to refine validity.

Validity aids the precursor generation task in a unique way. Validity can be used to identify situations in which a particular reaction is <u>not</u> applicable. (Most structures have validity = 100, but some, like Bredt's rule violators, would have a lower validity.) Reactions of very low (predicted or known) yield or impractical structures are called "negative instances". Mitchell uses negative instances in the learning process to rule out otherwise plausible generalizations.(12) We use validity to define a continuum from the most positive to the most negative instances. The mechanism for precursor generation then automatically uses these negative instances (structures or reactions of low validity) to block the use of good generalizations in specific invalidating situations.(13)

All the generalizations calculated from a set of known reactions are assigned a validity (reliability factor) based on (i) how much these subgraphs deviate from the known reactions from which they are derived and (ii) the known yields of these known reactions. This primitive method of predicting yields based on precedent serves to illustrate challenges to be met if machines are to acquire reliable chemical judgement independent of, but consistent with, an expert's evaluations. The validities calculated here are used to guide the precursor generation task and provide a means of evaluating proposed precursors.

The System in Action

Interactive Sessions. The system has been implemented in LISP (Franz Lisp) and is running on a Digital Corporation VAX 11/780 at the University of Texas.   Interactive sessions with the system are illustrated in Figures 4-7. (During the development stages of this project a linear notation was created for reaction input and output. A brief description of this notation is provided in the Appendix.)

The figures are annotated and little additional comment is required.   Figure 4 illustrates retrieval of a structure and its supergraphs and subgraphs. Figure 5 illustrates reaction retrieval.

The system is able to use its knowledge to generate precursors to a target molecule. Two examples are shown (Figures 6 and 7). At present, the program compares known reactions and generalizations based on known reactions to the target and chooses to apply reactions which have the most reactive bonds in common with the target. The result is that precursors are suggested with little sophistication. In fairness, it should be emphasized that the data base was generated from only about 230 reactions, and no generalizing concepts were provided by the operators. We look forward to testing the system when it has acquired more knowledge.

Conclusions

The system described in this paper stores and retrieves reactions and structures, creates generalizations which further organize the knowledge base, estimates the validity of these generalizations, and uses both specific reactions and machine derived generalizations to generate precursors. We have shown that the representation of reactions as single labeled graphs is possible based on the idea of a bond which changes during a reaction and this graph representation simplifies the machine driven act of induction.   Concepts are generated automatically and these concepts organize the data base, aid in the retrieval, and support the precursor-generation capabilities of the system. A method for calculating the validities of a given generalization has been devised and methods of refining these calculations have been identified.

This study examined some unexplored aspects of conceptualization in organic chemistry. How are classificatory concepts created? Can the value of a generalization be quantified? Although here these questions are presented in relation to organic chemistry, they are in fact basic questions of epistemology and go beyond organic chemistry.(9)

This program makes generalizations about real-world reactions and uses these generalizations to generate precursors. Mitchell's approach to conceptualization requires an "instance language" to represent observations, a "generalization language" to create concepts, and a "matching predicate" to associate observations with generalizations.(12,23) Our approach to generalization in organic chemistry relies on a bond-centered labeled graph representations of reactions and structures (observations). In this language "more-general-than" is defined as equivalent to "subgraph-of". We take advantage of the fact that in organic chemistry the instance language and the generalization language are identical, and matching predicates are based on graph comparisons.

■ ■

The system is now ready to process your requests.
There are 1230 concepts. |comments|

What would you like to do? |several options are available.
 1 = Change the database.
 2 = Ask a question.
 3 = Go to lisp level input. |the user is interested in asking
 4 = Save changes. |a question about the database.
 5 = Quit
 2
Which of the following do you need help with?
 1 = Structure or reaction retrieval.
 2 = Finding a precursor.
 3 = Finding a postcursor.
 4 = A multistep synthesis.
 5 = Return to previous menu.
 1 |structure retrieval or reaction
Initiating a query... |retrieval are possible.
Please enter the list of classes: |the system asks for a list of
                                                              |data types which will restrict
 (s) |the search. We choose to
                                                              |search only structures.

Please enter the legal substitutions:
nil |no substitutions are allowed
Type in the structure please: |The user enters the following
                                                              |structure using a mnemonic system:

(c1-c-c-c-c-c-),c1=o.

Searching data-base of graphs... |graphic input is not yet available.

Exact matches: (282) |the query structure is known.

Subgraphs: (4 7 16 49 63 69 86 88 102 137 282 306 539)

Supergraphs: (196 282 296 432 436 484 509 510 515 526 668 669 670 677 678 682 683 684 766
815 816 817 819 828 829 830 831 987 989 991 1164 1183 1192 1193 1194 1225 1226 1227)

Close matches: nil |by convention, since supergraphs were
                                                              |found, close matches are not sought.
Number of concepts searched: 16 |16 concepts were examined to
                                                              |find the 51 matches shown above.
Number of complete node-by-node searches required: 15
                                                              |a complete subgraph isomorphism
                                                              |test was required on 15 concepts.
 3
Going to lisp level input.
To return to this menu type '(hi)'

 -> (show 539) |The user asks to view two sub-
C1-C2-C3,. |graphs. Eventually, graphical output
 -> (show 306) |will be possible.
C1-C2-C3-C4-C5-C6,C3=O7,.
 -> (show 484) |a supergraph is viewed.
(C1-C2-C3-C4-C5-C6-),C7-C3-C2-C1=O12,C8-C2-C9-C10=C11,.

■ ■

Figure 4.   Structural retrieval.   Responses provided by the user
are in italics.   Annotations are inserted on the right.

■■■

What would you like to do?
1 = Change the database.
2 = Ask a question.
3 = Go to lisp level input.
4 = Save changes.
5 = Quit

2

Initiating a query...
Please enter the list of classes: | this time we are interested in
                                                               | only reactions.
(r)

Please enter the legal substitutions:

nil | no substitutions are allowed

Type in the structure please: | the user wants to see reactions
                                                               | which form c-c bonds at the alpha
(c1-c2-c-c-c-c-),c1=o,c2:-c. | carbon in cyclohexanone:

Searching data-base of graphs...

Exact matches: nil | the exact reaction is not in the data
                                                              | base.
Subgraphs: nil | no known subgraphs.

Supergraphs: (667 681 826 1224) | four known reactions are supergraphs
                                                              | of the query.
Close matches: ((508 7) (814 7) (676 7) (136 4) (1063 3) (105 3) (1057 2) (359 2))
                                                              | concept 508, for example, has a 7 bond
Number of concepts searched: 21 | subgraph in common with the query.

Number of complete node-by-node searches required: 19

Would you like to add the structure as a new concept? (y=yes)
no | this is one way in which the system
                                                             | can learn new concepts
3
Going to lisp level input.
To return to this menu type '(hi)'
-> *(show 826)* | the user now examines some
                                                             | supergraphs of the query.
(C1-C2-C3-C4-C5-C6-),(C12+C13+C14+C15+C16+C17+),07=C1-C6-C8,C6:-C9=:-C10-C11-C12,C1
1=018,.

-> *(com 826)* | comments include bibliographic
                                                             | information. Yields are stored
(House, H. O. "Modern Synthetic Methods", pp 595-6 | in a separate file.

-> *(show 1224)*
(C1-C2-C3-C4-C5-C6-),(C2-C3-C10-C11-N12:-C13:-),07=C1-C2:-C13:-N12-C15,C1-C6-C8,C6-C9
,C13=:014,.

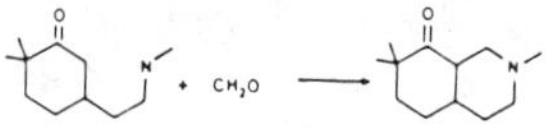

-> *(com 1224)*
(Corey , E. J., et al J. Amer. Chem. Soc. 1974, 96, 6516)

■■■

Figure 5.   Reaction retrieval.

■■■

What is the target molecule?
Type in the structure please: |the target:

(c1-c2-c3-c-c-c-c7-c-),c1=o,c2-c-c=c,c3-c7.

Adding concept... |the system temporarily adds the target
Searching data-base of graphs... |to the data base. In this way known
The concept is 1231. |subgraphs of the target are found together
                                                       |with the reactions that will produce them.
The following precursors are suggested: |these reactions are then used to generate
  #   reaction   validity   size |precursors.
  -    --------   --------   ----
                                                       |the table gives five precursors. the
  1   1216   83   13 |concept used to generate the precursor is
  2   11   61   13 |shown with the transform validity of this
  3   236   56   14 |application (see text). the last column
  4   193   56   13 |gives the number of bonds in a precursor.
  5   27   41   13

The precursors are on list 'pre'. |the user now views the first three
                                                       |precursors.

-> *(view pre 1)*
(C1=C2-C3-C7-C8-),(C3-C4-C5-C6-C7-),C3-C2-012-C11-C10=C9,.

-> *(view pre 2)*
(C1=C2-C3-C7-C8-),(C3-C4-C5-C6-C7-),C3-C2-C1=C12,C2-C9-C10=C11,.

-> *(view pre 3)*
(C1=C2-C3-C7-C8-),(C3-C4-C5-C6-C7-),C3-C2-C1=O9,C2-C10-C11-C12-I13,.

-> *(show 1216)* |this is the concept which was used
(C1-:=C2=:-C3:-C6=:-C5:=O4-:),C5-:07, |to suggest the first precursor:
 -> *(up 1216)*
(1219)
 -> *(up 1219)*
(1210)
 -> *(com 1210)*
(Dauben W G J Org Chem 1972 37 1212) |the reference and a reaction from which
 -> *(show 1210)* |the general concept were derived
(C1-C2-C3-:=C4=:-C5-C10-),(C5-C6-C7-C8-C9-C10-),C3-:012:=C13-:-C14:-C5-C10-C11,C13-:01
5-C16,. |can easily be found.

■■■

Figure 6. Precursor generation. Note that overall transforms
may be encoded and applied without restrictions as to the actual
mechanism.

****

What is the target molecule? |the target:
Type in the structure please:

(c1-c2-c3-c-c-c-),c1=o,c3=o,c2-c-c-c #n.

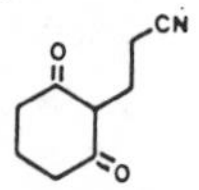

Adding concept... |as in Figure 7, the target is first added
Searching data-base of graphs... |to the data base. subgraphs of the target
The new concept is added: 1232 |are identified and reactions known to
                                                      |generate such subgraphs are applied in a
The following precursors are suggested: |retrosynthetic sense to the target.

#	reaction	validity	size
1	441	77	14
2	11	76	11
3	425	36	13
4	308	20	13

The precursors are on list 'pre'. |the user now views three of the
                                                      |suggested precursors.
3
Going to lisp level input.

-> *(view pre 1)*
O7=C3-C4-C5-C6-C2=C1-C8-C9-C10#N11,C2-N13-C12,C3-CL15,N13-C14,.

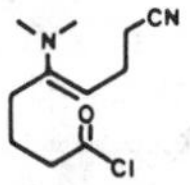

-> *(view pre 2)*
(C1-C2-C3=),(C1-C2-C3-C4-C5-C6-),C2-C7-C8-C9#N10,.   [a very naive answer]]

-> *(view pre 3)*
(C1-C2-C3-),(C1-C2-C3-C4-C5-C6-),O7-C1-C2-C3-O8,C2-C9-C10-C11#N12,.

****

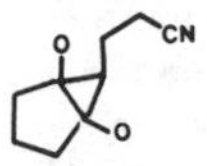

Figure 7.   The capacity to generalize from specific facts is
revealed by the systems ability to provide these precursors.

Following the seminal work of Corey and Wipke, elegant and powerful programs have been developed to aid the synthetic organic chemist. These programs use man-made generalizations and special heuristics to guide the computer to the solution of complex problems. This project complements these earlier and ongoing efforts. The limits and utility of machine-made generalizations are our central interest.

Acknowledgments

Enlightening conversations with Dr. Elaine Rich (Department of Computer Science, University of Texas) are gratefully acknowledged. Mr. James Wells wrote the Pascal programs which allow input and output via mnemonic strings of characters. This research was sponsored in part by the Robert A. Welch Foundation, Research Corporation, and NSF (MCS-8122039). Additional support was provided by a National Science Foundation Graduate Fellowship to RAL.

APPENDIX

Data Organization and Retrieval

A new data base organization for storing and retrieving organic structures was created for this project. Although this retrieval system is applied here to chemistry, it is written in a general manner and is applicable to other graph-based domains. The organization is based on a partial ordering of graphs by the ordering relation "subgraph-of". A simple yet powerful retrieval algorithm has been developed to accompany the partial ordering. These methods offer an alternative to the scheme used by most retrieval systems - the screen approach.

The Partial Ordering. Labeled graphs stored in this data base will be referred to as concepts because they represent structural features that are useful to consider when reasoning about molecules and reactions. Both molecules and reactions are represented as labeled graphs. Those graphs that represent known molecules and reactions sit near the top of the partial ordering. Primitives (the single node graphs that represent bonds) form the lowest level of the partial ordering. As the system evolves, intermediate concepts are created. These concepts usually represent partial structures (such as functional groups) or reaction generalizations. The intermediate concepts are discovered (constructed) by the system to improve retrieval efficiency and precursor generation. Figure 8 shows a simple partial ordering. Notice that the concepts in the partial ordering can be viewed as forming a continuum from general concepts to more specific concepts.

The Retrieval Algorithm. The retrieval algorithm efficiently tells the system user how a new concept relates to all other known concepts. The algorithm solves the following basic problem: Given an element G and a partial ordering return the following four sets:

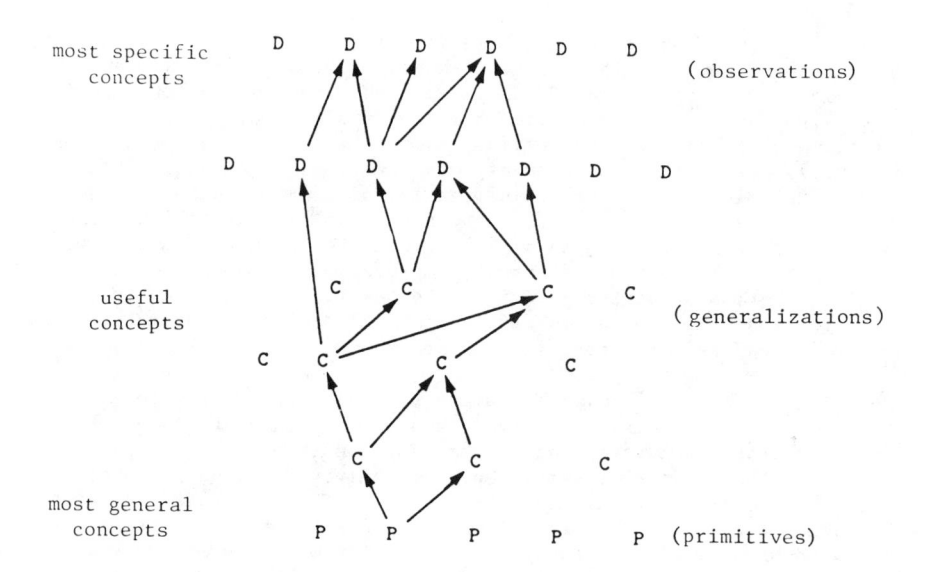

Figure 8. A simplified view of the partial ordering. A typical
upward chain is shown.

1. The set of elements that are the same as **G**. (exact matches)
2. The set of elements that are predecessors of **G**. (subgraphs)
3. The set of elements that are successors of **G**. (supergraphs)
4. The set of elements that have predecessors in common with **G**. (close matches)

The algorithm does something more powerful: It finds the immediate predecessors to **G** (the largest known subgraphs of **G**), and the immediate successors to **G** (the smallest known supergraphs of **G**). This is the key to the algorithm. By finding where **G** fits in the partial ordering we find the four desired sets. The algorithm must minimize the number of comparison operations required to find the four desired sets. This minimization of comparison operations is very important in a system that uses complex objects like graphs since the complexity of these comparisons varies exponentially with size.(14) The algorithm is easy to implement and searches nodes in a logical bottom-up order. This may be useful in domains where, for example, one may wish to apply general concepts or rules to a situation before more specific ones are found to be applicable.

Details of the Algorithm.   The algorithm has two phases. In Phase 1 the immediate predecessors (largest known subgraphs) are found and in Phase 2 the immediate successors (smallest known supergraphs) are found.   These two phases are enough to answer all four parts of the query.   To understand the algorithm, note that transitive edges between concepts in the partial ordering are not stored: if $a \leq b$ (a is a subgraph of b) and $b \leq c$, an edge from a to c is not stored. $IP(y)$ is the set of immediate predecessors of the data element y and $IS(y)$ is the set of its immediate successors. These sets are stored in files as LISP lists, one line per concept. Phase 1 determines $IP(G)$ where **G** is the query object.

$$S := \{ \}$$

-While there is an unmarked element **y** in the database such that each member of $IP(y)$ is marked T and **y** has fewer nodes than **G**:

    <u>If</u> $y \leq G$ (graph comparison needed)

      <u>Then</u> mark y as T

$$\{S\} := \{S - IP(y)\} \cup \{y\}$$

    Else mark **y** as F.

It can be shown that when this process terminates $S = IP(G)$, the set of largest known subgraphs of the query graph. When Phase 1 begins, all objects at the bottom of the partial ordering (the primitives) are compared to G since they have no immediate predecessors. This process is fast because the bottom of the partial ordering contains single node graphs for which the comparison operation is trivial.

Phase 2 may be informally described as follows: The goal of Phase 2 is to calculate $IS(G)$ - the immediate successors (smallest known supergraphs) of G:

$$S := \{\ \},$$

-Sequence through the elements of IP(G) in any order, chaining up the partial ordering for each element. Beginning with the last element of IP(G) a breadth-first search is required and if an unmarked element **y** is encountered which has been reached from all other elements of IP(G), execute:

<u>If</u> G ≤ **y** (comparison needed)
    Then$\{S\} := \{S\} \cup \{y\}$
        -(mark all concepts chaining up from **y** as successors without further comparison)
<u>Else</u> continue breadth-first search from **y**.

When Phase 2 terminates S = IS(G). All supergraphs of G have been identified by chaining up from each element of IS(G) as these are found.

Phase 1 and Phase 2 answer parts 1-4 of the query as follows:

1. Exact match: If G already exists in the database, then IP(G) = IS(G). G is the single element contained in these sets.
2. Subgraphs: The subgraphs are simply all nodes that were marked T in Phase 1.
3. Supergraphs: The supergraphs are marked in Phase 2. They are the union of the upward chains from each member is IS(G).
4. Close matches: The close matches are the union of the upward chains from each member of IP(G) (not including supergraphs).   In the most obvious implementation of Phase 2, a hash table is used to manage the breadth first search. It contains information about which nodes have been visited and which upward chains they are on.   The desired union can be found simply by collecting elements of the hash table.

<u>Other Chemical Structure Search Systems.</u> Many efficient systems have been designed to identify graphs in a file that contain a given substructure.   One system is the Cambridge Crystallographic Data Base.(15) In the Cambridge system the query structure is compared to every molecule of the database. This means that retrieval time for a query goes up linearly with the size of the database. Other search systems alleviate this problem.   These systems use a screen approach.(16-22)   The screen approach is an indexing scheme that includes, associated with each smaller concept of the database, a list of data items that contain the smaller concept (a list of upward pointers).

<u>Comparisons with the Screen Approach.</u>   The algorithm used by screen systems is a special case of our algorithm, the difference between the screen approach and this approach is in the number of levels allowed in the database organization and not in the retrieval algorithm.

Which organization supports more efficient retrieval in terms of number of concept comparisons?   No absolute conclusion can be reached, but there are reasons to believe that in general a multilevel aproach may be cheaper overall.(13) First, our approach tends to search smaller concepts than does the screen system. The cost of these searches will be much cheaper. Second, in Phase 2 of our system we ar‹ able to infer that some graphs are supergraphs of the query without d›ing further searching. Finding all subgraphs and all supergraphs of a query, with precision, is beyond the capabilities of most screen systems. Finally, experimental evidence supports our system.

To compare the performance of the multilevel organization against a two-leveled one we ran our retrieval algorithm on two data bases. The first contained molecular structures, discovered molecule concepts, and primitives, and had 630 concepts altogether. The second was a version of the first in which all intermediate levels between primitives and top-level structures have been removed, leaving just two levels. This database had 521 concepts in all. The algorithm ran more than twice as fast on the multi-leveled database, even though the two-level database contained fewer concepts. The algorithm produced 33% more answers (subgraphs and supergraphs) when running on database 1 than on database 2.

Linear Notation for Reactions and Structures.   To assist in the development of this program a new linear grammar was developed to describe reactions and structures (Figure 1). A program written by Mr. James Wells at the University of Texas accepts alphanumeric strings created by the chemist. From these strings which represent structures or reactions the Pascal program generates a connectivity table of the sort used in the Cambridge Crystallographic database. The connectivity files are transferred to the main LISP program which creates the LISP structure lists shown in Figure 1.

The grammar for reactions and structures is easily mastered by the organic chemist. The following symbols are used:

- ; single bond
= ; double bond
; triple bond
+ ; delocalized double bond

Other than these symbols, the chemist needs to remember only two rules: (i) rings are encoded in parentheses wherein the last atom is followed by a bond which connects it to the first atom in the parenthetical expression, and (ii) atoms at branching points must be numbered. Linear or cyclic strings are separated by commas. Hydrogens are ordinarily ignored. Thus cyclopentane is encoded as (c-c-c-c-c-) and sec-butanol as c-c1-c-c,c1-o. A menu is available which contains commonly used structures which can be used in an abbreviated form to define molecules. The t-butyldimethylsilyl ether derived from n-propanol can be represented as *tbs*-o-c-c-c. Further examples of representations based on this system are shown in Figures 4-7.

The chemist can encode a structure in many ways and, provided the representation follows the above rules, each alphanumeric string will generate a proper connectivity file. For example, "(c-c-c-c1-c-c-c-c2-),c1-c2" or "(c1-c2-c-c-c-),c1-c-c-c-c2" are both proper representations of 3.3.0-bicyclooctane. IUPAC numbering can be followed or the numbering can be arbitrary.

Reaction   graphs   are   encoded   in   the   same   way   as   static structures.   Bonds which change during the reaction are coded as "x:y" where **x** is the bond type before the reaction and **y** is the bond type after the reaction.   Thus "c-c=:-c" represents the reduction of propene   to   propane   and   "(c-o:-cl-c-c-),cl-:i"   represents   the formation   of   tetrahydrofuran   and   an   iodine   atom   from 4-iodobutan-1-ol.

A second program accomplishes the reverse process and will generate from a connectivity file an alphanumeric representation of molecules   or   reactions   based   on   this   linear   notation.   While we recognize   the need for a graphical interface for the main AI program we   are   enthusiastic   about   the   efficiency   of   this linear grammar. This linear notation should be adaptable to use in any application dealing with connected graphs.

Literature Cited

1. Lederberg, J.; Sutherland, G. L.; Buchanan, B. G.; Feigenbaum, E. A.; Robertson, A. V.; Duffield, A. M.; Djerassi, C. J. Amer. Chem. Soc. 1969, 91, 2973.
2. Corey, E. J.; Wipke, W. T. Science 1969, 166, 178.
3. Brandt, J.; Friedrich, J.; Gasteiger, J.; Jochum, C.; Schubert, W.; Ugi, I. In "Computer Assisted Organic Synthesis"; Wipke, W. T.; Howe, W. J., Eds.; ACS Symposium Series No. 61, American Chemical Society: Washington, D.C., 1977; pp. 33-59.
4. Wipke, W. T.; Braun, H.; Smith, G.; Choplin, F.; Sieber, W. In "Computer Assisted Organic Synthesis"; Wipke, W. T.; Howe, W. J., Eds.; ACS Symposium Series No. 61, American Chemical Society: Washington, D.C., 1977; pp. 97-125.
5. Hendrickson, J. B. J. Amer. Chem. Soc. 1971, 6844-6862.
6. "Computer Assisted Organic Synthesis"; Wipke, W. T.; Howe, W. J., Eds.; ACS Symposium Series No. 61, American Chemical Society: Washington, D.C., 1977.
7. Wipke, W. T.; Rogers, D. J. Chem. Info. Comp. Sci. 1984, 24, 71-80.
8. Lindsay, R. K.; Buchanan, B. G.; Feigenbaum, E. A.; Lederberg, J. "Applications of Artificial Intelligence for Organic Chemistry"; McGraw-Hill: New York, 1980.
9. Reidl, R. In "Biology of Knowledge"; John Wiley and Sons: New York, 1984.
10. "Computer   Representation   and   Manipulation   of   Chemical Information"; Wipke, W. T., Ed.; John Wiley and Sons: New York, 1974.
11. Michalski, R. S.; Stepp, R. E. "Learning from Observation: Conceptual Clustering" In "Machine Learning: An Artificial Intelligence Approach"; Michalski, R. S.; Carbonell, J. G.; Mitchell, T. M., Eds.; Tioga Press, 1983.
12. Mitchell, T. M.; Utgoff, P. E.; Banerji, R. "Learning by Experimentation: Acquiring and Refining problem Solving Heuristics" In "Machine Learning: An Artificial Intelligence Approach"; Michalski, R. S.; Carbonell, J. G.; Mitchell, T. M., Eds.; Tioga Press, 1983.
13. Levinson, R. A. Ph.D. Thesis, University of Texas at Austin, Austin, 1985.

14. Tarjan, R. E. In "Algorithms for Chemical Computations"; Christoffersen, R. E., Ed.; American Chemical Society: Washington, D.C., 1977; pp. 1-20.
15. Allen, F. H.; Bellard, S.; Brice, M. D.; Cartwright, B. A.; Doubleday, A.; Higgs, H.; Hummelink, T.; Kennard, O.; Motherwell, W. D. S.; Rodgers, J. R.; Watson, D. G. Appl. Cryst. 1979, 35, 2331-2339.
16. Adamson, G. W.; Cowell, J.; Lynch, M. F.; McLure, H. W.; Town, W. G.; Yapp, M. A. J. Chem. Doc. 1973, 13, 153-157.
17. Bawden, D. J. Chem. Inf. Comp. Sci. 1983, 23, 14-22.
18. Dittmar, P. G.; Farmer, N. A.; Fisanick, W.; Haines, R. C.; Mockus, J. ibid. 1983, 23, 93-102.
19. Feldman, A.; Hodes, L. ibid. 1975, 15, 147-151.
20. Fugmann, R.; Kusemann, G.; Winter, J. H. Info. Process. Mgmt. 1979, 15, 303-323.
21. O'Korn, L. J. In "Algorithms for Chemical Computations"; Christoffersen, R. E., Ed.; American Chemical Society: Washington, D.C., 1977; pp. 122-148.
22. Willett, P. J. Chem. Inf. Comp. Sci. 1980, 20, 93-96.
23. Mitchell, T. M. Artificial Intelligence 1982, 18, 203-226.

RECEIVED December 17, 1985

Expert-System Rules for Diels–Alder Reactions

C. Warren Moseley, William D. LaRoe, and Charles T. Hemphill

Texas Instruments Inc., Dallas, TX 75265

Expert systems of today are powerful when used in the proper domains. Unfortunately, the most difficult part of applying these systems is the structuring of knowledge into rule format. This paper describes methods developed which allow the capture of Diels-Alder reaction knowledge into simple and elegant expert system rule format. Essential components of the system include: a grammar for matching the input molecular structure expressed in Wiswesser Line Notation (WLN), the unification of many reactions into a single generalized mechanism using synthon template patterns, use of WLN rules to produce valid synthons, and use of frontier molecular orbital theory (FMO) to verify the disconnection. This system is implemented in Prolog, whose natural backtracking and generation capabilities easily express and produce the many structural combinations possible.

There have been attempts to apply formal methods to the representation of organic compounds [1],[2], some attempts to apply artificial intelligence to organic synthesis [3],[4], and numerous attempts to apply the use of molecular orbital calculations to the verification of the validity of compounds in the synthesis route. This effort was a moderate attempt to examine the representation issues involved in writing production rules for Diels-Alder disconnections.

The disconnection approach [5] is adopted in this work because it is amenable to backward chaining systems. The starting point is the *target compound*, which is, in this case, a Diels-Alder product. The target compound is broken or *disconnected* into two distinct parts called *synthons*. The synthons are the ideal representations of the actual reactants used to produce the target compound. Synthons embody the physical properties of the actual compounds they represent.

As an initial implementation approach, rules could consist of specific targets and a list of their synthons. No one uses this method because the naive approach of expressing every possible chemical disconnection is impracticable: the number of rules involved to express even trivial synthetic routes grows exponentially. Any expert system solution to

0097–6156/86/0306–0231$06.00/0

the synthesis problem must attack two fundamental problems: the variety of functional groups which may participate in a given reaction and the symmetry involved between function groups in a reactant (intra-synthon and inter-synthon functional group interaction, respectively). The thrust of this research has been to capture the reaction routes for a chemical disconnection in a clear, symbolic notation which accommodates qualitative reasoning with functional groups and which comprehends the symmetry of this problem.

Ideally, an implementation language would support symbolic and linguistic approaches to representation and manipulation, a qualitative approach to verification, and a deductive approach to disconnection. Prolog [6] is a symbolic language which directly supports backward chaining deduction. Viewed as a declarative language it naturally supports elegant grammar formalisms and its procedural aspects support qualitative reasoning. For these reasons, Prolog was chosen as the implementation language for this project.

In summary, the following research goals are addressed in this effort:

1. A linguistic approach to the representation of chemical information.

2. Use of molecular orbital theory to qualitatively validate derived synthons.

3. Unification of synthetic disconnections into a general form.

4. Use of symbolic structure rearrangement in WLN.

2 Grammar Rules for Structure Recognition

The Definite Clause Grammar (DCG) formalism [7] is utilized throughout this project. Grammar rules are used in the expert system rules to recognize the general class of the parent molecule in the disconnection (*e.g.*, cyclohexene). The class determines the patterns used to construct the resultant synthons (discussed in Section 4).

2.1 Background for WLN and DCG

Many researchers have recognized the importance of having an unambiguous grammar for chemical notation, but they have mainly applied WLN [8] to on-line compound search [9] and structural summary (identification of common structural features)[10]. Johns and Clare point out that it is a linguistic rather than merely a symbolic notation. This means that the symbols are represented and manipulated in well defined structures. This section relies on the unambiguousness of WLN to recognize parent molecules while Section 5 relies on the WLN rules to actually manipulate symbol structures.

The DCG formalism is based on first order predicate logic and provides a clear and powerful method for describing languages. The formalism generalizes the Context Free Grammar (CFG) formalism and DCG grammars may be efficiently executed. DCG is most often implemented through a translation process from the DCG notation to a top-down, left-to-right, backtracking Prolog program. This program becomes a parser for the language specified by the DCG.

The required amount of work at each step in a backtracking parser is exponential in the number of constituents already found, just for recognition. This occurs because intermediate effort, which could become useful later, is not saved. Of course, classes of grammars exist for which this behavior does not occur. Most programming language grammars are carefully written to avoid exponential behavior. However, parsing algorithms exist (e.g., the active chart parser [11]) where the worst case parsing time is $O(n^3)$ for *any* CFG grammar and $O(n^2)$ when the grammar is unambiguous (n is the sentence length). Nevertheless, Prolog provides an adequate DCG grammar parsing mechanism for the purposes of this work.

2.2   Grammar for Diels-Alder Reactions

This section examines grammars used to recognize parent molecules (carbocyclic rings for example).

The following regular expression [12] recognizes cyclohexene:

$$\text{L6UTJ } [\text{A}\sigma_A] \ [\text{B}\sigma_B] \ [\text{C}\sigma_C] \ [\text{D}\sigma_D] \ [\text{E}\sigma_E] \ [\text{F}\sigma_F]$$

where if r is any regular expression, $[r]$ is an abbreviation for $(\epsilon + r)$ (in other words, r is optional). ϵ is a regular expression that denotes the empty set and '+' is the union operator for the languages represented by the regular expression arguments. The symbol 'σ' represents an arbitrary substituent, with the subscripts indicating to which ring locant the substituent belongs.

Using DCG, the more general class of carbocyclic rings can be recognized. The grammar rule

carbocyclic(Substituents, Number) $\longrightarrow$ "L", number(Number), "U", "T", "J",
          substituents(Substituents, Number).

achieves the desired result. Within this rule the *logical variables* are denoted by a leading capital letter. This declaratively states that carbocyclic *rewrites* into the letter "L", followed by a number (which in turn is recognized by DCG grammar rules), followed by the letters "UTJ", followed by the substituents. The **substituents** rule recognizes the **Substituents** at each ring locant and uses the instantiated value for **Number** to verify that the ring locant values are within the proper range. Subsequent steps in the disconnection process utilize the variables mentioned in the head of the rule.

Finally, using the grammar rule described above (and related rules not presented), the goal

carbocyclic(S, N, "L6UTJ A1 BNW F3", [])

rewrites the string "L6UTJ A1 BNW F3" into the empty set [] (meaning that the entire string is recognized) and produces the result

$$\text{S} = [[\text{A},1],[\text{B},\text{N},\text{W}],[\text{F},3]], \text{N} = 6.$$

S is a list of ring locants and the corresponding substituents used in subsequent disconnection stages. N represents the number of ring locants.

2.3   Application to Other Reactions

The general grammars and the mixture of declarative and procedural Prolog code allows easy grammar rule writing for other reactions. As an additional example, consider heterocyclic rings. The grammar rule

heterocyclic(Substituents, Number, Heteroatom) $\longrightarrow$ "T", number(Number),
          heteroatom(Heteroatom), "J", substituents(Substituents, Number).

recognizes this class of molecules.

The following grammar rule recognizes the heteroatom:

heteroatom(Heteroatom) $\longrightarrow$ [Heteroatom], {member(Heteroatom, "NOS")}.

Curly braces allow direct inclusion of Prolog terms within DCG grammars (the terms are not translated). In this case, the **member** predicate tests the value of the **Heteroatom** variable for membership in a list of heteroatoms.

3   The Reaction Check

This system covers concerted reactions of the π electron systems on two reactants to form new σ bonds yielding carbocyclic rings with a single unsaturation. If the reaction follows the rule of maximum orbital overlap, then it is a suprafacial, suprafacial process and is termed a $[_\pi 4_s + _\pi 2_s]$ reaction. By the Woodward-Hoffmann rules this is a symmetry-allowed thermal reaction [13].

The theoretical underpinnings used in this program are derived from those used by Jorgensen *et. al.* in the CAMEO system [14], [15] with the exception that our system works backwards, going from a product to either the reactants which form it, or issuing a statement informing the user that a disconnection is not possible.

3.1   Basic Frontier Molecular Orbital Theory

It is known from molecular orbital theory that molecules possess sets of individual molecular orbitals (as long as the molecules are sufficiently far apart from each other). These are the basic unperturbed molecular orbitals used in the evaluation of the reaction. As the molecules move more closely together, their orbitals begin to overlap. This interaction between the orbitals on the different molecules results in the mixing of the orbitals on each molecule [13].

According to frontier molecular orbital theory, the strongest interactions are between those orbitals that have coefficients with similar magnitudes relative to the unperturbed molecules, *i.e.* the interaction is between the small coefficient on the dienophile and the small coefficient on the diene [16], [17].

If both of the molecular orbitals involved in the bonding are filled, the resulting orbital is not significantly reduced in energy [18]. The greatest reduction in energy arises in the interaction between a filled molecular orbital and an empty one. Since the interaction is strongest between the orbitals of like energy, the ideal combination of orbitals is between the highest occupied molecular orbital (HOMO) on one molecule and the lowest unoccupied molecular orbital (LUMO).

Although Diels-Alder reactions can occur in the unsubstituted case, the reaction is most successful when the diene and the dienophile contain substituents which exert a favorable electronic influence [19]. In the normal electron demand case, the most favorable interactions are between dienes with electron-donating groups and dienophiles with electron-withdrawing groups. Cases have been reported in which inverse electron demand occurs and the electronic nature of the diene and dienophile are reversed [20], [21], [22]. This case of inverse electron demand is accounted for in the system.

3.2   Structural Constraints on Reactants

It became necessary early on in the project to develop a method for quickly checking the reactants for structural features which would make them unsuitable for the Diels-Alder reaction. The constraints are integrated into the notation package, since they are most easily recognized in terms of the notation patterns resulting from the disconnection. The synthons produced by a Diels-Alder disconnection are checked for proper configuration. All synthons are checked before the FMO algorithm begins, resulting in the failure of program execution and the return of a "no" to indicate no reaction. This assures that synthons produced by the rules are actually reactive.

The following structural features of diene-synthons are considered unreactive in $[_\pi 4_s + _\pi 2_s]$ cycloadditions:

1. Any diene-synthon unable to have an s-*cis* conformation.

2. Diene-synthons in which an exocyclic double bond is conjugated to a double bond in the ring (*e.g.*, a double bonded substituent on the diene).

3. Diene-synthons in large (greater than 7-membered) rings.

4. Acyclic compounds that have bulky substituents at the central positions on the diene-synthon. The substituents at these positions are relatively close to each other, and bulk leads to steric hindrance.

5. Substitution at both terminal diene-synthon positions is allowed only if the substituent is a primary atom or a triply bonded functional group (such as a cyano group).

All double bonds are perceived as possible dienophile synthons by the notation package. The screening involves only the elimination of all double bonds in aromatic compounds (WLN symbol "R").

3.3   Basic HOMO–LUMO Calculations

From work performed in 1983 by Burnier and Jorgensen [15], the following *ab initio* calculations for the HOMO and LUMO energies of the synthons were developed. The function n(x, *parent*) returns the number of atoms of type x in the *parent*. This function is abbreviated below as simply n(x) where the *parent* is understood. The symbols UU, O, N, S represent triple bonds, oxygen, nitrogen, and sulfer, respectively. The subscripts 'c' and 't' denote central and terminal locations respectively in the parent for the elements which they modify. For brevity, the terms diene-synthon and dienophile-synthon will be replaced with diene and dienophile respectively.

For Dienes:

$$E_{\text{HOMO}} = -2n(O) - n(UU) - 0.2n(N_c) - 0.5n(S_t) - n(S_c) - 9.0 \qquad (1)$$

$$E_{\text{LUMO}} = -n(O) - 0.5n(N) - 2n(S_t) + 1.5n(S_c) + 0.6 \qquad (2)$$

For Dienophiles:

$$E_{\text{HOMO}} = -n(UU) - 4n(O) - 2n(N) - n(S) - 10.5 \qquad (3)$$

$$E_{\text{LUMO}} = n(UU) - n(O) - 0.5n(N) - 4n(S) + 1.8 \qquad (4)$$

In the carbocyclic ring case, the HOMO-LUMO values default to the constants at the end of the equations. The formulas above are used to compute the orbital energies (both HOMO and LUMO) of the unsubstituted parent compounds. In the case of substituted compounds, additional formulas account for the electronic effects of the substituents.

The explanation of the regiospecificity of Diels-Alder reactions requires knowledge of the effect of substituents on the coefficients of the HOMO and LUMO orbitals. In the case of normal electron demand, the important orbitals are the HOMO on the diene and the LUMO on the dienophile. It has been shown that the reaction occurs in a way which bonds together the terminal atoms with the coefficients of greatest magnitude and those with the coefficients of smaller magnitude [18]. The additions are almost exclusively *cis* and with only a few exceptions, the relative configurations of substituents in the components is kept in the products [19].

It is known that the effects of substituent groups on a diene or dienophile vary between different types of parents [23]. A function, $\tau(Y)$, has been determined for several functional groups, with Y corresponding to their electron donating or withdrawing capability such that a reasonable estimate of the HOMO energy could be obtained by use of the equation [15]:

$$E_{HOMO} = \gamma(P) + \tau(Y) + E_{HOMO}(P) \tag{5}$$

This equation yields a value for the substituted molecule where $\gamma(P)$ is the sensitivity of the parent P. Some initial values, called τ values, which describe the electronic effects of functional groups have been found and developed by Jorgensen *et. al.* Hydrogen was assigned a τ of 0.0 eV so that electron withdrawing substituents have negative τ values and electron donating groups have positive τ values. The values for τ were chosen so that a 0.5 eV change in the substituent gives a change of 10 in the τ value. This algorithm, when combined with the notation rules, yields useful results for many functional groups and gives reasonable estimates of the values for those not known. The factor $\gamma(P)$ for an ethylene analog is given by:

$$\gamma(P) = 0.01n(UU) + 0.06n(O) + 0.03n(N) + 0.03n(S) + 0.05 \tag{6}$$

For any given diene the value for $\gamma(P)$ can be adequately represented by the value 0.03 eV. This provides the proper value for correction in the calculation due to the sensitivity of the parent compound towards different types of functionality.

3.4   Determination of Substituent Effects

To determine substituent effects, substituent groups are built from primary recognized atoms and functional groups. A functional group is scanned one Wiswesser symbol at a time. A Wiswesser symbol can represent either an individual atom (*e.g.*, "G" for chlorine) or a functional group (*e.g.*, "Z" for the amino group). This allows us to adapt the "layer" method of Jorgensen to the scanning of the functional groups on the rings. These groups are provided as Prolog sublists as outlined in the previous section. Once the comparison between the functional group elements and the known values are compared, τ is calculated by the following method. The formula for the numeric calculation is:

$$\tau_{total} = \tau_{max} + 2\tau_{sum}/(1 + NFG) \tag{7}$$

Table 1. Example τ Table Entries

name	WLN	τ
tau_entry(p-methoxyaryl,	"R DO1",	51).
tau_entry(trimethylamino,	"N1&1",	44).
tau_entry(aryl,	"R",	42).
tau_entry('methyl sulfate',	"S1",	38).
tau_entry(amino,	"Z",	36).
tau_entry(olefinic,	"1U2",	36).
tau_entry(sulfate,	"SH",	32).

The legend for this equation is:

- τ_{max} - the largest calculated reference value of τ in either the positive or negative direction.

- τ_{sum} - the sum of the remaining τ values in the functional group.

- NFG - the number of functional groups attached to the parent system.

The above is based on the calculation of a collective τ for the whole molecule. This value changes the HOMO of either the diene or dienophile, as is necessary. This equation is accurate to about 0.5 eV on either side of the "known" values [15]. The value of τ_{total} is inserted into the HOMO-LUMO calculation as the parameter $\tau(Y)$. Note that in its pure form, this equation only yields values for the HOMO orbitals. Corrections are used for the calculation of the LUMO values. Table 1 contains examples of the Wiswesser Line Notation and the raw τ values used in the computation of orbital energies.

3.5 Determination of Permutated LUMO Coefficients

The following rules were used for the determination of the LUMO orbital coefficients from the values determined for the HOMO coefficients [15].

1. An electron donating functional group raises the energy of the HOMO orbital of a system about twice as much as it raises the LUMO.

2. In contrast, an electron withdrawing functional group lowers the HOMO energy about one third as much as it lowers the LUMO.

3. Groups which add conjugation such as olefinic, acetylenic and aromatic groups lower the LUMO orbital energy one third to one half as much as the HOMO energy.

The same equations are used to determine both the HOMO and LUMO values. This is consistent with the fact that the HOMO and LUMO orbitals are calculated from the same parent system, and that the difference between the orbital energies can be adequately covered by the two parameters $\gamma(P)$ which represents the sensitivity of the parent to substitution and $\tau(Y)$ which represents the electronic effect exerted by the functional group acting as a substituent.

To implement the rules mentioned above, only the $\tau(Y)$ values for the functional groups are changed. Thus, the $\tau(Y)$ values for the calculation of the LUMO orbitals on both the diene and dienophile are changed following these rules:

1. Positive τ values except those for conjugated hydrocarbons are divided by a factor of 2.

2. Negative τ values are multiplied by 3.

3. τ values for conjugated hydrocarbons are divided by a factor of 3 and their signs are reversed.

This method covers many combinations of functional groups that influence the orbital energies. A feature of this method is that it uses the same functional group τ values as in the HOMO energy calculation. The algorithm described above is used for the calculation of both the HOMO and LUMO atomic coefficients. The τ values of the substituents are permutated to give the proper values for the LUMO orbitals. The following steps are required:

1. τ values on terminal positions are taken from the list previously described.

2. Resultant τ values on the central diene positions are divided by a factor of two to accommodate the fact that the orbital coefficients at these positions are very small.

3.6   Algorithm for Regiochemical Selection

Any functional group attached to a terminal carbon on either a diene or dienophile increases the magnitude of the coefficient on the opposite terminal. Any functional group attached to a central position on the diene (there is no analogous case for the dienophile) increases the magnitude of the coefficient on the terminal farthest from the substituted position. For cyclohexene, the central locants are the **A** and **B** positions on the Diels-Alder adduct. Thus, if a functional group is on position **A** the magnitude of the coefficient at terminal **C** increases. One of the remarkable aspects of the Diels-Alder reaction is the specificity of the bonding between the carbon atoms [13]. The orientation of the addition can be accurately predicted by an extended form of the frontier molecular orbital theory as developed by Fukui and Fujimoto *et. al.* [16]. For dienes the coefficients are determined as follows: if the sum of the absolute values of τ on positions **F** and **B** is greater than the sum of τ on positions **A** and **C**, then the coefficient on position **C** has the greater magnitude, otherwise the coefficient of position **F** has the greater magnitude. On dienophiles, if the sum of the absolute values of τ is greater on position **D** than on position **E**, then **E** has the greater magnitude.

4   Reaction Unification Using a General Form

This section examines the notion of a *general form* for representing the possible synthons in a reaction. Derivation of this form is illustrated and examples of the general form are presented. Symmetry and the encoding of optional notation is discussed and some examples of the naive approach are presented.

Table 2. The Naive Approach

Parent	Synthons
discon("L6UTJ A1 B1",	["1UY1&Y1&U1", "1U1"]).
discon("L6UTJ D1Q",	["1U2U1", "Q2U1"]).
discon("L6UTJ A1 B1 DOV1",	["1UY1&Y1&U1", "1V01U1"]).
discon("L6UTJ A1 B1 D1 ENW",	["1UY1&Y1&U1", "WN1U2"]).

4.1 Motivation: the Naive Approach

In the naive approach, disconnections are simply listed as facts with the molecule to disconnect as the first argument and a list of the synthons as the second. Table 2 contains some examples. This approach suffers in many ways; primarily, the number of rules would become unmanageable (quite huge even for cyclohexene), slowing the inferencing speed of the expert system.

A sample inference mechanism using these facts (given the natural backward chaining of Prolog) might be

```
disconnect(Parent, Given_Synthons) :-
        discon(Parent, Synthons),
        disconnect(Synthons, Given_Synthons).
disconnect(Parent, [Parent]) :-
        given(Parent).
disconnect([First|Rest], [First_Disc|Rest_Disc]) :-
        disconnect(First, First_Disc),
        disconnect(Rest, Rest_Disc).
disconnect([], []).
```

This procedure recursively **disconnect**s synthons until the final synthons for the original parent are all available (or **given**) compounds. Upon successful completion, the variable **Given_Synthons** contains a tree (in list notation) which denotes the synthon combination order to reproduce the parent compound.

4.2 Derivation of the General Form

Consider the domain of a six-membered ring with single unsaturation. Table 3 expresses the synthetic route with one substituent. Again, the symbol 'σ' represents an arbitrary substituent. Square brackets surrounding a set of symbols indicates optionality of those symbols (as in regular expression notation). For example, the string "$\sigma[\&]$" may reduce to the string "σ" or "$\sigma\&$" depending on whether the substituent represented by σ ends in a terminal symbol or not (following the rules of WLN).

Symmetry in the patterns, however, hides many details in the diene and dienophile patterns. Table 4, with combinations of symmetric substituents, reveals more of the details. The order of the symmetric substituents may be chosen arbitrarily. Alphabetical ordering was chosen here for consistency.

Finally, for a full cyclohexene molecule, the patterns become

$$\sigma_C 1UY\sigma_A[\&]Y\sigma_B[\&]U1\sigma_F + \sigma_D 1U1\sigma_E \qquad (8)$$

Table 3. Patterns for a Six-Membered Ring with One Substituent

Substituent Position	Diene	Dienophile
A or B	$1UY\sigma[\&]U1$	1U1
C or F	$\sigma 1U2U1$	1U1
D or E	1U2U1	$\sigma 1U1$

Table 4: Patterns for a Six-Membered Ring with Two Substituents

Substituent Position	Diene	Dienophile
A and B	$1UY\sigma_A[\&]Y\sigma_B[\&]U1$	1U1
C and F	$\sigma_C 1U2U1\sigma_F$	1U1
D and E	1U2U1	$\sigma_D 1U1\sigma_E$

It should be clear that this notation applies to many different classes of reactions. Use and manipulation of this general form will be discussed in the next section. The following discussion outlines its use in expert system rules.

4.3   Use of the Mechanism in Rule Formation

Given the general form, it is possible to capture many disconnections of a given class with a single rule. The following example illustrates the approach advocated in this paper for cyclohexene.

```
discon(WLN, [Diene, Dienophile]) :-
        carbocyclic(Substituents, 6, WLN, []),
        collect_substs(Substituents, ''CABF'', Dn_substs),
        collect_substs(Substituents, ''DE'', Dl_substs),
        fmo(Dn_substs, Dl_substs),
        make_synthon(Dn_substs, "*1UY*&Y*&U1*", Diene),
        make_synthon(Dl_substs, "*1U1*", Dienophile).
```

This rule declaratively states that the compound represented by **WLN** disconnects to the **Diene** and **Dienophile** pair *if* the **WLN** matches the carbocyclic grammar rule with 6 substituents, the collected substituents for the **Diene** and **Dienophile** pass the **fmo** test, and the respective constituents may be successfully incorporated into the general form for the cyclohexene **Diene** and **Dienophile**.

The goal **make_synthon** instantiates the general form and rewrites the instantiated general form into a pseudo-WLN form. The pseudo-WLN form has adjacent number values combined and redundant ampersands eliminated, but the branch ordering does not necessarily follow all the WLN rules. The '*' symbol in the second argument represents a general substituent, 'σ', where the subscript is determined by the order mentioned in the **collect_substs** predicate (*e.g.*, "CABF" and "DE").

The following grammar rewrites the instantiated general form to the pseudo-WLN notation. The unit symbol '[]' in the following grammar represents the NIL symbol (or empty symbol) and arises when a substituent is not present in a particular position. This grammar captures the following conditions: the '[]' symbol next to a number disappears, adjacent numbers are summed (for a longer carbon chain), a three way branch reduces to a carbon when one of the branches is empty, optional ampersands are eliminated, and required ampersands are retained. The rules must be applied to the string repeatedly until no changes to the string occur.

$$N[] \quad \longrightarrow N.$$
$$[]N \quad \longrightarrow N.$$
$$\sigma N \quad \longrightarrow \{\text{number}(\sigma), \text{NN is } \sigma + N\}, \text{NN}.$$
$$N\sigma \quad \longrightarrow \{\text{number}(\sigma), \text{NN is } N + \sigma\}, \text{NN}.$$
$$N_1 N_2 \quad \longrightarrow \{\text{NN is } N_1 + N_2\}, \text{NN}.$$
$$Y[]\& \quad \longrightarrow 1.$$
$$Y\sigma\& \quad \longrightarrow \{\text{not}(\text{number}(\sigma)), \text{ends\_in\_terminal}(\sigma)\}, Y\sigma.$$
$$Y\sigma\& \quad \longrightarrow \{\text{not}(\text{number}(\sigma)), \text{not}(\text{ends\_in\_terminal}(\sigma))\}, Y\sigma\&.$$

For example, performing these transformations with an empty cyclohexene (σ_A = [] ... σ_F = []) yields the diene "1U2U1" and the dienophile "1U1". Once the synthons are in pseudo-WLN form, they are rearranged to conform to the standard WLN form (described in Section 5).

4.4   Application to Other Reactions

General forms are easily developed for other reactions. The machinery introduced in this section can then be utilized to write disconnection rules for other reactions. For example, consider the Diels-Alder adduct bicyclo[2.2.1]hept-2-ene. Using the regular expression notation described previously, the line notation for these types of compounds can be represented as

L55 CU ATJ $[A\sigma_A]$ $[B\sigma_B]$ $[C\sigma_C]$ $[D\sigma_D]$ $[E\sigma_E]$ $[F\sigma_F]$ $[G\sigma_G]$ $[-A\&(F+G)]$ $[-B\&(F+G)]$

The information following the hyphens describes the orientation of the substituents at locants where stereoisomerism can occur. **F** and **G** are the locants where the stereochemistry may occur.

This compound can be disconnected into a cyclopentadiene synthon and a dienophile synthon similar to the the the one previously described. The general form for the disconnection is then given in the notation by

$$\text{L5 AHJ } A\sigma_A \text{ } B\sigma_B \text{ } C\sigma_C \text{ } D\sigma_D \text{ } E\sigma_E + \sigma_F 1U1\sigma_G \tag{9}$$

Additional pseudo-WLN rewrite rules would eliminate ring locant symbols which are followed by an empty substituent.

5   Notation Rearrangement

The previous section illustrated the formation of diene and dienophiles and noted that the intermediate notation did not necessarily obey the WLN rules. This section describes the transformation from pseudo-WLN form to legal WLN notation.

A predicate called **wln_order** occurs within the **make_synthon** predicate. This predicate builds a graph from the pseudo-WLN (using WLN Rule 8(a)) and possibly reorders the graph as described below. The following Prolog code describes this manipulation:

```
wln_order(Pseudo_WLN, WLN) :-
      notation_graph(Pseudo_WLN),
      rule6(Chain),        % uses graph in database
      rule7and8(Chain, WLN).
```

Construction of the **notation_graph** requires general knowledge about terminal symbols and their interaction with branch symbols. The pseudo-WLN is parsed using this knowledge. Vertices are created when branch symbols are encountered and the edges are labeled with the notation which occurs between the branch vertices. An undirected graph results from this process and all vertices with outdegree one are considered roots.

Rule 6 orients the molecule, collecting the vertices and edges in the proper order. To accomplish this, all root nodes are collected. Starting from each root, the primary chain of the notation is chosen using the longest path of notation symbols, breaking any tie by choosing the chain which ends in the latest notation symbol (Rule 2).

Next, Rule 7 orients branch choices along the primary chain chosen above. This rule orders branches using the branches with the lowest branching factor and with the fewest notation symbols. Ties are again broken by Rule 2. Rule 8 guides the reassembly of the molecule in proper WLN form. It reintroduces ampersands and inserts hyphens where necessary. All of this was easily implemented in Prolog, using DCG to parse the pseudo-WLN form and the Prolog database to represent the graph.

Many additional rules are required for other reactions. Probably the entire complement of WLN rules must be implemented for even moderately sophisticated chemistry. It may be desirable at this point, however, to design a notation which encompasses WLN'S strong points, but is more computationally oriented.

6   Conclusions

Other systems have developed FMO reaction checks and used WLN for cataloging, but this system has relied heavily on a symbolic approach to chemistry, including application of grammar techniques to WLN strings. We feel that our system is very successful in the domain that it has been applied, eliminating hundreds of naive expert system rules. We also feel that our techniques are applicable to many other reactions as well.

This paper has primarily stressed concepts rather than implementation details. A prototype system based on these concepts has been implemented, with concentration in the cyclohexene domain. The entire system, including grammars, the FMO verification, and WLN manipulation required only 12 pages of Prolog code. Although execution speed was never considered a factor at this stage, the system performs the disconnection

$$L6UTJ \; A1 \; B1 \; D1 \; ENW \Longrightarrow 1UY1\&Y1\&U1 + WN1U2 \tag{10}$$

in four seconds with a 1K Logical Inferences Per Second (LIPS) interpreter.

There are several future research directions for this project. First, results from the FMO reaction check are not infallible due to the qualitative nature of this check. A more precise, yet computationally feasible model may be possible. Second, more work remains in the WLN rearranger; a full system based on our concepts would require knowledge of the entire complement of WLN rules. It may also be desirable to adopt or develop another, more computationally tractable line notation for the purpose of synthetic analysis. Finally, we would like to extend our work to more reaction classes to examine its potential in more detail.

Acknowledgments

We wish to express our appreciation for the Texas Instruments IDEA program which sponsored the majority of this research. This is a unique program within a large company which provides excellent research opportunities. Texas Instruments' unsurpassed computing facilities also deserve acknowledgment.

Literature Cited

[1] Blower, P. E., Jr., *An Application of Artificial Intelligence to Organic Synthesis*, PhD Thesis, University of Wisconsin, 1975.

[2] Gordon, John E., "Chemical Inference. 2 Formalization of Organic Chemistry: Generic Systematic Nomenclature," *J. Chem. Inf. Comput. Sci.*, 24, (1984), pp. 81-92.

[3] Rodgers, David and W. T. Wipke. "Artificial Intelligence in Organic Synthesis. SST: Starting Material Selection Strategies. An Application of Superstructure Search," *J. Chem. Inf. Comput. Sci.*, 24, (1984), pp. 71-81.

[4] Sridharan, N. S., PhD Thesis, State Univerisity of New York at Stonybrook, 1971.

[5] Warren, Stuart, *Organic Synthesis: the Disconnection Approach*, John Wiley & Sons, New York, 1982.

[6] Clocksin, W. F. and C. S. Mellish, *Programming in Prolog*, Springer-Verlag, Berlin, 1981.

[7] Pereira, F.C.N., D.H.D. Warren, "Definite Clause Grammars for Language Analysis – a Survey of the Formalism and a Comparison with Augmented Transition Networks", *Artificial Intelligence*, 13, (1980), pp. 231-278.

[8] Smith, Elbert G., ed., *The Wiswesser Line - Formula Chemical Notation*, McGraw-Hill, New York, 1968.

[9] Fritts, Lois E., Margaret Mary Schwind, "Using the Wiswesser Line Notation (WLN) for Online, Interactive Searching of Chemical Structures," *J. Chem. Inf. Comput. Sci.*, 22, (1982), pp. 106-109.

[10] Johns, Trisha M., Michael Clare, "Wiswesser Line Notation as a Structural Summary Medium," *J. Chem. Inf. Comput. Sci.*, 22, (1982), pp. 109-113.

[11] Winograd, Terry, *Language as a Cognitive Process*, Volume 1, Addison-Wesley, Reading, 1983.

[12] Hopcroft, John E., Jeffry D. Ullman, *Introduction to Automata Theory, Languages, and Computation*, Addison-Wesley, 1979.

[13] Woodward, R. B. and R. Hoffmann, *The Conservation of Orbital Symmetry*, Academic Press, New York, 1970.

[14] Jorgensen, W. L. and Timothy D. Salatin, *J. Org. Chem.*, 45, 2043, (1980).

[15] Jorgensen, W. L. and Julia Schmidt Burnier, *J. Org. Chem.*, 48, 3923, (1983).

[16] Fukui, K., *Top. Cur. Chem.*, 15, 1, (1970).

[17] Herndon, W. C., *Chem. Rev.*, 72, 157, (1972).

[18] Lowry, T. H., K. S. Richardson, *Mechanism and Theory in Organic Chemistry*, 2nd ed., Harper & Row, New York, 1981.

[19] Onishchenko, A. S., *Diene Synthesis*, Daniel Davey and Co., New York, 1964.

[20] Sustmann, R., *Tetrahedron Lett.*, 2717, (1971).

[21] Sustmann, R., *Tetrahedron Lett.*, 2721, (1971).

[22] Sustmann, R. and H. Trill, *Agnew. Chem. Int. Ed.*, 11, 838, (1972).

[23] Fleming I., *Frontier Molecular Orbitals and Organic Chemical Reactions*, Wiley, London, 1976, Chapter 4.

RECEIVED December 17, 1985

20

Using a Theorem Prover in the Design of Organic Syntheses

Tunghwa Wang, Ilene Burnstein, Michael Corbett, Steven Ehrlich, Martha Evens, Alice Gough, and Peter Johnson

Illinois Institute of Technology, Chicago, IL 60616

This paper describes an expert system for organic synthesis which uses a resolution based theorem prover as its reasoning component. This reasoning component is built upon LMA (Logic Machine Architecture), a collection of Pascal subroutines written by the theorem proving group at Argonne National Laboratory. The SYNLMA system (SYNthesis with LMA) represents the target compound as a theorem to be proved, while the starting materials and reaction rules become axioms. The main advantages of SYNLMA stem from the independence of the database from the inferencing mechanism. This separation makes it possible to experiment with different representations of knowledge and different data bases, like the large chemical databases made available by ISI and Chemical Abstracts, without reprogramming.

Using LMA (Logic Machine Architecture), a collection of Pascal programs written by the theorem proving group at Argonne National Laboratory (1-2), we have developed SYNLMA (SYNthesis with LMA), an expert system for organic synthesis that uses a resolution based theorem prover as the reasoning component. The major advantages of SYNLMA stem from the independence of the database and the inferencing. First, the database can be modified or an entirely different one used without reprogramming the decision making unit of the system. This conversion involves modifying a short program that translates a database representation for molecules into a molecular representation the theorem prover recognizes; SYNLMA is not changed at all. Second, the scheme for representing

0097–6156/86/0306–0244$06.00/0

a molecule can be changed without changing SYNLMA. Once again SYNLMA remains the same, only the interface between the database and SYNLMA will have to be altered. This flexibility makes SYNLMA an attractive alternative to other organic synthesis programs.

SYNLMA performs a retrosynthetic analysis using a special purpose theorem prover built from LMA components. The compound to be synthesized becomes a theorem to be proved. The reaction rules and starting materials become axioms. The choice of a knowledge representation has been one of our greatest problems.

Data for the theorem prover has to be translated into clauses, the only form the theorem prover recognizes. A clause is the "OR" of one or more literals where a literal is a predicate and its arguments. A predicate is a property or relationship that is true or false. Its arguments can encompass any number of functions. A function returns true, false or some other value. The statement "x + y > y + z" can be written as a clause using the function "Sum" and the Predicate "GreaterThan." The resulting one-literal clause looks like this:

$$GreaterThan(Sum(x,y), Sum(y,z))$$

(See <u>3</u> for a formal definition of a clause.)

Molecular Representations

The representation of molecular structure in clause form is crucial to this research as it is a major determinant of the theorem prover's efficiency. The clause representation affects the time it takes to retrieve reaction rules and starting materials and the time necessary to make comparisons between structures. The importance of the relationship between efficiency and the clause representation is illustrated by the difference in the run times between proving our first clauses and current ones. Our first representation scheme was a simple one with one predicate for each atom except hydrogen and one for each bond (Figure 1a). Using this clause form, a molecule with ten atoms took several hours to prove on an IBM mainframe. For SYNLMA to be a viable system for organic synthesis the "proving time" has to be reasonable and one key to this is the clause representation. By using a single predicate to describe each atom and its "bond environment," the proof of a molecule has been reduced to a few seconds. We will continue to experiment with the representation for molecules, trying to find the right balance between the number of clauses and their length. We currently represent starting materials and compounds that we want to synthesize (targets) by a clause list (Figure 1b). In this scheme:

1. A molecule is represented by a list of clauses, where each clause corresponds to one atom and describes its environment (i.e., its bonds, charge, etc.).
2. The number of atoms in a molecule does not correspond to the number of clauses in the clause list. An atom generates a clause only if it is bonded to two or more atoms; otherwise the atom will be ignored as all its information will be contained in a clause generated by another atom.
3. Each clause consists of the predicate called Fragment, a Bond function (Brr1, B211, B1111, etc.) listing the types of bonds, such as aromatic, resonant, triple, double, single, for the atom being described and an Atom function for this central atom of reference and for each atom bonded to it. A clause is terminated with a semicolon.
4. The arguments for the Atom function are: the chemical symbol for the element, a number assigned by our numbering scheme, the charge on the atom (-1, 0, +1, +2 etc.), a stereochemistry flag and a ring flag indicating whether or not the atom is a member of a ring. Default values for the last three arguments are zero.

```
        H(7)              O(3)          C(1);
          \               //            C(2);
      H(6)__ C(2) __ C(1)               O(3);
          /              \              O(5);
        O(5)             H(4)           Double(1,3);
          /                             Bond(2,1);
        H(8)                            Bond(2,5);
```

Figure 1a. Our First Clause Representation for a Simple Molecule. The numbers following the element symbols in the diagram are used to identify atoms in the clauses.

```
Fragment(B211(Atom(C,1,0,0,0),Atom(O,3,0,0,0),
          Atom(C,2,0,0,0),Atom(H,4,0,0,0)));
Fragment(B1111(Atom(C,2,0,0,0),Atom(C,1,0,0,0),
          Atom(O,5,0,0,0),Atom(H,6,0,0,0),
                         Atom(H,7,0,0,0))));
Fragment(B11(Atom(O,5,0,0,0),Atom(C,2,0,0,0),
                         Atom(H,8,0,0,0)));
```

Figure 1b. Our Current Clause Representation
for the Same Molecule

Figure 1b is a simple example of a clause list and the rules for constructing it. In actuality, there are no spaces between characters in a clause. They are included to make it easier to grasp the clause notation. Note, that although there are eight atoms in the molecule only three generated clauses. For example, O(3) does not generate a clause since it would be redundant. The clause for O(3) would be "Fragment(B2(Atom(O,3,0,0,0),

Atom(C,1,0,0,0)))" and all this information is contained
in the clause generated by C(1). The first Fragment
predicate in figure 1b is:

Fragment(B211(Atom(C,1,0,0,0),Atom(O,3,0,0,0),
         Atom(C,2,0,0,0),Atom(H,4,0,0,0)));

The Bond function describes a central atom of
reference and the atoms bonded to it. B211 states that
there is a central atom of reference bonded to one atom
by a double bond (2) and to two other atoms by single
bonds (1). The order of the Bond function arguments
corresponds to this Bond function notation. These
arguments are not simple atomic symbols, but Atom
functions that can relate considerable information about
the atom. In this Bond function, Atom(C,1,0,0,0) is the
central atom. The next three arguments are atoms that are
bonded to this central atom: the first, Atom(O,3,0,0,0)
by a double bond; the next two, Atom(C,2,0,0,0) and
Atom(H,4,0,0,0), by single bonds.

The first two arguments in the Atom function for a
particular Atom never change, as they identify the atom.
Atom(O,3,0,0,0) describes the oxygen atom numbered 3, as
opposed to the oxygen atom numbered 5, in the drawing in
Figure 1a. The number does not indicate position. If
some reaction resulted in the "O3" bond to "C1" being
broken and "O3" was replaced by some other atom, "O3"
remains "O3"; the new atom will have a new number.
Suppose "O3" were to become charged, then the function
describing it would become Atom(O,3,-1,0,0), reflecting
the change.

Reaction Rule Database

Our present reaction rule database is made up of
approximately one hundred rules adapted from a microfiche
generously sent to us by Gelernter (4). For a given
reaction, a rule specifies the reactants (subgoal) and
the product(s) (goal), in connection table format and any
constraints on their composition (Figure 2a). The rules
are identified by chapter and schema numbers. The
connection tables are organized as follows:

1.   A reaction rule connection table includes all the
     atoms in both the goal and subgoal molecules. The
     atoms are numbered uniquely and the numbering of the
     atoms (see the drawing of the molecules) corresponds
     to the row numbers in the tables. The same atom
     appearing in both a goal and subgoal keeps the same
     number. If an atom in the goal does not appear in the
     subgoal, the subgoal connection table will still
     include the atom as a row atom but all values to the
     right will be zero.
2.   The symbols in the first column (to the right of the
     row number) identify the atom or variable described

by the row. It will be referred to as the row atom.
There are three types of row atom symbols: periodic
table notation for elements; the variable #J which
represents a halide; and a variable composed of a
dollar sign followed by a number($1,$2...). The
"$/even numbered" variables can represent any
substructure or any atom. The "$/odd numbered"
variables can represent any substructure or any atom
except hydrogen. The following four structures could
be equivalent. The structures range from the very
specific on the left where the atom pointed to is
defined as a chlorine atom to the very general where
the atom or substructure can be anything.

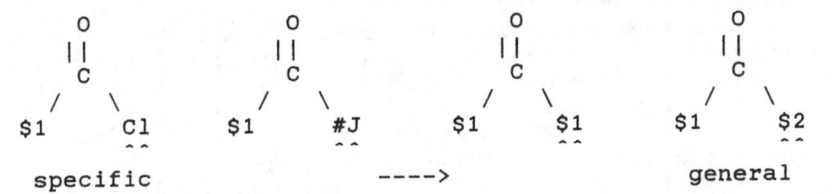

       specific   ---->   general

3. The next twelve columns (six pairs: up, down, left,
    right, in, out) describe the bonds of the atom in
    column one. The first number in each pair is the row
    index, identifying the atom bonded to the row atom.
    The second number is one of five bond types (1:
    single, 2: double, 3: triple, 16: resonant bond,
    8:single bond between an atom and a resonant
    structure). If the row atom does not appear in the
    goal or subgoal structures the default values are
    zero.
4. The last sixteen columns contain symmetry
    information.

    Figure 2a is the Gerlertner reaction rule for the
"reaction of magnesium with alkyl bromides". The number
of (six) and type of row atoms (Mg, Br, C, $2, $4, $6)
are identical for both the the goal and subgoal
connection tables and is a composite of all atoms in both
the product and reactants. Differences between goal and
subgoal structures are indicated by the numbers to the
right of row atoms and not their presence or absence in
the tables. For example, in the goal table Row Atom 1,
magnesium, is bonded to Row Atom 2 by a single bond
(index:bond = 2:1) and to Row Atom 3 by a single bond
(index:bond = 3:1). While magnesium does not appear in
the subgoal structure, it is still the first row atom in
the subgoal's table. But the values for bond indexes and
bond types are now zero; that is, Mg(1) is not bonded to
other atoms in the table. An example of an atom that
appears in both the goal and subgoal structures is Row
Atom 3. One of the atoms that C(3) is bonded to changes
(Br to Mg) but C(3) is considered the same throughout the
reaction and keeps the same index.

```
   $2(4)        $4(5)                         $2(4)        $4(5)
      \         /                                \         /
         C(3)            ------------>              C(3)
      /         \                                /         \
   Br(2)        $6(6)                         Mg(1)        $6(6)
                                                 /
                                              Br(2)
```

Diagram illustrating the following reaction rule

Schema 2

Schema name is reaction of magnesium with alkyl
bromides. The starting values for ease, yield and
confidence are: 90, 95, 100. The reagent class for
this schema is: 0. This is a single application
schema. The maximum no. of nonidentical subgoal
molecules allowed for this schema is 9.

```
┌─────────────────────────────────────────────────────────────────┐
│ The Transformation Patterns:                                      │
│                                                                   │
│ Goal TSD                                                          │
│                                                                   │
│ no elem   up  down   left   right   in    out                     │
│                                                                   │
│ 1   Mg   2:1  3:1   0:0    0:0    0:0   0:0   0000000000000000     │
│ 2   Br   1:1  0:0   0:0    0:0    0:0   0:0   0000000000000000     │
│ 3   C    1:1  4:1   5:1    6:1    0:0   0:0   0000000000000000     │
│ 4   $2   3:1  0:0   0:0    0:0    0:0   0:0   0000000000000000     │
│ 5   $4   3:1  0:0   0:0    0:0    0:0   0:0   0010000000000000     │
│ 6   $6   3:1  0:0   0:0    0:0    0:0   0:0   0010000000000000     │
│                                                                   │
│ Subgoal TSD                                                       │
│                                                                   │
│ no elem   up  down   left   right   in    out                     │
│                                                                   │
│ 1   Mg   0:0  0:0   0:0    0:0    0:0   0:0   0000000000000000     │
│ 2   Br   3:1  0:0   0:0    0:0    0:0   0:0   0000000000000000     │
│ 3   C    2:1  4:1   5:1    6:1    0:0   0:0   0000000000000000     │
│ 4   $2   3:1  0:0   0:0    0:0    0:0   0:0   0000000000000000     │
│ 5   $4   3:1  0:0   0:0    0:0    0:0   0:0   0000000000000000     │
│ 6   $6   3:1  0:0   0:0    0:0    0:0   0;0   0000000000000000     │
│                                                                   │
│ Schema Tests: Can't have any of the following attributes:         │
│                                                                   │
│                   136 Thiol                                       │
│                   126 Oxime                                       │
│                   122 Diazoketone                                 │
│                       (and others)                                │
└─────────────────────────────────────────────────────────────────┘
```

Figure 2a. Gerlernter reaction rule.

   The constraints listed under the schema tests give
limitations on the possible values of the variables in

column one, etc. The reaction rules can be characterized
as single or multistep, where a multistep reaction is
defined as a rule that can be rewritten as a series of
single step reactions. An example of a single and
multistep reaction rule for a malonic ester synthesis
follows.

Multistep:

```
          O
          ||              1) NaOET
    ET-O-C                2) RX                O            CH3
          \               3) OH-,H20           ||           /
           CH2   ---------------->     H-O-  C -CH2-C -H
          /               4) H+                            \
    ET-O-C                                                  CH3
          ||
          O
```

Single Step Equivalent:

```
          O                                     O
          ||                                    ||
    ET-O-C                               ET-O-C
          \                                     \
           CH2 + NaOET  ------>                  CH-
          /                                     /
    ET-O-C                               ET-O-C
          ||                                    ||
          O                                     O
```

```
    O                                          O
    ||                                         ||
ET-O-C           H    CH3               ET-O-C           CH3
    \  -          \ /                       \            /
     CH    +       C         ------->        CH-C-H
    /             / \                       /            \
ET-O-C          Br   CH3               ET-O-C           CH3
    ||                                     ||
    O                                      O
```

```
    O                                      O
    ||                                     ||
ET-O-C     CH3                         H-O-C     CH3
    \     /            OH-, H20            \     /
     CH-C-H    --------------->             CH-C-H
    /     \                                /     \
ET-O-C     CH3                         H-O-C     CH3
    ||                                     ||
    O                                      O
```

```
    O
    ||
H-O-C        CH3        1) H+                          O        CH3
     \      /           2) -CO2                         ||      /
      CH-C-H         ---------------->        H-O- C -CH2-C-H
     /      \                                                \
H-O-C        CH3                                              CH3
    ||
    O
```

In this example a four step synthesis is also
expressed as a very general one step reaction.

We have written a program that translates the
connection tables into clauses, a form that the theorem
prover can process, and stores them in files organized
first by goal or subgoal and then by the functional
groups in the molecule. The constraints are in another
set of files. SYNLMA uses these files; it does not use
the files of Gerlernter formatted rules. In addition to
the reaction rule database, we have functional group and
starting material databases (also in clause form).

The Translation of Reaction Rules into Clauses

Each atom in a target or starting material molecule is
defined. This is not true for a reaction rule or
functional group molecule where parts of the molecule are
represented by variables ($1, $J, etc.). SYNLMA treats a
reaction rule or functional group structure as a
molecule, even though some of its atoms are unknown, and
represents it in essentially the same form as known
molecules (Figure 2b). A molecule with a variable
substructure differs from a known molecule in the
following:

1.   The predicates are ORed for a molecule with variables
     (one clause per molecule) instead of ANDed (one list
     of clauses for each molecule).
2.   The sign of the predicate is negative instead of
     positive.
3.   Variable atoms or substructures are represented by
     the letter "y" followed by a number (y1, y2) or the
     letter "j" (yj). "Yj" represents a halide; the
     "y/even numbered" variables can represent any
     substructure or atom; and the "y/odd numbered" can
     represent any substructure or atom except hydrogen.
4.   The Atom functions have variables for arguments, not
     constants.
5.   Each goal or subgoal clause is terminated with the
     predicate Rxnrule whose first argument is a reaction
     rule identification number. After this number, the
     predicate uses the function LL (for linked list) to
     list all the atoms in the connection table.
     Functional group clauses are terminated with the

similar function Funcgr. The main difference between
these two functions is that Funcgr only lists atoms
or substructures that are in the molecule being
described, while the reaction rule lists all atoms in
the connection table regardless of whether they
appear in the structure(s) being described.

Reaction Rule Chapter 20, Schema 2: GOAL

```
-Fragment(B11(Atom(Mg,x1,s1,t1,u1),Atom(Br,x2,s2,t2,u2),
            Atom(C,x3,s3,t3,u3)))|
-Fragment(B1111(Atom(C,x3,s3,t3,u3),
            Atom(Mg,x1,s1,t1,u1),y2,y4,y6))|
Rxnrule(202,LL(Atom(Mg,x1,s1,t1,u1,),
            LL(Atom(Br,x2,s2,t2,u2),
            LL(Atom(C,x3,s3,t3,u3),
            LL(y2,LL(y4,LL(y6,NIL)))))));
```

Reaction Rule Chapter 20, Schema 2: SUBGOAL

```
-Fragment(B1111(Atom(C,x3,s3,t3,u3),
            Atom(Br,x2,s2,t2,u2),y2,y4,y6))|
Rxnrule(202,LL(Atom(Mg,x1,s1,t1,u1,),
            LL(Atom(Br,x2,s2,t2,u2),
            LL(Atom(C,x3,s3,t3,u3),
            LL(y2,LL(y4,LL(y6,NIL)))))));
```

Figure 2b. Clause representation of the goal and subgoal
in reaction rule Chapter 20, Schema 2.

A comparison between the connection tables in figure
2a and their clause representations in figure 2b
illustrates the conversion rules and some of the
differences between a known molecule's clause and a
reaction rule clause. Two row atoms, Mg(1) and C(3), in
the goal and only C(3) in the subgoal are bonded to two
or more atoms and therefore generate predicates. Unlike
the clause list (figure 1b) these predicates are not
separated by semicolons (implicitly ANDed one predicate
clauses) but are joined by a vertical bar, the symbol for
OR. The predicate, Fragment, is constructed in the same
way as for a known molecule with the exception that some
of the Atom functions arguments are variables (e.g. x1,
s1, t1, etc.). Variables are not written using Atom
functions (they are unknowns) but are simply listed in
the proper order in the bond function. The clause is
terminated with an identifying Rxnrule predicate that
lists the reaction rule chapter and schema (chapter
number * 1000 + schema number) and every row atom in the
connection table. Note that the Rxnrule predicate is
identical for the goal and subgoal, linking the two
clauses together.

The Synthetic Design Process

SYNLMA is currently capable of handling the synthetic design for compounds the size of the analagesic Darvon using an in-core database of approximately a hundred reactions. The synthesis process starts with the input of the structure of the compound (in clause form) that we are trying to synthesize. Next, an internal representation of the compound is generated. This becomes the target (the theorem to be proved). The theorem prover begins by identifying the target's major functional groups and uses them as keys into the database. As the search begins for reactions and compounds from which the target can be synthesized, the theorem prover only searches the goal functional group files corresponding to the functional groups it has already found in the target. For example, if the target contained an acid and a benzene ring, the theorem prover would search the goal files containing acids and benzene rings for a matching molecular structure. When a matching structure is found, its corresponding subgoal becomes the new goal and is translated into the internal representation for a molecule. Then the functional group identification and the search and match process is repeated. This process of examining alternative reaction paths and setting up intermediate compounds as new goals is repeated until all the possible reactions can be performed using the available compounds.

This process of working backward from the available starting materials, called "retrosynthetic analysis" by the organic chemist, is immediately recognized as an example of backward chaining by workers in artificial intelligence. This backward chaining process creates a large problem solving tree in which goals or nodes correspond to compounds while the branches correspond to possible reaction pathways. (A more detailed description of how SYNLMA handles this process can be found in 5-6.)

An example of a problem solving tree for the synthesis of Darvon appears in Figure 3. The tree contains both AND nodes and OR nodes (7). The AND branches, connected by double arcs, indicate that both compounds are required to make the compound above them. The OR branches (there are three OR paths to make compound II) indicate different routes for making the compound. The terminal nodes corresponding to starting materials are enclosed in boxes. At present, a branch is terminated when the number of clauses in the clause list, the internal representation of the goal, is less than or equal to six or the clause list matches the clause list of a starting material molecule.

Currently, SYNLMA generates one problem solving tree for each molecule that it synthesizes. Some of the tree's paths are viable synthetic routes, others are deadends. Unfortunately, good and bad paths are pursued with the

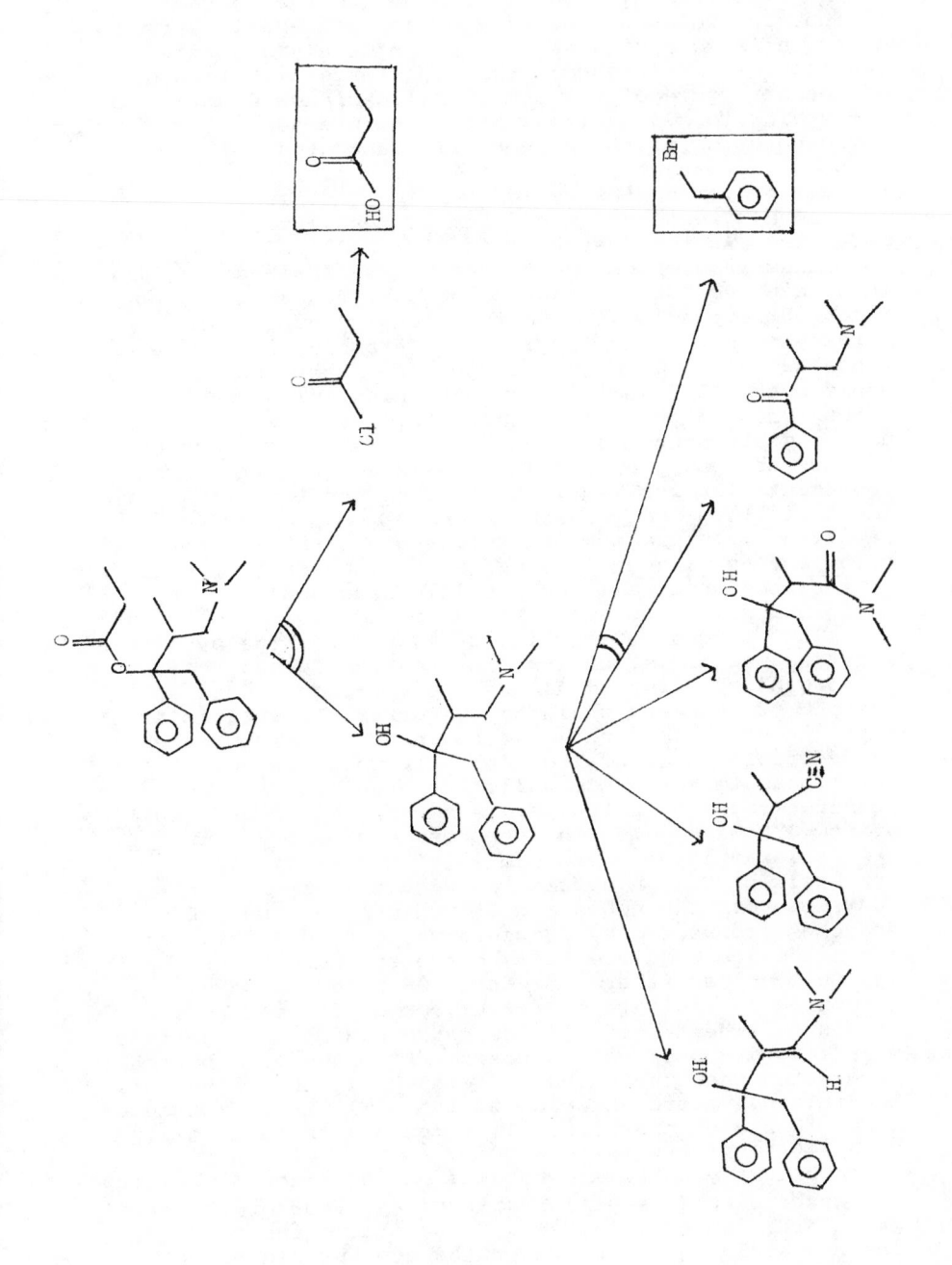

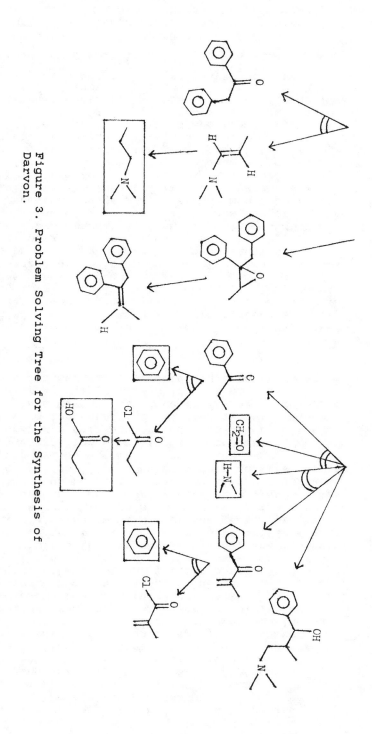

Figure 3. Problem Solving Tree for the Synthesis of Darvon.

same intensity and a lot of time is spent pursuing deadend paths. The program has the potential to be more clever in its approach. It can generate a number of trees of varying specificity. First, SYNLMA could generate a tree of multistep reaction rules. A tree built from multistep reaction rules would be quicker to build than one where each step is specified. Then a second, more specific tree could be generated using the knowledge gained from the first. For example, some synthetic pathways could be ruled out on the basis of the multistep rules. The more pathways that can be eliminated on the basis of one multistep rule as opposed to a series of single step rules, the faster the system can work. For paths that appear promising, the products and reactants in the first tree form pairs of targets and starting materials that will direct the growth of the second tree. A synthetic path that works in the more general tree does not necessarily work when SYNLMA tries to fill in the gaps between nodes with single step rules. Some other condition, like a substructure constraint, may block the pathway. So a combination of the two approaches, general and specific, is necessary. Two databases, one of single step rules, the other multistep, are necessary to implement the two tree system. Since our database is a mixture of these two types of rules, a two tree system is not yet possible. It will have to wait until we can separate our database into two parts.

Future Directions

After the two tree system is functioning, we would like to add a third tree definition layer that precedes the others and determines an overall synthetic strategy. The focus during this stage is on the recognition of cogent substructures, thus it requires a database of about 200 compounds instead of reaction rules. The target will be compared to these compounds rather than reaction rules and "matches" one of these compounds when a large substructure in the target is identified in a compound. This matching compound now becomes the new target and the process is repeated, resulting in a much more abstract problem solving tree. Then the two-tree system is applied to this tree to define targets and starting materials. The system moves from the general to the specific, using the information from the first tree to build the second tree and information from the second tree to build the third. The third and final tree describes the specific steps in the synthetic pathway.
    If an organic synthesis system is to be of practical use to chemists, it must be set up to interface with large chemical databases such as the databases made available by ISI (the Institute for Scientific Information) and by Chemical Abstracts. We have started
to convert our database to the CAS connection table format to simplify database interfaces. Fortunately, this does not require changing SYNLMA. We only need to write a new program to translate connection tables into

clauses, but this is a short program independent of SYNLMA.

From the user's point of view the most important step in improving SYNLMA is to rewrite and expand the user interface. Currently, the compound we want to synthesize is entered in clause form and the program is run in batch mode. This means that the user cannot affect SYNLMA's behavior once the system starts working on a synthesis. We plan to develop an interactive system where the user enters the initial target molecule by drawing it on the screen using a graphics package and is able to monitor the progress of the theorem prover. The user will be able control its actions by removing intermediate targets and suggesting starting materials.

Summary

The success of SYNLMA shows that it is possible to base an expert system on a theorem prover. The advantage of using a theorem prover as deductive component is that it allows us to experiment with a number of different representations for chemical information. The same flexibility makes it easy to add new starting materials and reaction rules from large commercial online databases.

Acknowledgments

This research was partially supported by the National Science Foundation under Grant MCS 82-16432.

Literature Cited

1. Lusk, E.; McCune, W.; Overbeek, R. Proc. Sixth International Conference on Automated Reasoning, Loveland, D., Ed.; Computer Science Lecture Notes #138, Springer-Verlag: New York, 1982, pp. 85-108.
2. Lusk, E.; McCune, W.; Overbeek, R. Proc. Sixth International Conference on Automated Reasoning, Loveland, D., Ed.; Computer Science Lecture Notes #138, Springer-Verlag: New York, 1982, pp. 70-84.
3. Wos, L.; Overbeek, R.; Lusk, E.; Boyle, J. "Automated Reasoning"; Prentice-Hall: Englewood Cliffs, New Jersey, 1984.
4. Gelernter, H.; Sanders, A.; Larsen, D.; Agarwal, K.; Boivie, R.; Spritzer, G.; Searleman, J. Science, 1977, 197, 1041.
5. Wang, T.; Ehrlich, S.; Evens, M.; Gough, A.; Johnson, P. Proc. Conference on Intelligent Systems and Machines, 1984, pp. 176-181.
6. Wang, T.; Burnstein, I.; Ehrlich, S.; Evens, M.; Gough, A.; Johnson, P. Proc. 1985 Conference on Intelligent Systems and Machines, 1985.
7. Nilsson, N. "Principles of Artificial Intelligence"; Tioga: Palo Alto, California, 1980.

RECEIVED December 17, 1985

21

Acquisition and Representation of Knowledge for Expert Systems in Organic Chemistry

J. Gasteiger, M. G. Hutchings[1], P. Löw, and H. Saller

Institute of Organic Chemistry, Technical University Munich, D-8046 Garching, West Germany

Many of the models used by the organic chemist to explain his observations provide a good basis for representing chemical knowledge in an expert system. Such
knowledge can be acquired by developing algorithms for
these models and parameterizing them with the aid of
physical or chemical data. This is demonstrated for
concepts such as electronegativity, polarizability, or
the inductive and resonance effects. Combination of
these models permits construction of systems which
make predictions worthy of an experienced chemist.
This is exemplified by EROS, a system that can predict
the course of chemical reactions or can design organic
syntheses.

Chemistry - as a scientific and technological discipline - has some
unique characteristics. In contrast to physics, where most of the
underlying laws can be given in explicit and sometimes simple mathematical form, many of the laws governing chemical phenomena are
either not explicitly known, or else have a mathematical form that
still eludes an exact solution. Still, chemistry does provide - and
rests on - quantitative data of physical or chemical properties of
high numerical precision. A search for quantitative relationships is
thus suggested, despite the lack of a tractable theoretical basis.
 Chemists have accumulated over the last two centuries an enormous
amount of information on compounds and reactions. However, this information appears largely as a collection of individual facts devoid
of any comprehensive structure or organization. This is most pain-
fully felt by the novice studying chemistry. However, the more he
progresses in his scientific discipline, the more concepts and rules
emerge that allow him to bring order into his knowledge. These concepts include partial atomic charges, electronegativity, inductive,
resonance, or steric effects, which have all been coined by the

[1] Current address: Organics Division, Imperial Chemical Industries plc, Blackley, Manchester M9 3DA, England

chemist do derive models for the principles governing chemical observations. The design of these models has involved the reduction of
collections of individual observations to general principles.

Throughout this paper we use the term model. It will refer to
concepts of varying degrees of sophistication and specification. A
model can be a notion developed by the chemist to classify an observation, it can be an explicit procedure for the calculation of a
value for a physico-chemical concept, or, it can refer to a mathematical equation for the prediction of an observation. We intention-
ally do not distinguish between these different uses in order to
stress the point that the development of a model to further understanding is quite a common approach in science.

The huge amount of information available in chemistry early on
invited the use of the computer for storing and retrieving information. Documentation systems have been developed, and are being
maintained, that contain a sizeable amount of the known chemical
information. Thus, they have gained importance as a knowledge base
for assisting the chemist in solving his problems. Clearly, the construction of a large chemical information retrieval system is an
enormous endeavor. Furthermore, the work will never be complete as
new information is constantly being gathered and should be incorporated into the system. Beyond that, pure retrieval can only give
access to known information. Without appropriate structuring of
information no predictions can be made of new information.
Thus, some of the most important and interesting problems of a
chemist could not be tackled.
These are:
1. What will be the properties of an unknown compound?
2. What is the structure of a new compound?
3. How can a compound with a new structure be synthesized?

These questions fall into the domains of structure-activity relationships, structure elucidation, and synthesis design, respectively. They all ask for new information not yet known explicitly. That
is, they require predictions.

It would be highly desirable to reduce the individual facts in
an information retrieval system to general principles just as the
chemist has done in devising his empirical concepts mentioned previously. Such a reduction of information to its essential contents
asks for insights, to transform information to knowledge.

We have not attempted to make the computer do the job of automatically finding the fundamental laws of chemistry from a compilation
of individual facts. Rather, we have explicitly built into the
computer specific models that we believe can represent the structure
of chemical information. We were guided in this endeavor by concepts
derived by the chemist and have tried to develop models and procedures that quantify these concepts. In doing so we have put more
emphasis on the acquisition and representation of knowledge than on
problem-solving techniques. In any expert system the quality of the
knowledge base is of primary and desicive importance.

We are mainly concerned with the development of EROS (Elaboration of Reactions for Organic Synthesis), a program system for the
prediction of chemical reactions and the design of organic syntheses
(1-3). This system does not rely on a database of known reactions.
Instead, reactions are generated in a formal manner by breaking and

making bonds and shifting electrons. In Figure 1 one of those reaction schemes contained in the program is shown. This scheme, breaking
two bonds and making two new ones is quite important; many rather
diverse reactions follow that scheme. Such a scheme can be applied in
both a forward search (reaction prediction; 1a and 1b) as well as in
a retrosynthetic search (synthesis design; 1c).

Clearly, not all reactions obtained by such a formal scheme can
be realistic ones. In fact, many have no chemical reality (cf. 1d).
A major task in program development is therefore, to find ways of
automatically extracting the chemically feasible reactions from
amongst the formally conceivable ones. To this end a modelling of
chemical reactivity seems indispensable.

Finding the Pieces

The high quality numerical data on physical and chemical properties
of atoms, molecules, and compounds present a good starting point for
the development of a knowledge base. The task is to condense the information contained in a series of individual data into a quantitative parametric model which will reproduce the primary data with a
certain accuracy. If this is successful it can be used to predict
new, as yet unknown data for which the same kind of accuracy can be
expected. Furthermore, the parameters could also be of use in other
models which in turn give new types of data.

In developing models for treating chemical reactivity we have
been guided by the concepts used by the organic chemist in discussing
the causes of organic reactions and their mechanisms. Examples of
the more prominent effects are shown in Figure 2.

Our intention has been to derive models that can quantify these
various effects and thereby build a basis for a quantitative treatment of chemical reactivity. The following simple models that enable
calculations to be performed rapidly on large molecules and big data
sets have been developed.

Heats of Reaction and Bond Dissociation Energies. The simplest form
of a model is an additivity scheme that derives a molecular property
through summation over increments assigned to atoms, bonds or groups
(4). We have explored such an approach by assuming that heats of
formation can be estimated from values assigned to direct (1,2) and
next nearest (1,3) atom-atom interactions (5). Values for these parameters have been derived from experimental heats of formation through
multi-linear regression analyses (6). As an example, the heats of
formation of 49 alkanes have been condensed into four fundamental
parameters that reproduce the data with a standard error of 0.87 kcal
/mol (6).

This amounts to a sizeable reduction of the information that has
to be stored, while conserving a rather good accuracy in the data.
With these four parameters unknown heats of formation of alkanes can
be estimated by the additivity scheme with a similarly high accuracy.
This approach has been extended to other series of compounds.

Using these parameters for the estimation of the heats of formation of starting materials and products of a reaction and then
taking the difference in these two numbers provides values for reaction enthalpies. Only parameters of those substructures that are
changed in a reaction need be considered.

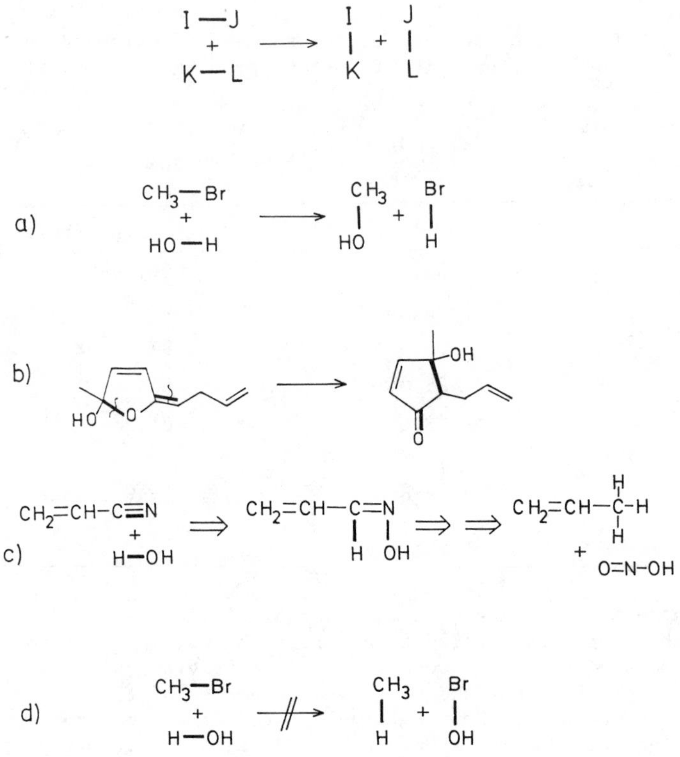

Figure 1. Formal reaction scheme with examples

Furthermore, the effects of strained rings and of aromatic compounds must be considered (7), and algorithms that perform these tasks have been developed (8,9). Values on bond dissociation energies can be calculated by extending the parametrization to radicals (10). Table I gives results obtained for methyl propionate; experimental values are from compounds containing similar structural situations around the bond being considered (11).

Table I. Comparison between calculated and experimental bond dissociation energies in methyl propionate (in kcal/mol)

$$\underset{1}{CH_3} - \underset{2}{CH_2} - \underset{3}{C} \overset{\overset{4}{O}}{\underset{\underset{5}{O} - \underset{6}{CH_3}}{<}}$$

bond	BDE (calc)	BDE (exp.)(ref. 11)
C^1-H	98.7	98.2 ± 1
C^2-H	93.4	92.3 ± 1.4
C^6-H	93.8	94 ± 2
C^1-C^2	85.0	86.4 ± 1
C^2-C^3	83.6	81.2 ± 1
C^3-O^4	123.4	–
C^3-O^5	96.9	95.5 ± 1.5
C^6-O^5	86.4	83.6 ± 1.5

An additivity scheme is a rather simple model, but despite this, such schemes can be applied to a variety of physical data of molecules. Benson and Buss have classified additivity rules into successive approximations and have given examples of their applicability (4). According to their terminology the zero-order approximation of a molecular property is given by additivity of atomic properties, first-order approximation by additivity of bond properties, and second-order approximation by additivity of group properties. More recent widespread use of additivity schemes is found in methods for estimating spectroscopic data, in particular those for deriving ^{1}H- or ^{13}C-NMR chemical shifts of organic molecules.

Polarizability Effects. The next model demonstrates that an additivity scheme can be combined with other forms of mathematical relations to extract the fundamental parameters of a model from primary information. And furthermore, it shows than an additivity scheme useful for the estimation of a global molecular proparty can be modified to obtain a local, site specific property.

Miller and Savchik (12) have given Equation 1 for estimating the mean polarizability, $\overline{\alpha}$, of a molecule, where N is the total number of electrons in the molecule, and τ_i is a polarizability contribution for each atom i, characteristic of the atom type and its hybridization state.

$$\overline{\alpha} = \frac{4}{N} \left(\sum_i \tau_i \right)^2 \tag{1}$$

Mean molecular polarizability can be calculated through the Lorenz-Lorentz- Equation from refractive index, n_D, molecular weight, MW, and density, d, of a compound, demonstrating that the parameters τ_i can be derived from these elementary molecular properties (Figure 3).

Polarizability is a measure of the relative ease of distortion of a dipolar system when exposed to an external field. The stabilization energy due to the interaction between an external charge and the induced dipole is highly distance-dependent and can be calculated through classical electrostatics. The situation is, however, less clearly defined when the charge resides <u>within</u> the molecule that is being polarized. To model the stabilization resulting from polarizability in these situations, we have modified Equation 1 by introducing a damping factor d^{n_i-1}, where $0 < d < 1$, and n_i gives the smallest number of bonds between an atom i and the charge center (Equation 2)(13).

$$\alpha_d = \frac{4}{N} \left(\sum_i d^{n_i-1} \cdot \tau_i \right)^2 \tag{2}$$

α_d is called <u>effective</u> polarizability, as the damping factor models the distance dependent attenuation of the stabilization effect. Furthermore, this factor gives different values for α_d for the same molecule depending on where the charge center is located. An alternative additivity scheme (14) for estimating mean molecular polarizability can be similarly modified to obtain values of effective polarizability (15). The significance of these values has been demonstrated by correlation with physical data (13).

<u>Charge Distribution, Inductive and Resonance Effects</u>. Until now, the discussion has been concerned with models based on additivity schemes and their modifications. However, we have also explored other types of models that can be put into algorithms that are fast, albeit less convenient for pencil and paper application.

This is true for our procedure for calculating partial atomic charges in σ-bonded molecules (16). The method starts from Mulliken's definition of electronegativity, χ, derived from atomic ionization potentials, IP, and electronegativities, EA (Equation 3)(17).

$$\chi = 0.5 \ (IP + EA) \tag{3}$$

Electronegativity was considered to be dependent both on orbital type, and on the occupation number of an orbital (or, equivalently, the charge on an atom). On bond formation, negative charge is transferred from the less to the more electronegative atom. Because of the charge dependence, the electronegativities change in the sense that they tend to equalize. The problem of the mutual dependence of electronegativity on charge and of charge transfer on electronegativity was solved by an interative procedure that takes explicit account of the molecular topology (16). This gives access to a self-consistent set of values of partial charges and associated residual

— heat of reaction and bond dissociation energy

$$H_3C-H + X-X \longrightarrow H_3C-X + H-X$$

— charge distribution

$$H_3C-\overset{\overset{\displaystyle O}{\parallel}}{\underset{H}{C}} \overset{\delta+}{} + :Nu \longrightarrow H_3C-\overset{\overset{\displaystyle O^{\ominus}}{\mid}}{\underset{\underset{H}{\mid}}{C}}-Nu$$

— inductive effect

$$Cl-CH_2-COOH \longrightarrow Cl-CH_2-COO^{\ominus} + H^{\oplus}$$

— polarizability

$$Br-CH_2CH_2-Cl + OH^{\ominus} \longrightarrow HO-CH_2CH_2-Cl + Br^{\ominus}$$

— resonance effect

$$H_2C=CH-\overset{\overset{\displaystyle O}{\diagup}}{\underset{H}{C}} + H-CN \longrightarrow NC-CH_2-CH_2-\overset{\overset{\displaystyle O}{\diagup}}{\underset{H}{C}}$$

$$\updownarrow$$

$$\overset{\oplus}{H_2C}-CH=\overset{\overset{\displaystyle O^{\ominus}}{\diagup}}{\underset{H}{C}}$$

Figure 2.   Concepts used in discussing the causes of organic reactions

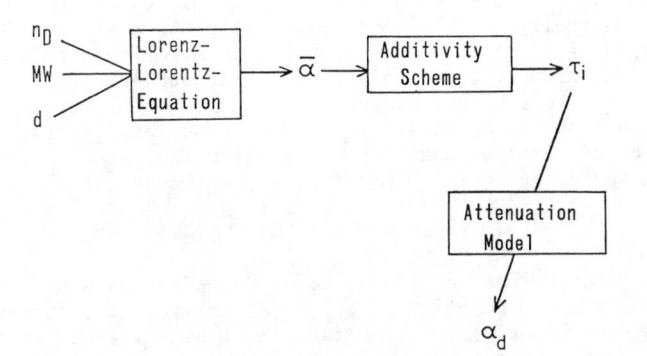

Figure 3.   Deriving values for effective polarizability, α_d, from refractive index, n_D, molecular weight, MW, and density, d

electronegativity values for each atom of a molecule, reflecting atomic type as well as the influence of the molecular environment.

The charge values have been used to correlate or calculate a variety of physical data including dipole moments (18), ESCA chemical shifts (16), ^{1}H-NMR chemical shifts (19), and $^1J_{C-H}$coupling constants (20), thereby relating these physical data to the fundamental values of IP and EA, in concert with a proper consideration of the network of bonds in molecules.

An extension of the method has been developed for conjugated π-systems which arrives at charge distribution in these systems by generating the various resonance structures and assigning weights to them (21, 22). Again, the significance of the charge values was established by reproducing physical data of molecules.

It was found that the residual electronegativity values calculated for σ-bonded molecules can be taken as a quantitative measure of the inductive effect (23). In a similar manner, the values of π-electronegativities can be used for quantifying the resonance effect.

Hyperconjugation.   Empty or partially filled p-orbitals can be stabilized through overlap with adjacent C-H and C-C bonds of appropriate symmetry. Following a previous suggestion (24), we have taken the number of such bonds as a measure of this stabilization through hyperconjugation.

Putting the Pieces Together

The previous chapter has briefly presented methods that quantify the various effects used by the organic chemist to rationalize his observations on reactivity, reaction mechanisms, and the course of organic reactions. Physical data were chosen to demonstrate the significance of the calculated values.

But are the values calculated by the above methods also useful for understanding and prediction of chemical reactivity data? Here, the situation is less well-defined than with physical properties. In many cases our knowledge of chemical reactivity is more of a semi-quantitative nature. Furthermore, in many reactions the various effects operate simultaneously, and they do so to varying degrees.

Several statistical and pattern recognition techniques were used to unravel the relationships between chemical reactivity data and the previously described effects which influence them.

Multilinear Regression Analysis.   As an entry to the problem we have selected simple gas phase reactions involving proton or hydride ion transfer which are influenced by only a few effects and for which reactivity data of high accuracy are available. In these situations where a larger set of numerial data are available multilinear regression analysis (MLRA) was applied. Thus, the simplest mathematical form, a linear equation is chosen to describe the relationship between reactivity data and physicochemical factor. The number of parameters (factors) simultaneously applied was always kept to a minimum, and a particular parameter was only included in a MLRA study if a definite indication of its relevance existed.

The proton affinity (PA) of alkylamines can be described by

only a single parameter, the effective polarizability α_d (13). For 49
unsubstituted alkylamines a method was found to calculate proton af-
fintity from the refractive index, molecular weight, and density (cf.
Figure 3). For alkylamines carrying heteroatom substitution a
measure of the inductive effect had to be included. This could be a-
chieved by using residual electronegativity values, $\overline{\chi}_{12}$ in the two
parameter equation 4 (23).

$$PA = c_o + c_1\alpha_d - c_2\overline{\chi}_{12} \tag{4}$$

The signs of the coefficients in this equation are entirely consistent with intuition, polarizability stabilizing, and electronegativity destabilizing the protonated form of the amine. Similar equations could be developed for proton affinity data of alcohols and
ethers, as well as of thiols and thioethers (Figure 4b and 4c) (25).
Furthermore, α_d and $\overline{\chi}_{12}$ parameters were also sufficient to describe
quantitatively gas phase acidity data of alcohols (Figure 4d) (25). In
this case, the coefficients for the two parameters had the same sign
as both effects provide sources of stabilization for the alkoxide
ion. Figure 5 shows the results obtained by MLRA.

Simple linear equations could also be developed for the other
three systems of Figure 4, PA of aldehydes and ketones(4e), and their
hydride ion affinities, both of the neutral (4f) and protonated forms
(4g). However, in addition to effective polarizability and electronegativity, hyperconjugation had also to be used as a parameter, as
p-orbitals carrying a partial positive charge are involved in the
reactions 4e to 4g (26).

Multiparameter equations, such as Equation 4, obtained through
MLRA are the simplest form of parallel connection of several models.
Each model has been parameterized from its own source of primary
data. Combined application can reproduce new types of data and lead
to new information and knowledge.

The correlations with data on gas phase reactions have served
to establish that the parameters calculated by our methods are indeed
useful for the prediction of chemical reactivity data. Their application is, however, not restricted to data obtained in the gas phase.
This has been shown through a correlation of pK_a values (in H_2O) of
alcohols with residual electronegativity and polarizability parameters, by including a parameter that is interpreted to reflect
steric hindrance of solvation (27).

The Reactivity Space. In many reaction types the situation is not as
well defined as in the chemical reactions so far investigated. If
either fewer and less accurate reactivity data are available, or the
chemical system is under the influence of many effects, then MLRA is
no longer the appropriate analytical method.

For such situations we have developed a different approach. The
parameters calculated by our methods are taken as coordinates in a
space, the reactivity space. A bond of a molecule is represented in
such a space as a specific point, having characteristic values for
the parameters taken as coordinates. Figure 6 shows a three-dimensional reactivity space spanned by bond polarity, bond dissociation
energy, and the value for the resonance effect as coordinates.

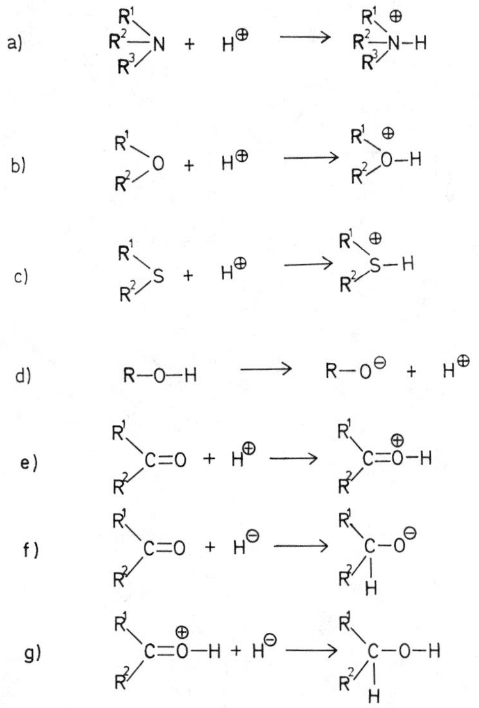

Figure 4. Gas phase reactions for which linear equations have been developed using polarizability, electronegativity, and hyperconjugation parameters. Reaction a) ref. 13, 23; b)-d) ref. 25; e)-g) ref. 26.

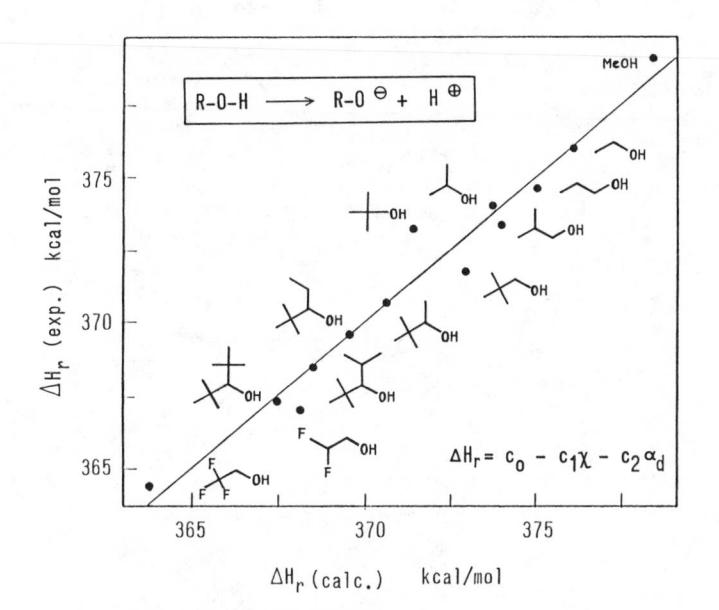

Figure 5. Experimental gas phase acidity data of alcohols plotted against values calculated from electronegativity and polarizability parameters. (Reprinted from: Gasteiger, J.; Hutchings, M.G. J. Am. Chem. Soc. 1984, 106, 6489).

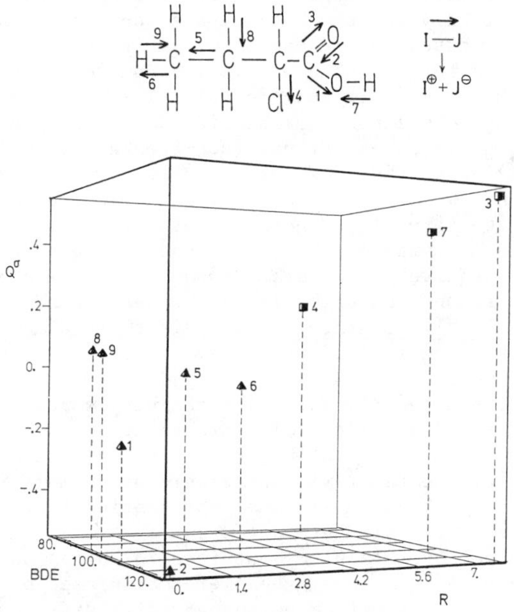

Figure 6. Reactivity space having bond polarity, Q_σ, bond dissociation energy, BDE, and resonance effect parameter, R, as coordinates.

In this space various bonds of 2-chlorobutyric acid are indicated. In fact, in this case we have investigated heterolytic bond cleavages. Thus, each bond will give rise to two points in this space, depending on which direction the charges are shifted on heterolysis, in the direction of bond polarity, or against it. For example, points 2 and 3 of Figure 6 both refer to the carbonyl double bond. In the case of point 2, its heterolysis against the preformed polarization of that bond (Figure 7), and therefore the bond polarity parameter Q_σ has a negative sign.

Points 2 and 3 are characterized by the same value for the (homo-lytic) bond dissociation energy. However, resonance stabilization of charges can occur only for the heterolysis represented by point 3. Therefore in this case, the resonance parameter R has a high value, whereas it is zero for the heterolysis represented by point 2.

Figure 6 shows an additional feature, The points are distin-guished according to whether the associated bond is considered react-ive (breakable; small cubes) or not (non-breakable; small pyramides). Any chemist will agree that the most reactive bonds of 2-chlorobutyric acid are the O-H, the C=O, and the C-Cl bonds, where the negative charge goes to the more electronegative atom (O or Cl) on heterolysis. This heterolytic cleavage of the three bonds is represented by points 7, 3, and 4, respectively. The other bonds are considered as much less reactive, or non-breakable. It can be seen that reactive and non-reactive bonds clearly separate. Thus, this three-dimensional space already represents the ease of breaking a bond, a chemical reactivity phenomen, quite well.

With Figure 6 a three-dimensional reactivity space is shown. Where-as this is the limit for pictorial representation, statistical methods can deal with spaces of higher dimensionalities. In a study aimed at modelling the reactivity of single bonds in aliphatic chemistry a data set of 28 molecules representing that field was chosen. Table II gives this data set.

The entire set of molecules contained 782 bonds out of which 111 σ-bonds were selected. The parameters were calculated by our methods to build a reactivity space with electronegativity difference, resonance effect parameter, bond polarizability, bond polarity, σ-charge di-stribution, and bond dissociation energy as six coordinates.

First, unsupervised-learning pattern recognition methods were applied. A principal component analysis showed that the dimension-ality of the space could be reduced without much loss of information. With three factors, instead of six, 85.9% of the variance of the data set could still be reproduced. The first factor can be identified as containing the σ-electron distribution, the second factor is highly loaded with the bond dissociation energy and bond polarizability. The third factor contains a mixture of effects. Cluster analysis was applied as a second unsupervised learning technique. In this case it was applied to a reactivity space of reduced dimensionality using – for reasons that become clearer below – the resonance effect, bond polarity, and the bond dissociation energies as coordinates. The re-sults are shown as a dendrogram in Figure 8.

It is probably not too surprising that the same bond types cluster together, as they are characterized by similar values for the respective parameters. However, the interrelationships between dif-ferent bond types indicated by the overall structure of the dendro-

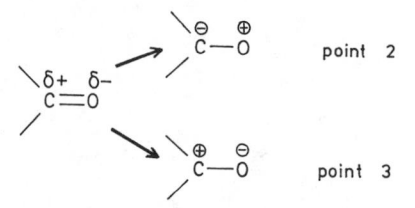

Figure 7. The two choices for heterolysis of the carbonyl double bond, and their representation as points in Figure 8.

Table II. List of compounds used in deriving a reactivity function

---

 1) cyclopropane
 2) cyclobutane
 3) cyclopentene
 4) cyclopentadiene
 5) ethyl bromide
 6) ethyl iodide
 7) methylene chloride
 8) allyl chloride
 9) neopentyl chloride
10) 1-methyl-1-cyclopropyl-ethyl bromide
11) 1-methyl-1-cyclobutyl-ethyl iodide
12) 2,2,4,4,-tetramethylcyclobutanol
13) acetaldehyde
14) acetone
15) trimethylacetaldehyde hydrate
16) choral hydrate
17) aldol
18) methyl propionate
19) ethyl acetoacetate
20) α-chloropropionic acid
21) 5-hydroxy-nona-3,5,8-triene-2-one
22) 2-oxocyclopentane carboxylic acid
23) 5-hydroxy-5-methyl-butylrolactone
24) 1-dimethylamino-propene
25) 4-amino-2,4-dimethyl-2-pentanole
26) succinimide
27) α-picoline
28) 6-chloro-6-methoxy-bicyclo[3.1.0]hex-2-ene

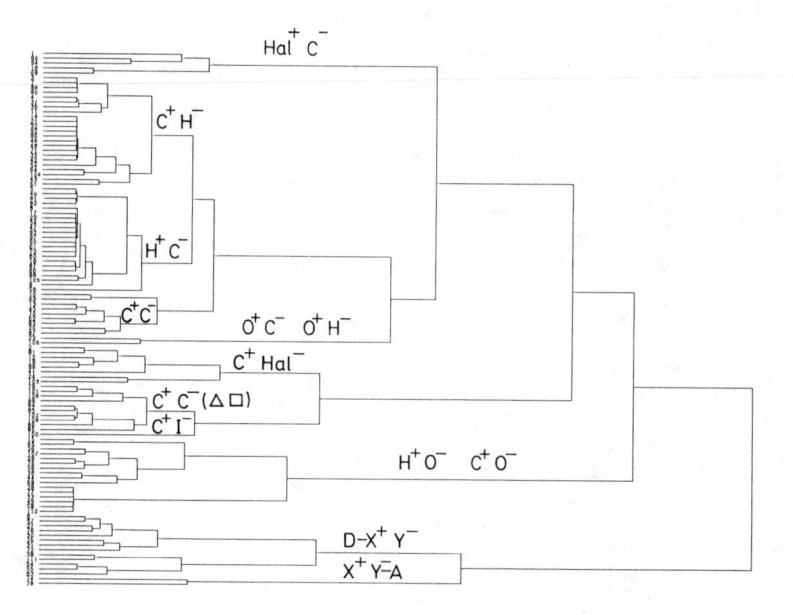

Figure 8. Dendrogram of the relationship between the various
bonds on heterolysis as obtained by a cluster analysis (A= accep-
tor; D= donor).

gram, and some of the smaller details of the dendrogram give interesting information. To name just one: the C-C bonds of the three-
and four-membered carbocycles are found to be rather closely related
to the carbon-halogen bonds (in both cases carbenium ions can be obtained, either directly as with the C-Hal bond, or after attack of an
electrophile (H^+, Lewis acid) as both with halocarbons and with cyclo-
propanes and cyclobutanes).

Next, supervised-learning pattern recognition methods were applied to the data set. The 111 bonds from these 28 molecules were
classified as either breakable (36) or non-breakable (75), and a stepwise discriminant analysis showed that three variables, out of the
six mentioned above, were particularly significant: resonance effect,
R, bond polarity, Q_σ, and bond dissociation energy, BDE. With these
three variables 97.3% of the non-breakable bonds, and 86.1% of the
breakable bonds could be correctly classified. This says that chemical reactivity as given by the ease of heterolysis of a bond is well
defined in the space determined by just those three parameters. The
same conclusion can be drawn from the results of a K-nearest neighbor analysis: with k assuming any value between one and ten, 87 to
92% of the bonds could be correctly classified.

One method that we have found particularly useful for our purposes is logistic regression analysis (LoRA). In this method, a
binary classification is taken as a probability, P_0, (given the value
0 or 1) and modelled by the two coupled equations 5 and 6.

$$P = 1/(1 + e^{-f}) \tag{5}$$

$$f = c_0 + c_1 x_1 + c_2 x_2 + \dots \tag{6}$$

In the linear function f, the x_i are the parameters considered
relevant to the problem. The coefficients c_i are determined to maximize the fit of the calculated probability P as closely as possible
to the initial classification P_0.

The method applied to the problem of chemical reactivity translates into the following. A data set of molecules is chosen and bonds
in these molecules are selected and specified either breakable or
non-breakable (P_0 = 0/1). Then, the physicochemical parameters deemed
important for the reactivity of the bonds under investigation are
calculated and used as variables x_i in Equation 6. LoRA is applied
to model the initial classification of bonds into breakable or non-
breakable classes.

In this process, a function f is obtained that can be used as a
numerical estimate for the ease of breaking of a bond. We therefore
call it a reactivity function. The all-important point is that
through LoRA the qualitative information of whether a bond is breakable or not is used to construct a function that predicts chemical
reactivity quantitatively.

A reactivity function (Equation 7) applicable to single bonds in
aliphatic species was obtained with the data set of 111 bonds from
the 28 molecules mentioned above.

$$f = 2.87 + 0.162 \cdot R + 32.9 \cdot Q_\sigma - 0.084 \cdot BDE \tag{7}$$

In a similar manner, a function quantifying the reactivity of
bonds in charged species was developed. These functions are of quite

general validity. The numerical values calculated with them permit
prediction of which bonds and combinations of bonds will react preferentially. Inferences on the course of complex organic reactions
can be drawn from this information.

As an example: What will be the product of heating 1,2:1,4-di-
epoxy-p-menthane, 1. (Figure 9) with alumina in toluene? A chemist
would assume initial breaking of an epoxide-ring. But which one of
the two? Or will both break? Furthermore, for each epoxide ring there
are two possible choices of C-O bonds.

Figure 10 shows the sequence of bond breaking obtained by application of the reactivity function for neutral aliphatic molecules and
the one for charged species. The consecutive bond breakings that are
explored lead to the conclusion that the pattern of breaking and
making bonds as indicated in structure 2 should be the most favored
one. Thus, it is predicted that both oxirane-rings are broken, one
even in the direction leading to the seemingly less stable carbenium
ion. Furthermore, even a bond in the saturated six-membered ring is
found to be breakable. The mechanistic pattern of structure 2 permits
to make the inference that compound 3 is the most likely product of
this reaction. This is indeed the observed course and product of the
rearrangement of 1 (28).

Examples of other cases of prediction of complex organic reactions have been given elsewhere (3). Functions applicable to the
reactivity of multiple bonds and of aromatic systems have been developed in an analogous manner.

Conclusion. It has been demonstrated that the methods developed for
the calculation of physicochemical effects can form the foundation
for a general quantitative treatment of chemical reactivity. Based on
the factors calculated with these various methods, reactivity functions can be elaborated that are able to assign a numerical reactivity to bonds and combinations of bonds in a molecule. In this manner
the course and outcome of organic reactions can be predicted. A
quantitative treatment of chemical reactivity is also an essential
component in synthesis design since it allows evaluation of the
feasibility of various synthetic reactions and pathways.

The knowledge base of that part of the EROS system that predicts
chemical reactivity consists of the procedures for calculating the
physicochemical effects and the way in which they are connected.
These methods can be part of a series connection (Figure 3) or of a
parallel connection (Equation 4). In other words, the knowledge base
consists of the chemical models that form the building blocks and the
statistical models that form the network of connections.

As the chemical models mentioned here refer to some fundamental
thermochemical and electronic effects of molecules, their application
is not restricted to the prediction of chemical reactivity data. In
fact, in the development of the models extensive comparisons were
made with physical data, and thus such data can also be predicted
from our models. Furthermore, some of the mechanisms responsible for
binding substrates to receptors are naturally enough founded on
quite similar electronic effects to those responsible for chemical
reactivity. This suggest the use of the models developed here to calculate parameters for quantitative structure-activity relationships
(QSAR).

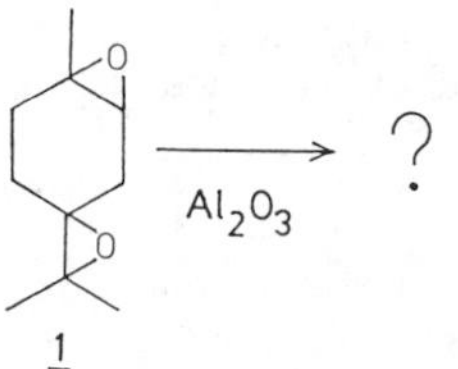

Figure 9. Example of a problem for reaction prediction

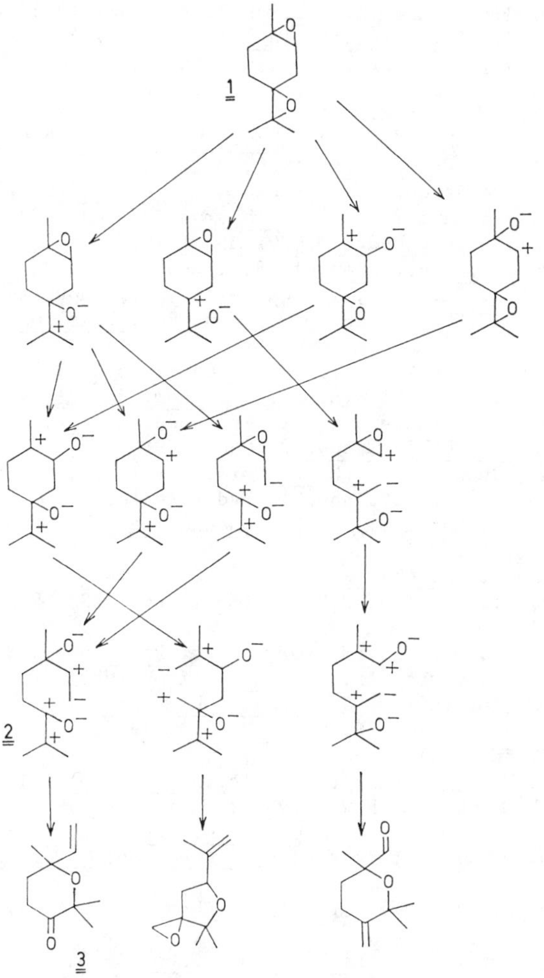

Figure 10. Network of bond-breaking and -making patterns explored by the reactivity functions leading to the correct prediction of product 3 from 1.

In this sense, expert systems for the prediction of chemcial reactions, for the design of organic syntheses, for the prediction of physical data, for structure elucidation , and for QSAR can be founded on the knowledge base comprized by the models presented here.

Acknowledgments

Support of this work by the Deutsche Forschungsgemeinschaft and by Imperial Chemical Industries, plc, United Kingdom, is gratefully appreciated.

Literature Cited

1. Gasteiger, J.; Jochum, C. Topics Curr. Chem. 1978, 74, 93.
2. Gasteiger, J. Chim. Ind. (Milan) 1982, 64, 714.
3. Gasteiger, J.; Hutchings, M.G.; Christoph, B.; Gann, L.; Hiller, C.; Löw, P.; Marsili, M.; Saller, H.; Yuki, K. Topics Curr. Chem., submitted.
4. Benson, S.W.; Buss, J.H. J. Chem. Phys. 1958, 29, 546.
5. Allen, T.L. J. Chem. Phys. 1959, 31, 1039
6. Gasteiger, J.; Jacob, P.; Strauss, U. Tetrahedron 1979, 35, 139.
7. Gasteiger, J.; Dammer, O. Tetrahedron 1978, 34, 2939.
8. Gasteiger, J. Tetrahedron 1979, 35, 1419.
9. Gasteiger, J. Comput. Chem. 1978, 2, 85.
10. Gann, L.; Löw, P.; Yuki, K.; Gasteiger, J., unpublished results.
11. a) McMillen, D.F.; Golden, D.M. Ann. Rev. Phys. Chem. 1982, 22, 493.
    b) Egger, K.W.; Cocks, A.T. Helv. Chim. Acta 1973, 56, 1516.
12. Miller, K.J.; Savchik, J.A. J. Am. Chem. Soc. 1979, 101, 7206.
13. Gasteiger, J.; Hutchings, M.G. Tetrahedron Lett. 1983, 24, 2537; J. Chem. Soc., Perkin Trans. 2 1984, 559.
14. Kang, Y.K.; Jhon, M.S. Theor. Chim. Acta 1982, 61, 41.
15. Löw, P.; Gasteiger, J., unpublished results.
16. Gasteiger, J.; Marsili, M. Tetrahedron Lett. 1978, 3181; Tetrahedron 1980, 36, 3219.
17. Mulliken, R.S. J. Chem Phys. 1934, 2, 782.
18. Gasteiger, J.; Guillen, M.D. J. Chem. Res. (S) 1983, 304; (M) 1983, 2611.
19. Gasteiger, J.; Marsili, M. Org. Magn. Resonance 1981, 15, 353.
20. Guillen, M.D.; Gasteiger, J. Tetrahedron 1983, 39, 1331.
21. Marsili, M.; Gasteiger, J. Croat. Chem. Acta 1980, 53, 601.
22. Gasteiger, J.; Saller, H. Angew. Chem. 1985, 97, 699; Angew. Chem. Int. Ed. Engl., 1985, 24, 687.
23. Hutchings, M.G.; Gasteiger, J. Tetrahedron Lett. 1983, 24, 2541.
24. Kreevoy, M.M.; Taft, R.W. J. Am. Chem. Soc. 1955, 77, 5590.
25. Gasteiger, J.; Hutchings, M.G. J. Am. Chem. Soc. 1984, 106, 6489.
26. Hutchings, M.G.; Gasteiger, J. J. Chem. Soc., Perkin Trans. 2, in press.
27. Hutchings, M.G.; Gasteiger, J. J. Chem. Soc., Perkin Trans. 2, in press
28. Ho, T.L.; Stark, C.J. Liebigs Ann. Chem. 1983, 1446.

RECEIVED December 17, 1985

ANALYTICAL CHEMISTRY

22

An Expert System for High Performance Liquid Chromatography Methods Development

René Bach[1], Joe Karnicky[1], and Seth Abbott[2]

[1] Varian Research Center, Varian Associates, Inc., Palo Alto, CA 94303
[2] Varian Instrument Group, Varian Associates, Inc., Walnut Creek, CA 94598

ECAT (Expert Chromatographic Assistance Team) is an expert system being developed at Varian Associates. The goal of our project is to create a computer program that performs, at the human expert level, the tasks of designing, analyzing, optimizing, and trouble-shooting a high performance liquid chromatography (HPLC) separation method. The program is successfully reaching conclusions relating to a number of probes that test the design and trouble-shooting capabilities. This paper describes the development of ECAT in terms of the overall strategy of the program, the hardware and software used, and the development of the knowledge bases. Current results and future plans are discussed.

The goal of our current research is to apply Artificial Intelligence (AI) techniques to the writing of an expert system for High Performance Liquid Chromatography (HPLC) methods development; that is, to produce a computer program capable of developing HPLC separation methods in a manner comparable to that of an expert chromatographer. The expert system program is named ECAT (an acronym for Expert Chromatographic Assistance Team).

Creating a machine chromatographer is a highly ambitious goal. Because it will involve a very large effort to complete the ECAT program as envisioned, we are developing the system as a set of (eventually interacting) modules whose functionality can be separately specified and implemented.

Once one has built or acquired an expert system shell, an expert system is usable and useful at an early stage of development. Subsequent development consists of increasing and refining the knowledge, expanding the functionality and improving the efficiency of the system.

The reader of this paper should be aware that the overall design of ECAT (described in the sections on SYSTEM DESIGN and FUTURE WORK) has only been implemented to the extent of the running modules that are described under CURRENT STATUS.

Background

The original design for ECAT was described in an article by Dessy (1). The development of the system to date has deviated considerably from this plan both in implementation methodology and rate of progress. This is primarily attributable to the fact that the skill and experience of our group in applying AI programming techniques has grown with time.

Project Motivation. Chromatography, in general, and methods development, in particular, exhibit characteristics which indicate that writing an expert system is worthwhile: while chromatography is used by a large and diverse technical group (i.e., biologists, engineers), the number of skilled chromatographers is in finite supply.

HPLC is characterized by a dynamic, expanding knowledge base, which should benefit from a systematic reorganization of knowledge in a common repository (the expert system). Writing an expert system for HPLC would make available to users of chromatographic techniques an automatic, reliable and fast application of existing chromatographic expertise. It could communicate this expertise in an instructional manner, and provide for the convenient construction and manipulation of data (rule) bases containing structured representations of chromatographic knowledge.

AI research in the last decade has demonstrated that it is possible to capture and apply the human expertise related to a specialized field by means of an expert system computer program.

Limitations of Conventional Programming. It is clear that truly intelligent and comprehensive methods development is sufficiently complex to be beyond what a conventional computer program can manage. Conventional programming methods are inadequate because of the difficulty of writing, and subsequently debugging and modifying, a procedural algorithm which could perform the complex task of HPLC method development. In addition, conventional programming methods don't support efficiently the ability to represent and manipulate information which is non-numeric, judgmental, uncertain, and incomplete. Research in AI over the last two decades has yielded programming languages and programming methods for writing expert systems which do not suffer from the above limitations. We have applied some of these methods, described below, to create ECAT.

Expert System Programming. Many of the concepts and terms which will be used in the description of this work are unique to the fields of AI and computer science. The reader should refer to the article by Dessy (2) or to the introductory article of this symposium for a more detailed description of these concepts.

There is some disagreement within the AI community as to what qualifies a computer program to be called an "expert system". We use the term to describe a program which has the following characteristics: 1) The program performs some task (e.g., HPLC methods design) which requires specialized human expertise. This human expertise often takes the form of heuristics (empirical rules of

thumb);   2) The "domain knowledge" (i.e., the knowledge in the
program specific to the task at hand, for example: methods design)
is explicitly encoded in a computer readable form and is segre-
gated from the mechanisms for its application (collectively called
the "inference engine");   3) The system has the potential for
explaining its reasoning;   4) The amount of knowledge encoded in
the system is non-trivial (i.e., for a rule-based system there are
hundreds or thousands of rules).

Figure 1 illustrates the elements and individuals involved in
developing and using the ECAT expert system.   The domain knowledge
(including heuristic knowledge) is elicited from the domain expert
by the knowledge engineer who uses software tools to convert the
knowledge into computer processable form (i.e., facts and rules in
knowledge bases).   An individual uses the program by communicating
with the system via the user interface.   In response to the user's
requests, the inference machinery makes logical deductions and
performs tasks by processing the appropriate knowledge base.
Results are communicated back to the user via the user interface.

Related Work

Algorithmic Methods Development.   The recent development of sta-
tistically-based HPLC solvent optimization computer programs (3-9)
have achieved useful behavior in experimental design by optimizing
separations with respect to specific performance criteria.   However, AI programming techniques were not applied in these programs.

Expert Systems for Chemistry.   At this time, there are a very
large number of expert systems for chemistry in various stages of
development.   (See, for example, some of the other papers in this
symposium.)   Some of the more successful systems developed in the
past include: The DENDRAL series of programs from Stanford- These
include DENDRAL (started in 1965), CONGEN, and META-DENDRAL
(10-11).   These programs elucidate chemical structures from mass-
spectral information.   Similar programs have been used to compu-
terize C13 NMR spectral analysis (12-13).   Most recently the
PROTEAN project aims at computing the three-dimensional structure
of proteins in solution using NMR data (14); The CRYSALIS program
interprets a three-dimensional image of the electron density map
obtained by X-ray crystallography of proteins (15); SYNCHEM and
SYNCHEM2 (16-17), LHASA (18), and SECS (19) are examples of computerized or computer-assisted organic synthesis.

System Strategy

The goal of ECAT is to provide assistance to the user of a chromatograph in the development of an HPLC method.   To do this, one
must specify the tasks performed in developing an analytical
method.   The computer performs these tasks by processing information.   In ECAT we are calling the collection of information specific to a task a Module.   The modules and information flow which
will be needed for the completely implemented ECAT are shown in
Figure 2.

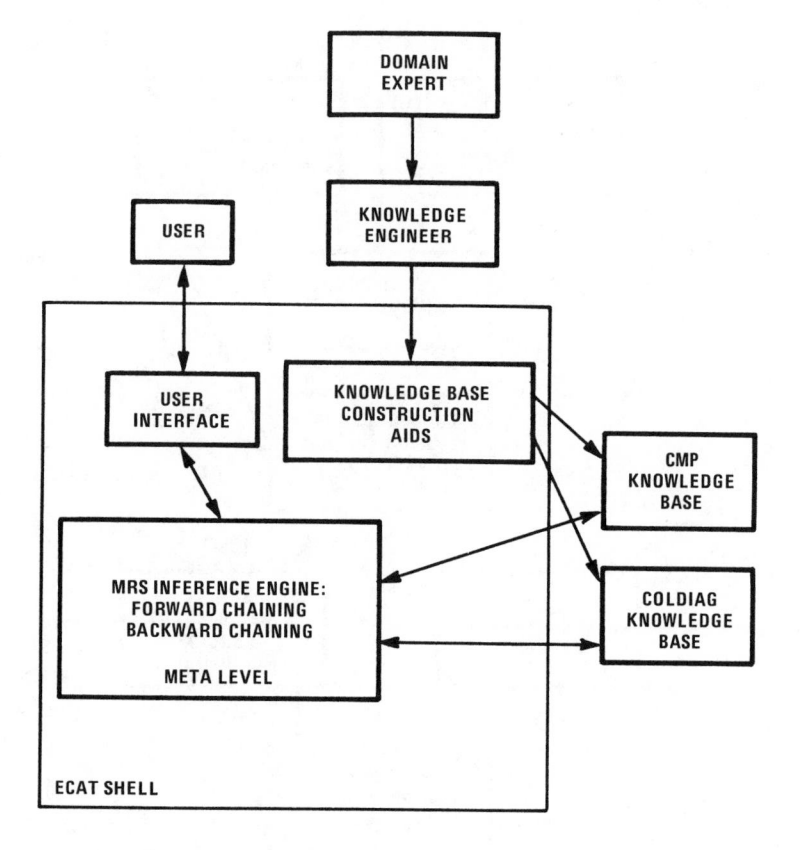

Figure 1. Elements involved in development and application of the ECAT expert system.

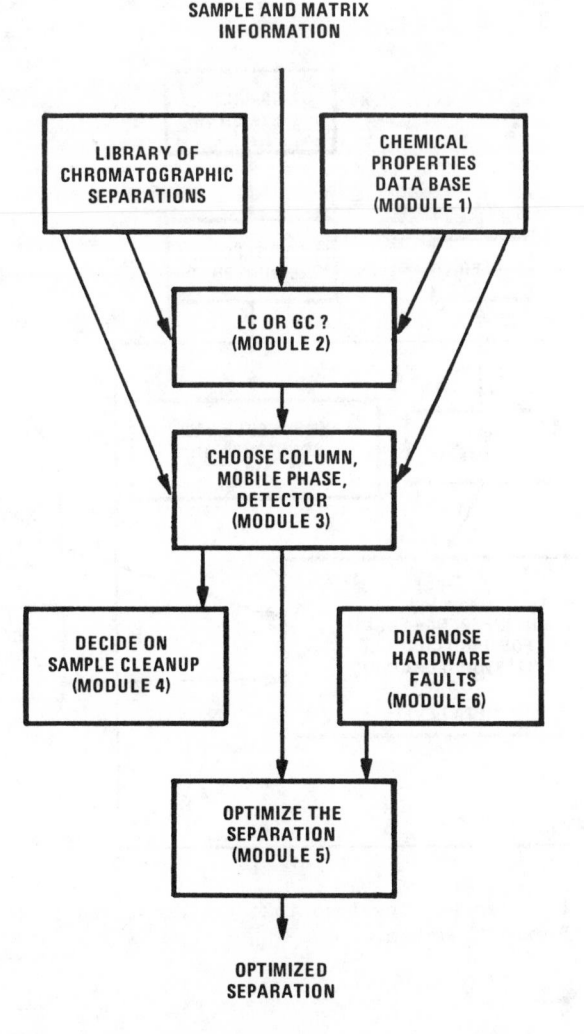

Figure 2. ECAT task modules: flow of information in the method design process.

The complete ECAT system as envisioned in Figure 2 will take as input the user's specification of the sample to be analyzed (analytes, matrix) and will ultimately produce a separation method that satisfies the user's requirements for resolution and analysis time. The strategy adopted is that the system will first decide whether Gas Chromatography (GC) or Liquid Chromatography (LC) is the best separation method.   If LC is the method of choice, and a qualified separation is not found in the program's library, it will design and optimize a separation, also specifying pre-column sample treatment where applicable.   The design of this separation will ultimately include cycles of designing an initial separation, performing the experiment, analyzing the results, and redesigning until a satisfactory separation is achieved.   During the optimization step it may be necessary to diagnose for column and hardware failure.

The chemical information which the program will need will be stored in data bases or input by the user.   The details of these modules are discussed in the sections on current results or future plans.

To summarize, a complete methods development program must be able to: 1.   provide chemical information,
2.   choose between GC and LC,
3.   specify column, mobile phase constituents and detector,
4.   decide on sample cleanup,
5.   optimize (or redesign) the separation,
6.   diagnose hardware problems.

Implementation

Our program is being implemented as a knowledge-based system.   The knowledge about chromatography which is imbedded in the program is in the form of facts and rules.   These facts and rules are represented within the computer by statements in predicate logic.   In predicate logic a statement is represented by a list of symbols, where the first symbol (the predicate) represents a relationship among the objects which are represented by the other symbols in the list.   Complex facts are expressed using what are called "logical connectives" (e.g., AND, OR, NOT, IF).   We distinguish statements starting with IF and call them rules.   A rule is also referred to as an IF-THEN statement.   A rule asserts that the statements in the left hand side imply the statements in the right hand side.   An inference engine is used to interpret those rules to generate new facts or to answer questions.

Figure 3 shows examples of some ECAT facts and rules.   A rule's components are a name, a type declaration, an English language description of the rule, an English language translation of the rule, and the actual form that is processed by the program during inferencing.   Variables are bound during inferencing.

Development Environment.   Hardware: the hardware currently consists of a Symbolics 3670 workstation, a Symbolics 3640 workstation, and a VAX 750, all connected by Chaosnet (an Ethernet proto-

```
(user1
  (type fact)
  (pform (largest-mw 500 daltons)))

(user2
  (type fact)
  (pform (analyte-class phenols)))

(user3
  (type fact)
  (pform (asked (analyte-class $class))))

(cmpgen1
  (descr nil)
  (type rule)
  (text  (if sample molecular-weight is > 100 then there are more
             than three carbons in the molecule))
  (pform (if (and (largest-mw $mw daltons)
                  (> $mw 100))
             then (more-than-three-carbons))))

(cmpgen7
  (type rule)
  (text (if the analyte class is not a protein and not a peptide,
            then use the specified analyte class for further
            inferencing))
  (pform (if (and (analyte-class $class)
                  (asked (analyte-class $class))
                  (unknown (analyte-class protein))
                  (unknown (analyte-class peptide)))
             then (consider (analyte-class $class)))))

(cmp1
  (descr (a default rule for selecting separation mode))
  (type rule)
  (text  (if the chemical class of the analyte is not a protein,
             and the analyte has more than three carbons, and the
             analyte does not belong to a class for which straight
             phase is recommended, then use a reverse phase sepa-
             ration mode))
  (pform (if (and (consider (analyte-class $class))
                  (more-than-three-carbons)
                  (unknown (consider (analyte-class protein)))
                  (unknown (straight-phase-packing $class $x $y)))
             then (separation-mode reverse-phase))))
```

Figure 3. Examples of facts and rules in ECAT. $... are
variables.

col). Software: the Symbolics machines run Zetalisp, and we are using Franzlisp in Eunice (a UNIX emulator running under VMS) on the VAX (see Figure 4).

To develop the expert system we are using a first-order logic programming system called MRS (20). It is a general inference engine providing for forward chaining, backward chaining and control of the inferencing by a meta-level reasoning system. Reasoning at the meta-level refers to reasoning about reasoning, that is, reasoning about what needs to be done next, or what is the best way to solve the problem at hand.

Forward chaining is reasoning from known facts via rules to conclusions. For example, if a user asserted the three facts listed at the top of Figure 3 the program would conclude, by forward chaining, that the separation mode should be reverse phase. We use forward chaining to process the Column and Mobile Phase (CMP) design knowledge base. Backward chaining proves given hypotheses by testing whether the "if" parts of relevant rules are known or provable using other rules. For example, if the program was asked the equivalent of "What separation mode should I use?" it could use backward chaining through the rules in Figure 3 to infer that it should ask the user about molecular weight and analyte classes to provide the answer to the question. We use backward chaining for the column diagnosis. MRS runs in Zetalisp, Maclisp and Franzlisp. We have made some modifications to the MRS inferencing capability and provided a better user interface.

We selected MRS for the following reasons: The domain expertise of the column troubleshooting and of the CMP design is readily expressed in IF-THEN rules that MRS is designed to handle. Previous users of MRS had indicated that it was a versatile tool for reasoning with various forms of domain expertise and that the meta level reasoning could be used to solve particularly difficult problems. MRS doesn't require, although it runs well on, specialized hardware such as a Lisp machine supporting high resolution graphics. Because the source code is provided, it is easy to write extensions to MRS directly in Lisp (such as the user interface). Finally, since MRS is academic software, it is inexpensive.

Results

Development of ECAT Knowledge Bases. The extent of an expert's domain knowledge typically exceeds that which he or she realizes, or is capable of immediately articulating. Our experience has shown that an expert asked to begin with a "tabula rasa" and perform an instantaneous brain dump of domain knowledge will yield only a small portion of that knowledge. The technique we are using to facilitate transfer of human expertise to the expert system program involves an iterative process which incrementally improves program functionality. Incorrect or incomplete conclusions reached by the program are presented to the human expert who is asked to provide the information necessary for the program to yield the expert's recommended solution. This process is repeated, expanding the knowledge base and hence the frequency with which successful problem solving occurs.

This strategy is illustrated by our construction of the knowledge base of Module 3 of the ECAT program. Module 3 specifies the HPLC analytical column, mobile phase constituents, and detector to be used. The design problem given to the program is termed a "sample probe". A sample probe consists of a specification of a user's sample (input) and the recommendations which the program SHOULD compute (output). Sample probes are prepared by colleagues outside the program (i.e., by chromatographers other than the domain expert) by selection from new separations appearing in refereed chromatographic journals, and from standard, qualified HPLC methods.

First Rules. The first probe tested was the trace analysis of phenols in wastewater. At this point, the knowledge base contained no rules and thus no answer was given as to column, mobile phase or detector specification. The expert stated that the separation should be run in a reverse phase mode on a C18-silica column, with a water-acetonitrile mobile phase containing 0.1% acetic acid as a competing acid additive (to reduce peak tailing of weakly acidic phenols). At this point, the expert was asked to explicate his reasoning as a series of rules which concluded the correct design recommendation. Nine rules were specified.

It should be noted here that in specifying the rules for the first probe (phenols), it became clear that rules for choosing the column and mobile phase interact significantly with detector rules. 0.1% acetic acid works well as a competing acid additive in terms of chromatography of the phenols. However, carboxylate ions are known to quench the fluorescence of phenols. Thus, if one were to use a fluorescence detector for trace phenol detection, an alternative competing acid, such as 0.1% phosphoric acid should be substituted. It was decided that mobile phase/detector interaction rules would be the first detector rules to be added to the knowledge base.

More Rules. Figure 5 tracks the number of IF/THEN rules added to the knowledge base to specify column and mobile phase constituents. Detector rules other than those relating to mobile phase compatibility were not entered at this time. As the knowledge base expanded, subsequent probes of similar molecular structure (and hence similar chromatographic properties) were solved with addition of few or no rules. Solution without requiring incrementing of the knowledge base is termed a "direct hit." Spikes in the graph of Figure 5 occur for new sample probes having major structural differences from those already tested - for example, the sample probe "LDH isoenzymes" required special rules regarding protein chromatography.

It should also be noted that new sample probes can generate additions to the sample information queries asked of the user at the beginning of the "probe session." Thus, protein probes required the addition of queries regarding molecular weight, isoelectric point and whether biological activity is to be preserved in the chromatographic step. These questions are triggered only if the user specifies the sample as a peptide or protein in answer to the initial sample questions. Also, once the sample is specified as a protein, the question as to the pKa or pKb of the

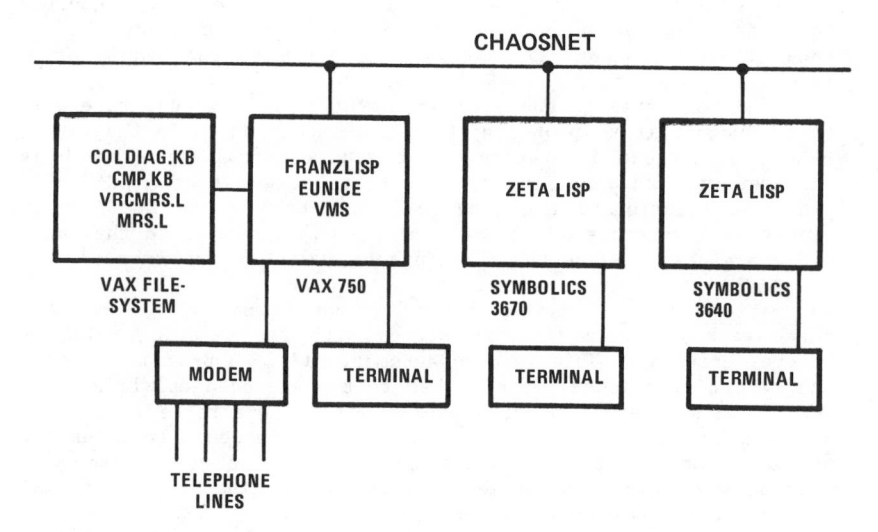

Figure 4.   Hardware and Software for the development of ECAT.

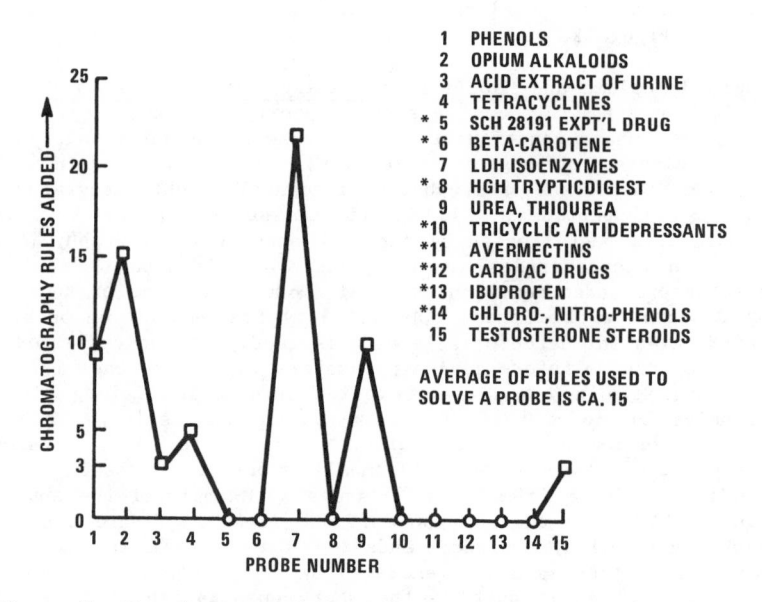

1   PHENOLS
2   OPIUM ALKALOIDS
3   ACID EXTRACT OF URINE
4   TETRACYCLINES
* 5   SCH 28191 EXPT'L DRUG
* 6   BETA-CAROTENE
7   LDH ISOENZYMES
* 8   HGH TRYPTICDIGEST
9   UREA, THIOUREA
*10   TRICYCLIC ANTIDEPRESSANTS
*11   AVERMECTINS
*12   CARDIAC DRUGS
*13   IBUPROFEN
*14   CHLORO-, NITRO-PHENOLS
15   TESTOSTERONE STEROIDS

AVERAGE OF RULES USED TO
SOLVE A PROBE IS CA. 15

Figure 5.   Development of knowledge base rules to select column
and mobile phase constituents.   * indicates "direct hit".

sample will not be asked, and the isoelectric point is requested instead. Thus the initial query session is "intelligent" in the sense that the questions asked are specific to the sample probe itself.

The next step in the development of this knowledge base will be to subject it to probe input by chromatographers in the Varian HPLC Applications laboratory. For each external probe which is answered with an incorrect or incomplete answer by the ECAT program, an interrogation of the probe creator by the ECAT domain expert will generate additional rules to be conveyed to the knowledge engineers. Thus the knowledge base will be incremented.

Automatic Testing. As the knowledge base expands, the need to check each new rule for consistency with the existing rule set becomes critical. This is done automatically. A program subjects the file of previous sample probes to the expanded knowledge base and checks to see if previous solutions are unaffected. If previous solutions have been affected, one must proceed to debug the new additions to the knowledge base. This often requires rewriting some rules. Sometimes it provokes rethinking and reformulating part of the knowledge base.

Current Performance - Modules

The performance of ECAT is primarily determined by the correctness and extent of the knowledge and data bases, that is, the modules shown in Figure 2.

Module 3, Column and Mobile Phase Design (CMP). This is the core module for ECAT. It can currently specify i) analytical column and mobile phase constituents for reverse phase chromatography of common classes of organic molecules; ii) reverse phase, ion exchange phase and hydrophobic interaction chromatography of proteins and peptides; iii) a limited set of specialty classes of molecules best treated by straight phase chromatography (e.g., mono- and disaccharides). The rules for selection of the HPLC detector are under development within Module 3. Some of the rules for detector mobile phase compatibility are already encoded. A set of rules for detector selection is ready but not yet encoded.

The program infers design parameters using data base information from Module 1 and user-supplied information, along with an extensive knowledge base of chromatography heuristics. Module 3 currently contains ca. 160 rules, generated to cover 15 sample probes which represent some commonly separated classes of compounds (see Table I). Figure 6 shows an example of the application of ECAT to a design problem. The items in Figure 6 are the user inputs and system recommendations in the form in which they are actually processed and generated by the program.

Figure 7 shows part of the user consultation that elicited the inputs listed in Figure 6. The current user interface provides on-line help as well as a menu of numbered valid responses. The user may either type in the number or the listed item. In answer to the user typing "?", the system rephrases the question, redisplays acceptable values, and specifies what other characters are recognized. If this is not enough information, the

Table I.   Sample probes used to develop knowledge base
of Module 3 for specification of analytical
HPLC column, and mobile phase constituents.

Probe 1.	phenols	moderately polar, weakly acidic molecules
Probe 2.	opium alkaloids	polar, basic nitrogen hetero cycles, typical of many drugs
Probe 3.	acid extract of urine	carboxylic acids
Probe 4.	tetracyclines	molecules with significant metal-complexation character
Probe 5.	SCH 28191 (experimental drug)	same as opium alkaloids (Probe 2)
Probe 6.	beta-carotene	non-polar, neutral molecules
Probe 7.	LDH isoenzymes	proteins
Probe 8.	HGH tryptic digest	peptide fragments
Probe 9.	urea, thiourea	small, polar molecules
Probe 10.	tricyclic anti-depressants	same as opium alkaloids (Probe 2)
Probe 11.	avermectins	moderately polar, neutral molecules
Probe 12.	cardiac drugs	same as opium alkaloids (Probe 2)
Probe 13.	ibuprofen	moderately polar carboxylic acid
Probe 14.	chlorophenols nitrophenols	non-fluorescent phenols (see Probe 1)
Probe 15.	testosterone steroids	complex mixture of compounds sharing same hydrocarbon backbone and differing in functional group

USER ENTRIES

(analyte-class phenols)
(specific-analyte phenol)
(pka-of phenol 11)
(largest-mw 400 daltons)
(detector-type fluorescence)
(smallest-analyte-amount 10 ng)
(class-of sample-matrix river-water)

RECOMMENDATIONS

Guard column
(additional-column guard-column)
(packing-of guard-column pellicular)
(packing-of guard-column silica-based)
(diameter-of pellicular 25 micron)

Analytical column
(separation-mode reverse-phase)
(restrict (diameter-of particle $value micron)
        (<= $value 5))
(packing-of $column silica-based)
(prefer (bonded-phase $column C18)
        (bonded-phase $column C8) 0.2)

Mobile phase
(prefer (liquid-of solventb acetonitrile)
        (liquid-of solventb methanol) 0.4)
(liquid-of solventb methanol)
(liquid-of solventb acetonitrile)
(liquid-of solventa water)
(additive-of solventb competing-acid phosphoric-acid)
(additive-of solventa competing-acid phosphoric-acid)
(restrict (ph-of $3 $4) (>= $4 2) (<= $4 7.5))
(concentration-of phosphoric-acid solventb 0.1%)
(concentration-of phosphoric-acid solventa 0.1%)

Figure 6. Example of user inputs and system recommendations for
a CMP probe. $... are variables.

```
You are running the Column and Mobile Phase Selection module.

You should type ? or H any time you require help.
Some prompts require an additional <CR> to terminate input acqui-
sition.
Be careful not to type ahead.


Valid values:
      1.  amino-acid-hydrolysate        17.  oligonucleotides
      2.  amino-acid-                    18.  oligosaccharides
          physiological-fluids          19.  oligosaccharides
      3.  citric-acid-cycle-acids       20.  peptide
      4.  diastereomers                 21.  phenols
      5.  carboxylic-acid               22.  phospholipids
      6.  disaccharides                 23.  porphyrins
      7.  glucosamines                  24.  porphyrins
      8.  glycolipids                   25.  prostaglandins
      9.  glycosphingolipids            26.  protein
     10.  hydroxyvitaminsd2+d3          27.  sphingolipids
     11.  lipids                        28.  stereoisomers
     12.  methylated-nucleosides        29.  stereoisomers
     13.  monosaccharides               30.  sugar-alcohols
     14.  monosaccharidesl              31.  tricarboxylic-acids
     15.  nucleosides+nucleotides       32.  other
     16.  nucleotides
Analyte class: phenols <CR>
Analyte class: <CR>

Valid value is a number.
phenol pKas: ? <CR>
Enter the pKa values for phenol.

Valid value is a number.
phenol pKas: 11 <CR>

Valid value is a number (unit: daltons).
Largest molecular weight: H <CR>
You are asked to enter the molecular weight of the largest mol-
ecule you are interested in analyzing.  Typical values are ranging
from the low hundreds to a few hundred thousand (in the case of
proteins).

Any of the following is a valid response: <number> unknown
Largest molecular weight: 400 <CR>

Are you using a fluorescence detector ? [y]: <CR>

Valid value is a number (unit: Nanograms)
Smallest analyte amount: 10 <CR>
```

Figure 7. Excerpts from a user/expert-system consultation.
Underlined items are user input. <CR> indicates user typed a
carriage return.

user can type "h" for help. The help text is not yet completely written, but is readily extensible. "How" and "why" queries are not yet recognized.

Module 6, Column Diagnosis (COLDIAG). This module uses chromatographic parameters such as efficiency, asymmetry, retention time, selectivity and operating pressure, to detect failures of the column or other chromatographic hardware. Table II lists the types of column failure which the module can currently handle. Note that the module will also correctly diagnose some problems which are NOT column malfunctions but which might be interpreted as such by a user.

Table II. Types of Column Failure Diagnosed and Treated by
Module 6

Column Failures	plugged column bed or frits
	dissolution of column bed at high pH
	physical compression of column bed
	hydrolytic cleavage of bonded phase
	chemical alteration of cn bonded phase
	reaction of NH_2 bonded phase with C=O
	deactivation of Si adsorbtion sites by trace H_2O
	loss of packing material from column
	irreversible adsorbtion of sample matrix components
Non-column Failures	too large an increase in injection volume
	inadvertent change to strongly eluting injection solvent
	inadvertent overloading of column

Module 1, Determination of Chemical and Structural Information on the Sample. The task of Module 1 is to provide non-chromatographic data for analytes prior to specification of the chromatographic method. Data bases have been developed for pK values of organic molecules, isoelectric points of proteins, and fluorescence spectral properties of organic molecules.

Other Modules. Modules 2, 4 and 5 are currently in a design stage.

Future Plans

Module (Knowledge Base) Development. Future development of the ECAT system will involve, in chronological order:
    1) Incrementing Module 3 (CMP) according to sample probe testing by chromatographers not directly associated with the project and adding detector selection rules.
    2) Development of the knowledge base for Module 5 (optimization of mobile phase composition and program). The column and mobile phase constituents having been specified by Module 3, the knowledge base of Module 5 will be used along with inputs of required analysis time, and desired degree of resolution to guide

optimization of the mobile phase composition and program (if gradient elution is required). This module will require a great deal of effort as one must write knowledge base rules to determine which parameters (pH, ionic strength, ion-pair reagent concentration, organic modifier concentration, etc.) to optimize by either simplex or factorial design techniques. One must also develop algorithms to analyze the quality of chromatographic separations with respect to user input requirements.

    3) Integration of Module 6 (diagnosis of hardware problems) into Module 5. It is sometimes necessary to detect and troubleshoot hardware and column failures during the optimization step. When abnormal changes in separation parameters occur during an optimization, this detection can halt the series of optimization experiments and notify the user of the system failure, preventing a useless optimization of a "broken" HPLC system. An example of this coupling of troubleshooting to optimization would be a situation in which one is optimizing a reverse phase separation of opium alkaloids, with respect to the parameters pH and % acetonitrile in the mobile phase. Suppose one has run a series of experiments varying the pH between 2 and 3 and the concentration of acetonitrile between 40% and 50%. Assume the efficiency has remained between 6000-7000 plates and asymmetry between 1.2-1.3. On the next experiment, with pH 3 and 55% acetonitrile, the retention time decreases as expected, but the peak efficiency drops to 500 plates, and asymmetry increases to 9.0. Simultaneously, a slight pressure increase occurs. The column troubleshooting module would flag the abnormal change in chromatographic parameters occuring for a very slight change in mobile phase character. It would then go back and repeat a previous experiment such as pH 3 and 50% acetonitrile. If the previous efficiency cannot be reproduced, it is certain a malfunction has occurred. The module could then halt the optimization, troubleshoot the fault (collapse of column bed with formation of a void at head of column), and recommend corrective action to the user.

    4) Development of Module 4 (knowledge base for sorbent cartridge-based sample cleanup prior to the analytical chromatography step). The column and mobile phase constituents specified by Module 3 will be fed into Module 4 in order to determine the procedure for sample cleanup, isolation and elution steps prior to the analytical chromatography step. In developing Module 4, we will use the recently developed techniques based on sorbent cleanup and isolation of sample analytes rather than the classical liquid-liquid extraction techniques. This decision was based on the ability to automate the sorbent technique by using short chromatographic sorbent cartridges and on technical advantages discussed in detail elsewhere (21).

    5) Expansion of Module 3 to include rules for selection of detectors and detector parameters. The rules will handle optical absorbance and fluorescence (including pre- and post-column derivatization) and electrochemical detection.

    6) Expansion of the data bases in Module 1 to include spectroscopic and electrochemical data to be used by the detector selection rules of Module 3. (This would include UV absorbance spectral properties of organic molecules, fluorescence quenching and activating properties of solvent environments, and electro-

chemical activity of organic molecules.)   Additional data which
will eventually be needed in ECAT includes solubility characteristics, boiling points, and melting points of organic molecules.

7) Development of Module 2: Knowledge base for screening out
samples which are best done by GC, and development of a library of
standard, qualified HPLC and GC methods. The system will decide
whether the analytical demands of the separation are best served
by gas chromatography or by liquid chromatography. Information
available from Module 1 is needed (boiling and melting point data,
molecular weight), along with information on analyte levels,
matrix properties, sample complexity and resolution requirements. As the ECAT program evolves, one might eventually consider
adding decision capability regarding other important separation
techniques, such as gel electrophoresis, to the knowledge base of
step 2.   It will be quite useful here to include a "library" of
standard methods used in GC and LC. The experts must specify
those known analytes that are best separated by GC. For example,
trace analysis of volatile pesticides at sub-picogram levels is
best performed by GC, and the user should have access to the
recommended GC method before considering an LC development.   On
the other hand, analysis of relatively non-volatile ionic drugs is
best done by HPLC, and here the user should be provided with a
standard, qualified HPLC method if such a method exists.   Thus,
Module 2 will include both a knowledge base guiding the decision
as to GC versus LC and will provide a library of standard, qualified chromatographic methods.

Knowledge Representation

We are currently investigating the expansion of the ECAT capability to represent and process knowledge by including a representation scheme based on heirarchically structured descriptions of
object properties (so-called "frame-based").   It is sometimes
awkward to express subtle or indirect knowledge in the simple form
of forward chaining reasoning we are currently using. A specialized planning software architecture such as SPEX (22) might be
required to handle the full fledged design module.

We are looking into ways of expressing uncertainty.   For
example, uncertainty occurs in ECAT when there are alternate suspected causes of a separation malfunction or alternate choices of
bonded phase for some sample classes. In ECAT, representation of
uncertain information within causal reasoning is currently handled
by predicates such as "prefer" or "consider".   There is a long-
standing discussion of reasoning about uncertain, inexact or
unreliable information (23). Certainty factors (24), statistics,
fuzzy sets (25) and explicit reasoning are methods that can be
applied to solve this problem. MYCIN-type certainty factor handling can be revised to fit entirely into the realm of statistics
(26). Gordon and Shortliffe have proposed a computable method for
using the Dempster-Shafer theory of evidence (27).   Cohen and
Grinberg (28) have argued that it is best to reason explicitly
about uncertainty in ways similar to human thought processing.
However, the latter method requires extensive computing which we
feel is hardly justifiable in our case. We will thus investigate
further   application   of   simple   statistics   and   Gordon   and

Shortliffe's proposal.   In addition, we will experiment with a simplified, domain restricted form of explicit reasoning about uncertainty.

User Interface

This, of course, is a very important part of the program. We are developing it on an as-needed basis in response to feedback from users.   In particular, we still have not implemented a complete explanation facility.   The user interface currently provides online help and a menu based selection of valid responses whenever applicable.

Conclusion

We have presented the development of an expert system in HPLC. The overall project goals are fairly ambitious and will require continuous work for many years to come.   However, the CMP design module (the key module of the project) and the COLDIAG module are beyond a prototype stage of development.   Since those two modules cover the two modes of reasoning we consider using, we will be able to complete the entire ECAT program using the approach described in this paper.

Acknowledgments

The authors wish to thank Steve Rosenblum for critical reviews of the manuscript and June Shelley for help in the preparation of the manuscript.

Literature Cited

1. Expert Systems Part 2, Dessy, R. E., Ed.; Anal. Chem. 56, 1984, 1312A
2. Expert Systems Part 1, Dessy, R. E., Ed.; Anal. Chem. 56, 1984, 1200A
3. Glajch, J. L.; et al. J. Chrom. 1980; 199, 57.
4. Glajch, J. L.; et al. J. Chrom. 1982; 238, 269.
5. Kirkland, J. J.; Glajch, J. L. J. Chrom. 1983; 255, 27.
6. Debets, H. G.; et al. Anal. Chim. Acta 1983; 150, 259.
7. Schoenmakers, P. J.; Drouen, A. C. J. H.; Billiet, H. A. H.; de Galan, L. Chromatographia 1982; 15, 688.
8. Haddad, P. R.; Drouen, A. C. J. H.; Billiet, H. A. H.; de Galan, L. J. Chrom. 1983; 282, 71.
9. Billiet, H. A. H.; Drouen, A. C. J. H.; de Galan, L. J. Chrom. 1984; 316, 231.
10. Barr, A.; Feigenbaum, E. "Handbook of AI"; William Kaufman Inc., 1982; Vol. II, Chap. VIIB.
11. Lindsay, R.; Buchanen, B. G.; Feigenbaum, E. A.; Lederberg, J. "DENDRAL"; McGraw Hill; New York, 1980.
12. Crandell, C. W.; Gray, N. A. B.; Smith, D. H. J. Chem. Inf. and Comp. Sci. 1952; 22, 48.
13. Gray, N. A. B. Artificial Intelligence (1984); 22, 1-21.
14. Jardetzky, O. Proc. Int. Conf. on Frontiers of Biochemistry and Molecular Biology 1984.

15. Terry, A. Stanford Heuristic Programming Project, Report
     No. HPP-83-19, May 1983.
16. Agarwal, K. K.; Larsen, D. L.; Gelernter, H. L. Computers in
     Chemistry 1978; 2, 75.
17. Gelernter, H. L.; et al. Science 1977; 197, 1041.
18. Corey, E. J.; Long, A. K.; Rubenstein, S. D. Science 1985;
     228, 408-418.
19. Wipke, W. T.; Ouchi, G. I.; Krishnan, S. Artificial
     Intelligence 1978; 11, 173.
20. Russell, S. Stanford Knowledge Systems Laboratory, Report
     No. KSL-85-12, 1985.
21. Van Horne K.; Good, T. American Laboratory 1983; 15, 116.
22. Bach, R.; Iwasaki, Y.; Friedland, P. Nucleic Acids Research
     1984; 12, 11-29.
23. Panel on Reasoning with Uncertainty for Expert Systems,
     International Joint Conference on Artificial Intelligence,
     Los Angeles, California, 1985.
24. Buchanan, B. G.; Shortliffe, E. H. "Rule Based Expert-
     Systems"; Addison-Wesley, 1984, Chap. 10.
25. Zadeh, L. A. in "Machine Intelligence"; Hayes, J.;
     Michie, D.; Mikulich, L.I. Eds.; John Wiley and Sons:
     New York, 1979; pp. 149-194.
26. Heckerman, D. Proceedings of the Workshop on Uncertainty and
     Probability in Artificial Intelligence; American Association
     for Artificial Intelligence, 1985; pp. 9-20.
27. Gordon, J.; and Shortliffe, E.H., Artificial Intelligence
     1985; 26, pp. 323-357.
28. Cohen, R.; Grinberg, M. R. AI magazine 1983; 4, 17-24.

RECEIVED January 16, 1986

An Expert System for Optimizing Ultracentrifugation Runs

Philip R. Martz, Matt Heffron, and Owen Mitch Griffith

Beckman Instruments, Inc., Fullerton, CA 92634

The SpinPro Ultracentrifugation Expert System is a
computer program that designs optimal ultracentrifugation procedures to satisfy the investigator's
research requirements. SpinPro runs on the IBM PC/XT.
Ultracentrifugation is a common method in the separation of biological materials. Its capabilities,
however, are too often under-utilized. SpinPro
addresses this problem by employing Artificial
Intelligence (AI) techniques to design efficient and
accurate ultracentrifugation procedures. To use
SpinPro, the investigator describes the centrifugation
problem in a question and answer dialogue. SpinPro
then offers detailed advice on optimal and alternative
procedures for performing the run. This advice
results in cleaner and faster separations and improves
the efficiency of the ultracentrifugation laboratory.

Ultracentrifugation is a common and powerful method in the separation of biological materials. Despite its widespread use, however,
few investigators fully exploit its capabilities. As a result, run
times are unnecessarily long and separations are indistinct. In the
long run, the efficiency and performance of the laboratory suffer.

The fundamental cause of this situation is the increasing
complexity of the ultracentrifugation environment; the investigator
must select the run parameters from a growing list of rotors,
gradient materials, and literature references. Knowing which rotor
to use and at what run speed and run time is a difficult matter.
Furthermore, the selection of one parameter complexly limits the
available choices for the remaining parameters.

Reliance on procedures reported in the literature has com-
pounded the problem. Often these procedures, perhaps initiated by
investigators with a limited set of rotors, are inefficient by
today's standards: the rotor is inappropriate, the run speed is too
slow, or the run time is too long. A new investigator applying this
procedure does not take full advantage of the potential of ultracentrifugation.

0097–6156/86/0306–0297$06.00/0

One solution to the problem is to provide the investigator with technical advice. Good advice should yield several immediate benefits: 1) Reliance on inappropriate or outdated techniques can be eliminated. 2) Better use can be made of the available equipment; shorter run times and improved separations will result. 3) The advice can be specific to the research requirements of the investigator. 4) The time usually wasted in performing standardization runs, designing an ultracentrifuge procedure, or researching ultracentrifugation techniques can be minimized. In general, good advice will improve the procedures, and thereby, improve the efficiency of most laboratories.

Designing efficient ultracentrifugation procedures and providing good advice, however, is a complex task; the knowledge and experience of an ultracentrifugation expert are often required. In this paper we describe a computer program, the SpinPro Ultracentrifugation Expert System, that designs ultracentrifugation procedures in response to the requirements of the investigator. SpinPro runs on the IBM PC/XT. The program is based on techniques from the field of Artificial Intelligence (AI) and expert systems: the powerful capabilities of the Lisp programming language; an inferencing procedure capable of drawing conclusions from a complex knowledge base; and a knowledge base derived from the expertise of ultracentrifugation experts. Indeed, SpinPro's use can be compared to the advice any person might seek from an expert. The investigator and SpinPro enter into a question and answer dialogue in which the investigator describes the research goals and sample characteristics. At the conclusion of the dialogue, SpinPro produces the following reports:

1. The Design Inputs Report is a summary of the SpinPro–investigator dialogue.
2. The Optimal Plan Report describes an optimal ultracentrifugation procedure designed to solve the problem described in the dialogue. It uses the most appropriate rotor from the entire line of Beckman rotors.
3. The Lab Plan Report is similar to the Optimal Plan, but it describes a procedure based exclusively on the ultracentrifuges and rotors available in the investigator's laboratory.
4. The Plan Comparisons Report compares the Optimal Plan and Lab Plan, identifying significant differences and trade-offs between the two plans.

The reports constitute a complete set of recommendations for the ultracentrifugation problem posed to SpinPro. Thus, SpinPro performs the advisory role of an ultracentrifugation expert: interviewing the investigator for the problem description, offering expert advice on the most appropriate centrifugation procedure, and finally, comparing alternative procedures.

Major Functions

SpinPro has four major functions: CONSULTATION, INFORMATION, CALCULATION, and CONFIGURATION. The CONSULTATION function performs the role of expert advisor. It is the main topic of this paper. The INFORMATION function provides a database of ultracentrifugation

techniques, centrifuges, rotors, and literature references. The
CALCULATION function performs a variety of routine calculations
including rotor speed reductions, k factors, and pelleting time.
The CONFIGURATION function records the ultracentrifuges and rotors
in the investigator's laboratory. This information is used by the
CONSULTATION function when designing a run using the equipment from
the laboratory.

User Interface

All user inputs are made by pointing at text on the computer screen
with a "mouse" controlled cursor. The mouse is a hand-held pointing
device, which when moved by the investigator over a flat surface,
controls the movement of a cursor or pointer on the computer screen.
To run the CONSULTATION function, the user points at the text
"CONSULTATION" on the screen and clicks the mouse button. When
using SpinPro, the keyboard is not required. In our observations,
novice users of the program have been able to design ultracentrifugation procedures within minutes of using the program.

The CONSULTATION Function

The primary goal of the CONSULTATION function is to provide the best
advice possible on precisely how to set up and run an ultracentrifugation procedure that is specifically designed for the investigator's research. SpinPro addresses virtually all problems in the
ultracentrifugation of biological samples excluding whole cells. To
this end, SpinPro is "knowledgeable" about differential, rate-zonal,
and isopycnic methods. It addresses the separation of proteins,
glycoproteins, proteoglycans, lipoproteins, subcellular fractions,
nucleic acids, and viruses. SpinPro's rotor knowledge includes
swinging bucket, fixed angle, vertical tube, zonal, and continuous
flow rotors.

Operation

The CONSULTATION function is run by using the mouse to select the
text "CONSULTATION" from the computer screen. The first question of
the dialogue, "Please enter the class of your sample of interest",
appears on the screen. The pop-up menu lists the sample types to
chose from. The investigator then uses the mouse to select the
appropriate response from the pop-up menu. This question and answer
procedure continues until SpinPro has enough information, typically
10 to 15 questions, from which to infer all of the relevant parameters. The dialogue is directed by SpinPro in response to answers
to previous questions. Thus, if the sample is a protein, SpinPro
requests the sedimentation coefficient; if the sample is a nucleic
acid, SpinPro requests the type of nucleic acid. At the conclusion
of the dialogue, the reports are written to the disk. Using the
pop-up menu, the reports can be read or saved.
      The dialogue includes capabilities to increase its flexibility.
First, the investigator can change an answer to a previous question
without disrupting the course of the dialogue. This capability is
useful when describing a problem that differs only slightly from a
previously described problem. Second, the investigator can ask why

the current question is being asked. The "Why?" function informs
the user what SpinPro is attempting to infer (i.e., the line of
reasoning) at any particular step, and it describes the affect that
different answers will have on the line of reasoning. Third, when
the answer to a question is not known, the investigator can answer
the question with "unknown". Depending on the question, SpinPro
responds either by asking a related question or by assuming a
reasonable answer and designing the procedures based on this
assumption. Any assumptions that have been made are noted in the
reports. Finally, for the experienced users of SpinPro, there is
the option to request that, during the dialogue, a short form of the
question be used.

Optimization Criteria

Two of the dialogue questions are of unique importance and are
particularly representative of SpinPro's capabilities. The first is
a question of research requirements. Every ultracentrifugation procedure should reflect the investigator's concern for purity of the
separation or short run time, goals that often run counter to each
other. Rarely does any procedure state this trade-off explicitly.
The optimization criteria question, "Select one of the following
optimizations:", not only identifies the trade-offs involved when
designing a procedure, but allows the investigator to control them.
The investigator can select the criterion which satisfies the
specialized requirements of the research. The criteria are: 1)
purity, 2) minimize run time, 3) minimize cumulative run time, 4)
minimize number of runs, 5) continuous flow rotor procedures, and 6)
procedures for processing many samples of small volume.
    Based on the optimization criterion, SpinPro can select the
most appropriate rotor. For example, suppose the investigator has a
relatively large sample volume, all of which needs to be processed
as soon as possible. The "minimize cumulative run time" criterion
would be the appropriate choice. SpinPro would then initiate the
following rotor selection procedure: SpinPro determines the total
sample volume based on inputs of the sample volume, the current
concentration of the sample, and a correction for any pre-run
dilutions of the sample. Next, consideration is made for whether
tubes or bottles will be used. The program then evaluates rotors
for the number of tube positions and the amount of sample per tube.
At this point, SpinPro will have estimated for each rotor the number
of runs required to process the sample. SpinPro then estimates the
run time for each rotor to perform a single run. Based on these
estimates, SpinPro selects the rotor that will give the shortest
total run time when the run time is summed over the total number of
runs. Similarly, the investigator can select any of the
optimization criteria and initiate a variety of precise rotor
selection procedures.

Lab Rotors

The second question of unique importance concerns the investigator's
selection of a rotor for the Lab Plan. Whereas, in the Optimal
Plan, SpinPro selects the rotor; in the Lab Plan, the investigator
selects the rotor. The investigator, however, is not required to

select the rotor blindly from those available in the lab. SpinPro
assists in the selection by assigning each of the rotors in the lab
to a category based on how well the rotor satisfies the requirements
of the problem. The categories are as follows:

1. Optimal rotors – the rotors that are both best suited to per-
    form the run and to achieve the stated optimization criterion.
2. Alternate rotors – other rotors that are not optimal but can
    perform the run.
3. Not qualifying rotors – rotors that are not recommended for the
    problem usually because they are too large or too small for the
    sample volume, or because they do not generate sufficiently
    high centrifugal forces.
4. Not compatible rotors – rotors that are not classified, as part
    of the rotor safety program, for running in the ultracentrifuge
    chosen from the lab.

The investigator can select any rotor from categories 1 and 2 above.
This allows the investigator to experiment with the rotors in the
lab and to design procedures as variations on the theme established
in the Optimal Plan. Ultimately, the rotor selected in the Optimal
Plan by SpinPro and in the Lab Plan by the investigator are the
major source of difference in the run parameters, purity, and
overall effectiveness of the two plans.

The Design Inputs Report

As noted earlier, SpinPro writes four reports regarding the recommended procedures. The Design Inputs Report summarizes the questions posed by SpinPro and the answers provided by the investigator.
A Design Inputs Report is shown in figure 1. The pop-up menu on the
right allows the user to switch between reports, print the reports,
or perform other functions. The report summarizes the problem that
is addressed by the Optimal Plan (Figure 2) and the Plan Comparisons
Report (Figure 3).
     A summary of the report follows: The problem is to separate
proteins. Furthermore, SpinPro should pay particular attention to
the purity of the separation. The sample is not negatively affected
by sucrose, has a sedimentation coefficient of 16 Svedbergs, and is
in liquid form of 3 mL and a concentration of 1% w/w. The protein
of interest should be placed 45% from the top of the gradient at the
end of the run. Of the gradient concentrations 10–40% and 5–20%,
the 10–40% is preferred by the investigator. There are no solvents
in the sample that are harmful to the tubes. Finally, from the lab,
SpinPro should use the L2–75B ultracentrifuge and the SW 41 Ti
rotor, which does not require a speed derating due to its age.

The Optimal Plan Report

The Optimal Plan is SpinPro's recommendation of how best to perform
the run. The Optimal Plan of figure 2 is underlined and annotated
below:

SpinPro Ultracentrifugation Expert System
Design Inputs

Experiment: SpinPro Consultation 11-Sept-1985 9:30:00

Particle class: Protein
Separation vs Concentration: Separation
Optimization criterion: Purity
Assoc/Dissoc in sucrose: No
Sedimentation coefficient: 16.0
10-40% or 5-20% gradient?: 10-40
Sample form: liquid/semi-solid
Total sample volume (mL): 3.0
Sample concentration % w/w: 1.0
Selected final location: 45.0
Solvents: No
Selected lab centrifuge: L2-75B
Selected lab rotor: SW 41 Ti
Rotor derated?: No

Page Forward
Page Backward
Optimal Plan
Lab Plan
Comparisons
Design Inputs
Change Answer
Save Reports
SpinPro Top
Exit to DOS

Figure 1. The Design Inputs Report for the problem described to
SpinPro. The Optimal Plan and the Lab Plan are based on this
problem. The pop-up menu on the right allows switching to the
other reports or performing other functions.

SpinPro Ultracentrifugation Expert System
Optimal Plan

Experiment: SpinPro Consultation 11-Sept-1985 9:30:00

This is a complete plan for a protein sample separation
Optimization criterion: Purity

Method: Density gradient, Rate-zonal
Gradient: 10-40% continuous sucrose
Rotor/run conditions: SW 55 Ti rotor at 55000 rpm
  for approximately 6 hours
Potential tube materials: Polyallomer, Ultra-Clear

Centrifuge: L8-80M set at 4 degrees C
Omega-squared t: 7.132x10^11
Acceleration/deceleration: fast/fast

Prior to the run prepare sample as follows:
  No special sample preparation is required.
Load 0.3 mL of the Protein sample in full tubes at
  the top position of the gradient.
At the end of the run the 16 S particles will be
  approximately 45% from the top of the gradient.
To process the entire sample volume requires approximately
  2 centrifuge run(s) with an estimated total run time of
  12 hours, 5 minutes.

> Page Forward
> Page Backward
> Optimal Plan
> Lab Plan
> Comparisons
> Design Inputs
> Change Answer
> Save Reports
> SpinPro Top
> Exit to DOS

Figure 2.   The Optimal Plan Report for the problem described in
the Design Inputs Report of figure 1.   The plan gives the recommended procedure for doing the run.

SpinPro Ultracentrifugation Expert System
Plan Comparisons

Experiment: SpinPro Consultation 11-Sept-1985 9:30:00

Run summaries:
  Optimal: SW 55 Ti at 55000 rpm for 6 hours per run
  in 2 run(s). Requiring a total of approximately
  12 hours, 5 minutes

  Lab: SW 41 Ti at 41000 rpm for 15 hours,
  45 minutes per run in 2 run(s). Requiring a total of
  approximately 31 hours, 30 minutes

Comparisons:
  The Optimal Plan requires 38% of the Lab Plan run
    time for a single run. It requires 38% of the Lab
    Plan run time when processing the entire sample.

> Page Forward
> Page Backward
> Optimal Plan
> Lab Plan
> Comparisons
> Design Inputs
> Change Answer
> Save Reports
> SpinPro Top
> Exit to DOS

Figure 3.   The Plan Comparisons Report compares the Optimal and
Lab Plans.   The comparison shows that, because the Lab Plan uses
the SW 41 Ti rotor, the run times are dramatically different.

This is a complete plan for a protein sample separation. All of the
relevant parameters have been inferred in a "complete plan".
"Partial plans" indicate that one or more parameters could not be
determined.

Optimization criterion: Purity. The report restates the optimization criterion chosen by the investigator.

Method: Density gradient, Rate-zonal. The rate-zonal method is one
of six addressed by SpinPro. The other methods are differential,
differential-flotation, discontinuous, isopycnic, and 2-step
isopycnic. These methods differ dramatically in their set up,
principles of operation, and expected results. The rate-zonal
method is described here briefly so that the recommendations to
follow can be appreciated. Prior to the run in a rate-zonal method,
a gradient material is introduced to the rotor tubes in steps of
increasing density from the top to the bottom of the tube. The
sample to be separated is layered, as a thin band, on the top of the
gradient. As the run begins, each component in the sample moves
toward the bottom of the tube. Some components sediment faster than
others. This fact is the basis for the separation. If the run
parameters are appropriate, the components will form separate bands
within the gradient. At the conclusion of the run, the band
representing the component of interest can be removed from the tube.

Gradient: 10 - 40% continuous sucrose. SpinPro usually selects the
gradient concentration and the gradient material. Here, SpinPro
narrowed the choices to the 5-20% or 10-40% gradient, noting in the
dialogue that a trade-off between purity and run time exists between
the 5-20% and the 10-40% gradient, but either will work. The investigator selected the 10-40% gradient. The investigator could, if
desired, finish the plan based on the 10-40% gradient, and then
using the change answer function, try the 5-20% gradient to find out
how the recommendations differ. Sucrose is the gradient material of
choice here. SpinPro considers a wide variety of gradient materials
including cesium chloride, Nycodenz, Metrizamide, glycerol, and
potassium tartrate.

Rotor/run conditions: SW 55 Ti rotor at 55000 rpm for approximately
6 hours. These recommendations form the core of any procedure.
SpinPro usually considers more factors in the rotor selection
process than does the expert. In determining the run speed, SpinPro
considers every possible reason to reduce the run speed. If there
are none, the rotor is run at full speed. When there are reasons
(e.g., when using salt gradients, bottles, differential pelleting,
or discontinuous runs), the run speed may have to be reduced dramatically, from 80,000 rpm to 40,000 rpm, for example. There are many
cases of rotors being run too slow for the application or too fast
for safety. Accurate determination of the run time is a complex
problem based on the gradient characteristics, calculations, inter-
polations from numerical tables, and experience. SpinPro employs
all of these methods in order to infer run times for many special
cases.

Potential tube materials: Polyallomer, Ultra-Clear. SpinPro checks
that all gradient materials, samples, and solvents are compatible
with the tube materials. The affects of acids, bases, oils, organic
solvents, and salts on the tube materials are considered.

Centrifuge: L8-80M set at 4 degrees C. The Optimal Plan recommends
the L8-80M ultracentrifuge. SpinPro selects a temperature that will
protect the integrity of the sample.

Omega-squared t: 7.132x10E11. SpinPro calculates this measure of
the total force applied to the gradient and sample during the run.

Acceleration/deceleration: fast/fast. Many investigators overlook
the affect that improper acceleration or deceleration can have on
disrupting the separation, especially when re-orientation of the
gradient occurs in fixed angle or vertical tube rotors. SpinPro
addresses many special cases.

Prior to the run prepare sample as follows: No special sample
preparation is required. Proper sample preparation is important to
prevent "overloading" the gradient. A sample that is too concentrated will drift through the gradient before the run is started.
If the sample is in a proper form, as it is here, then no
preparation will be recommended.

Load 0.3 mL of the Protein sample in full tubes at the top position
of the gradient. Applying the correct amount of sample is important
to prevent "overloading" the gradient. The rotor tubes can be run
full or half full, or bottles can be used in place of tubes.
SpinPro determines which option is most appropriate. A number of
parameters are affected by this option, including the run time.
Knowing where to load the sample is important. Samples can be
loaded at the top, middle, or bottom of gradients, or mixed
homogeneously with them.

At the end of the run the 16 S particles will be approximately 45%
from the top of the gradient. In the rate-zonal method, common
practice is to have the component of interest at the 50% position in
the gradient when the run is over. SpinPro allows the final
position to be specified, giving the investigator the opportunity to
adjust the procedure so that components not of interest are widely
separated from the component of interest.

To process the entire sample volume requires approximately 2
centrifuge run(s) with an estimated total run time of 12
hours, 5 minutes. SpinPro determines how many runs are required to
process the entire sample volume. The total run time is estimated.
When large sample volumes are involved, and thus many runs are
required, the investigator can change the optimization criterion to
"minimize number of runs" or "minimize cumulative run time" in order
to more efficiently process the sample. Since two runs are required
here, the investigator may want to select a larger rotor for use in
the Lab Plan.

The Lab Plan Report

The Lab Plan provides information similar to that of the Optimal
Plan except that there is the additional constraint of using only
the ultracentrifuges and rotors available in the laboratory. This
requirement can result in dramatic differences between the Optimal
Plan and the Lab Plan. The run times can differ by hours, for
example, or the purity of the separation can be significantly
affected. A completely different gradient can be recommended as a
function of the rotor selected from the lab. If there are no rotors
in the lab capable of doing the separation, SpinPro reports that the
run cannot be done with the available rotors.

The Plan Comparisons Report

The Plan Comparisons report summarizes the differences between the
plans in terms of run time and number of runs required to process
the sample (figure 3). In the figure the Optimal Plan uses the
SW 55 Ti rotor and the Lab Plan uses the SW 41 Ti rotor. The
different run times resulting from these rotors are compared on a
percentage basis. A similar comparison is made for the total run
time required to process the entire sample. Each of the rotors
requires two runs to process the entire sample. The comparison of
the total run times can help in identifying the slower, but larger
capacity, rotors that are more efficient for handling large sample
volumes. If warranted, SpinPro makes qualitative comparisons
between the two plans.

Expert System Details

SpinPro is a typical backward chaining, rule-based expert system.
Rule-based systems are systems in which the expert's knowledge is
encoded primarily in the form of if-then rules, i.e., if a set of
conditions are found to be true then draw a conclusion or perform an
action. "Backward chaining" refers to the procedure for finding a
solution to a problem. In a backward chaining system, the inference
engine works backwards from a hypothesized solution to find facts
that support the hypothesis. Alternative hypotheses are tried until
one is found that is supported by the facts.
   SpinPro's backward chaining inference engine is called "MP".
"MP" has been developed by Beckman to support the development of
expert systems. It has several features that have been designed
specifically in response to the requirements of the SpinPro project.
Two of these requirements are that SpinPro run on an IBM PC/XT and
that the program-user interface be advanced and easy to use. The
report generator and the pop-up menu/mouse interaction provide the
advanced user interface. To be able to run the program on the IBM
PC/XT and still address the ultracentrifugation problem required the
development of fact tables, "why responses", rule functions, rule
groups, and "constraints". Development of these features has
greatly improved the ability of "MP" to make complex inferences.
   Some of these features are demonstrated in the rule example of
figure 4. The rule, one of approximately 800 rules in SpinPro, is
assigned to the rule group 2-STEP.ISOPYCNIC.DNA.RULES. Only those
rules, identified by the rule group name and pertinent to the solu-

tion of a particular problem, are applied to that problem. This
breakdown of rules into rule groups is one of the methods used to
facilitate putting a complex expert system on a microcomputer with
relatively limited memory and processing power.

The overall effect of the rule in figure 4 is to select, from a
set of rotors, those rotors that are best for minimizing the run
time when using the 2-step isopycnic method to separate DNA. The
initial set of rotors is called USERS.MATCHED.ROTORS. The final set
of rotors is called the MINIMIZE.RUN.TIME.ROTORS. The body of the
rule applies tests to the initial set of rotors and concludes that
the rotors passing the tests are the MINIMIZE.RUN.TIME.ROTORS. In
greater detail, Clause 1 of the rule tests the value of the parameter VERTICAL.TUBE.ROTORS. The value of this parameter tells
SpinPro whether vertical tube rotors should be considered for the
run. Often this can be deduced by SpinPro, but when it can't, the
question "Do you want to consider using vertical tube rotors in this
run" is posed to the user. The parameter VERTICAL.TUBE.ROTORS has a
set of properties that define its characteristics including the
prompt used to request the information, the "expect" property used
to specify the acceptable responses to the prompt, and the "Why
Response" property used in response to the investigator's input of
"Why?".

If the value of VERTICAL.TUBE.ROTORS is found to be true (or
"yes") then clause 2 of the rule is evaluated. The references to
"fact" in clause 2 cause the system to refer to a table that
contains the facts for particular rotors. References to the facts
ROTOR.DESIGN, TUBE.VOLUME, and K.FACTOR are applications of particular constraints to the rotors. For example, two constraints are
that the rotor must have a tube volume greater than 1 mL and a k
factor less than 50. Clause 3 further pares the set of rotors on
the basis of k factor by taking only the best rotor and any rotor
with a k factor within 50% of the k factor of the best rotor.

The Other Functions

SpinPro includes two other functions that enhance its role as an
expert advisor. This is in recognition that an expert provides more
than expert advice. An ultracentrifugation expert serves in many
roles: a teacher of centrifugation principles, a describer of
standard procedures, and a source of literature references.

The INFORMATION function contains an extensive database of
ultracentrifugation information organized in a hierarchical fashion
(Figure 5). The primary purpose of the INFORMATION function is to
provide an on-line reference to separation techniques, gradient
materials, rotors, tubes, and centrifuges. For example, INFORMATION
can be used to get information on the Type 70.1 Ti rotor, the
compatibility of polyallomer tubes with certain chemicals, a
description of rate-zonal separations, and references to isopycnic
methods. The subjects in the information hierarchy can be expanded
to give a more detailed breakdown of the subject. For example,
expanding the "Fixed Angle" subject yields a detailed breakdown of
the fixed angle rotors. The investigator could now select one of
the rotor names on the screen and get information about that rotor.
The INFORMATION function includes the subject "SpinPro", which is a
complete on-line manual of the SpinPro system.

RULE 2667: (Rulegroup: 2-STEP.ISOPYCNIC.DNA.RULES)

If: 1) VERTICAL.TUBE.ROTORS, and

2) Find all instances of THAT.ROTOR among the value of USERS.MATCHED.ROTORS
such that:

   1) the ROTOR.DESIGN fact of THAT.ROTOR = one of: SWINGING.BUCKET,
FIXED.ANGLE, or VERTICAL.TUBE, and

   2) the TUBE.VOLUME fact of THAT.ROTOR > 1, and

   3) the K.FACTOR fact of THAT.ROTOR <= 50
(saving those in COLLECTED.ROTORS), and

3) Find all instances of THAT.ROTOR among COLLECTED.ROTORS for
which: the K.FACTOR fact of THAT.ROTOR is within 50% of the smallest value
so computed (saving those in COLLECTED.ROTORS)

Then: 1) Conclude that MINIMIZE.RUN.TIME.ROTORS is each of COLLECTED.ROTORS.

Figure 4. A rule that selects rotors to minimize the run time
in a plasmid DNA separation. The rule examines a set of rotors
called USERS.MATCHED.ROTORS, selecting those rotors that satisfy
criteria based on the rotor design, tube volume, and k factor.

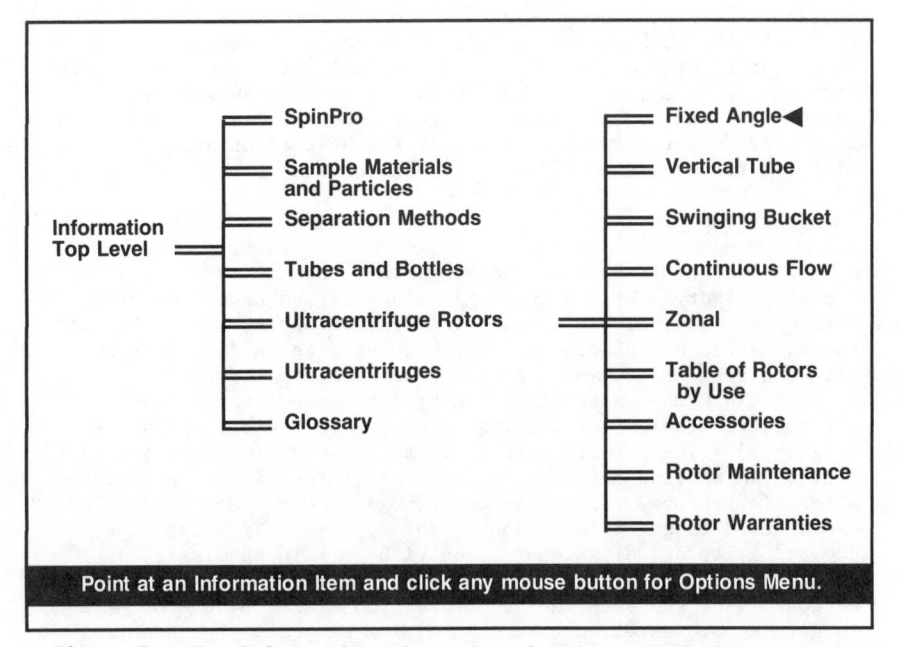

Point at an Information Item and click any mouse button for Options Menu.

Figure 5. The information hierarchy of SpinPro showing the categories of information available. The positions in the hierarchy can be expanded to give a more detailed breakdown of each
subject.

The CALCULATION function provides a variety of routine calculations performed in most ultracentrifugation laboratories. Included
are dilution calculations for sucrose, a pelleting time calculation,
and a calculation for determining rotor speed reductions for salt
gradients. As with the INFORMATION function, the CALCULATION function is a support tool in the effort to efficiently design and carry
out a separation.

Development of SpinPro

There is much concern about the length of time required to develop
expert systems, particularly since so many have achieved various
stages of prototype, but few have been completed. Our experience
with SpinPro has led to many insights, more than can be fully discussed here. Nevertheless, a few major points are worth mentioning.
    It is not particularly clear to us why SpinPro has succeeded in
achieving product status and other expert systems have not, although
we suspect that an early decision to produce a product rather than
to do AI research has been important. The problem domain of ultracentrifugation appears to have been a good choice. The domain has
proven to be fairly well bounded, even though the 800 rules required
has exceeded early estimates by a factor of four. When considering
the various stages of prototyping, debugging, and refinement, over
25,000 rules have been written, and tossed out. Perseverance,
sustained by having a concrete goal of "completeness" rather than a
more indeterminate goal of "demonstrating feasibility" or
"prototyping", was crucial to the success of the project.
    In some ways expert systems programming is little different
from more "traditional" programming. For example, similar to most
software programs, about 50% of the code in SpinPro is for the user
interface; debugging has been very time consuming; and miscommuni-
cation was the source of a great deal of additional effort. Since
these problems are a part of traditional programming as well, techniques designed to assist traditional programmers, such as organization principles, specification, and effective communication also
apply to expert systems.
    In other ways expert systems programming is much different.
Traditional principles of specification and organization are tested,
in part, because the program undergoes evolutionary and sometimes
revolutionary revisions as an understanding of the problem domain
grows. Despite early detailed specification, the tendency of the
specification and the project to evolve toward its final definition
seems to be unavoidable.
    From its inception to completion, the development of SpinPro
has taken about six person years. The development team has included
a manager, two knowledge engineers, one primary expert, four experts
for review, and two people responsible for the content of the
INFORMATION function. During this time, we have completed the
following major activities:

1.   specification and prototyping
2.   knowledge acquisition from the expert
3.   knowledge coding into rules and debugging of rules
4.   design and implementation of the "MP" inference engine
5.   design and implementation of the user interface

6. collecting and writing the contents of the INFORMATION function
7. converting from Interlisp-D on the Xerox 1108 AI workstation to
    Gold Hill Common Lisp (GCLISP) on the IBM PC/XT

Of these activities, task 3 (knowledge coding) and task 4 (inference
engine) were the major efforts. Knowledge coding and debugging
required at least five times as much effort as task 2, the knowledge
acquisition from the expert. Task 7, converting from the development environment to the product proved to be one of the major
hurdles.

There are two notable AI enhancements that are not a part of
SpinPro. First, the "MP" inference engine does not include uncertainty reasoning. The problem domain has only a limited use for it,
and where it is required, uncertainty is handled within the capabilities of "MP". Second, "MP" does not include an ability to explain
its reasoning beyond the "Why?" function discussed earlier. An
explanation capability was not implemented because the usual form of
presenting a trace of the rules that have fired is inadequate and
potentially confusing to the user. Why? Because rules typically
encode "shallow" knowledge (the expert's experience and rules of
thumb) and in a rule trace, are inadequate for communicating the
real, "deep" knowledge, reasons for making a decision.

SpinPro and the Expert

How does SpinPro compare to the expert in solving ultracentrifugation problems? For most problems, SpinPro designs procedures as
good as the expert, if not better. The inherent capabilities of
computers are responsible for this achievement; they are consistent,
they don't forget, and they are precise. For example, SpinPro
contains a vast amount of knowledge that is not a part of the
expert's active memory. Many of the rules are an integration of the
expert's knowledge and procedures reported in the literature. Other
rules are derived from literature references only. This vast amount
of knowledge is immediately available to SpinPro, but not to the
expert. For the new problems, the ones never described to SpinPro,
the expert is far superior. The expert has intelligence,
creativity, common sense, and an understanding of the principles of
ultracentrifugation. These are human tools that the expert can
bring to bear on new problems. At this stage in AI applications,
and despite the goal of AI to recreate these human abilities,
SpinPro, like other expert systems, is lacking.

From the SpinPro project emerged a strong SpinPro-expert
relationship. Early in the project the expert was doubtful about
the prospects of capturing years of education and experience in a
software program. Also the expert felt threatened by the expectation that his role would be subsumed by a computer. These problems
soon disappeared as the challenge of creating SpinPro became more
important. As the project neared completion, the expert took
personal responsibility for the accuracy of SpinPro and pride in its
level of achievement. SpinPro's future development remains closely
tied to the expert.

SpinPro required that the expert critically review the science
of ultracentrifugation and his knowledge of it. For example,
SpinPro sometimes designed a procedure using a rotor that was not

expected or recommended an exceptionally short run time that was shorter than thought possible. These procedures required careful review. Sometimes they were accepted as valid improvements to existing procedures. Isopycnic runs are one example, where SpinPro found that procedures typically requiring 12-16 hours, could be run for 7-9 hours with the same results. Thus, SpinPro is indirectly responsible for advancing the expert's understanding of ultracentrifugation and for improving ultracentrifugation techniques. SpinPro promoted a degree of rigorousness that had never before been applied to ultracentrifugation.

Updates to SpinPro continue as new rotors and new techniques are developed or as inadequacies are found. New expert systems techniques, such as the ability to incorporate the principles of a problem domain, rather than just the experience of the expert, should give SpinPro the ability to design procedures for novel problems and to explain its reasoning. The updates insure that SpinPro will be a repository of knowledge about the current state of ultracentrifugation; SpinPro's expertise should continue to improve. Furthermore, the expert remains gainfully employed as a final arbitrator on the inclusion or exclusion of any new knowledge.

Conclusion

The SpinPro Ultracentrifugation Expert System provides an integrated package of expert advice, information, and calculation functions. Its purpose is to allow investigators to fully exploit the capabilities of ultracentrifugation, thereby improving the efficiency of the ultracentrifugation laboratory. It uses AI techniques to provide the ability to advise on the best selection of run parameters that satisfy the investigator's requirements. Our experience with SpinPro has shown that it effectively performs the role of an expert advisor: designing efficient ultracentrifugation procedures that can reduce run times and improve the quality of separations.

Acknowledgments

For their contributions to the SpinPro Ultracentrifugation Expert System, the authors thank Gertrude Burguieres, Mike Brown, Phyllis Browning, Marsha Chase, Judy Cummings, Manny Gordon, Mary Jane MacDwyer, Edna Podhayny and Bruce Wintrode.

RECEIVED January 14, 1986

Elucidation of Structural Fragments by Computer-Assisted Interpretation of IR Spectra

Hugh B. Woodruff, Sterling A. Tomellini[1], and Graham M. Smith

Merck Sharp & Dohme Research Laboratories, Rahway, NJ 07065

Since its introduction to the scientific community in late 1980, PAIRS (Program for the Analysis of IR Spectra) has been used successfully by a large number of researchers. Recent improvements to PAIRS have made this package incorporate most of the aspects of expert systems. The improvement highlighted in this paper is the capability for scientists to inquire of the system why a particular interpretation result was achieved. This capability enhances the ability for scientists to learn from the knowledge base of interpretation rules present in PAIRS. It also simplifies the process by which the PAIRS knowledge base can be refined through incorporation of improved rules from expert spectroscopists.

One of the more interesting areas available for development in analytical spectroscopy is the generation of algorithms and software capable of interpreting IR spectra. A number of papers have been published recently on computerized interpretation of vibrational spectra (1-22). The generation of such software requires the analytical chemist to understand the interpretation process and be able to translate the process into an algorithm which the computer can perform. While generating the actual computer code is by no means trivial, the chemical knowledge required to solve the interpretation problem makes a chemist and not a computer scientist the likely producer of such a program.

Among the most widely distributed of these interpretation programs is a package called PAIRS Program for the Analysis of IR Spectra) which has been distributed by the authors and the Quantum Chemistry Program Exchange to nearly 100 researchers. The program is available in both IBM mainframe and DEC VAX versions. A simplified schematic of the information flow in PAIRS is shown in Figure 1. Spectral information in the form of a digitized IR spectrum including peak location, width and intensity values may be entered either

[1] Current address: University of New Hampshire, Durham, NH 03824

interactively or from a file created previously, perhaps with the
aid of a digitizing tablet.

Since the introduction of PAIRS in 1980, considerable effort
has been expended on improving various aspects of the package to
make it more valuable to researchers. A version of PAIRS capable of
running on a Nicolet FTIR instrument-based minicomputer was developed to eliminate the time required to digitize spectra and to
make the program available to the practicing analytical spectroscopist (12). Recently, versions of PAIRS capable of running on other
FTIR systems have been reported (23, 24).

Generating Interpretation Rules

The generation of interpretation rules for PAIRS has proven to be a
time-consuming and often inexact process. Many man-years were required to generate the first set of rules. Trulson and Munk (18)
emphasized the massive effort required for rule development in
their report on their promising work on a table-driven approach to
infrared spectral interpretation. Rule development and subsequent
testing are generally much more time consuming than either acquiring
test spectra or programming the interpretation routines.

One of the strengths of PAIRS is the accessibility of the interpretation rules in a form that is easily understandable and modi-
fiable by the scientist. To accomplish this feat, a special
English-like language known as CONCISE (Computer Oriented Notation
Concerning Infrared Spectral Evaluation) was developed (19).
CONCISE has a very small (62 words) and well-defined vocabulary
which can be mastered by non-computer-oriented scientists. It consists of if-then-else logic and begin-done blocking. Once the vocabulary and structure of CONCISE are known, the scientist is free to
create or change interpretation rules at will.

In order to expand the usefulness of the PAIRS package, an
automated rule generation program has been developed. An advantage
of automated rule generation is that a more mathematical and uniform
method of determining expectation values can be developed and used.
(An expectation value is a measure of the likelihood of occurrence
for the presence of a particular functionality in the unknown compound.) A detailed description of the algorithms used for the automated rule generator is presented elsewhere (21).

The simplicity and clarity of CONCISE has been retained in the
automated rule generator which creates CONCISE interpretation rules
for PAIRS based on a representative set of IR spectra. The rule
generator uses peak position, intensity, and width tables produced
by an automated peak picking routine. This method reduces the de-
pendency on published frequency correlation data and enhances the
usefulness of data already available. All work was done using the
version of PAIRS running on a Nicolet 1180 minicomputer and programs
generated have been optimized for this system.

CONCISE rules are generated based on the frequency of occurrence of peaks in compounds in a spectral database. Good interpretation rules have been created using a relatively small number of
spectra in the database. To recreate interpretation rules for the
168 classes of compounds currently addressed by PAIRS in an automated manner would require a substantial effort and a better

spectral database than currently exists. However, the automated
rule generator provides the tools to accomplish this task and to
expand the current rule base.

Tracing Interpretation Rules

The discussion thus far has centered on input to the interpreter;
however, the scientist is perhaps most interested in the information returned by PAIRS. The results were previously limited to a
numerical indication of the likelihood that any particular functionality or sub-functionality is present. While the rules upon which
interpretations are based are available in an English-like language,
CONCISE, it is normally a rather difficult process to determine why
a given functionality was assigned a given value. The usefulness of
PAIRS would be greatly enhanced, especially as a research tool, if
the program were able to provide the user with a clear trace of the
decision making process. Very recent efforts have resulted in an
improved version of PAIRS which not only allows the user to question
which functionalities may be present, but also why they are thought
to be present (22).
Major changes and additions were required to make PAIRS capable
of providing an easily understandable trace of the interpretation.
The interpreter required the vast majority of these modifications,
including the addition of a number of new subroutines. Full use was
made of the decompiling features already present in the interpreter.
Therefore, input-output and decision controlling routines make up
the majority of the subroutines added. The decision controlling
routines actually serve a dual purpose. Not only do these routines
decide which data should be printed during a trace, but they also
keep track of the progress of the interpreter as it makes its way
through the interpretation rules. Thus, the controlling routines
know at any given moment which rules have already been interpreted
and which rules remain to be interpreted. The rule compiler was
modified to create a file containing the "header" names, which are
the names of the major functionalities. The CONCISE interpretation
rules were not changed during this process. Now the user is pres-
sented with three options for interpreting a spectrum: 1.) trace
the decision making process for all functionalities; 2.) trace the
decision making process for any of the major functionalities (e.g.,
acid) and its corresponding sub-functionalities (e.g., acid-satur-
ated); or 3.) interpret the spectrum without any tracing as was
done previous to the modifications described in this paper. In any
case, an entire interpretation takes place and, therefore, a numerical indication is available for the likelihood that each functionality and sub-functionality is present.
It is important to remember that the if-then-else logic of the
CONCISE language forces the interpreter to follow one unique path
through the interpretation rules, a path dictated by the spectral
data entered. A very important consequence of being able to follow
only one path is that a trace of the decision making process can
give information about what decisions were made but cannot give any
information about what decisions might have been made had the spectral data been different. Knowing what decisions were made can,
however, give a good indication why a given functionality might have
been reported at a lower value than expected.

The best way to demonstrate the added capability and increased versatility of the interpreter due to the tracing feature is through example. Since the interpreter generally bases a good deal of importance on peak intensity information, it is obvious that mixtures and larger molecular weight compounds will often cause the interpreter to return less than desirable results. In cases where the intensities are lower than would normally be expected for a given functionality, a valuable feature of the modified program is the ability to see quickly what decisions have been made and why these decisions were made.

The antibiotic actinospectacin, the structure of which is given below, was chosen to demonstrate the improved interpreter. A spectrum of actinospectacin (published in Volume 10 of the Bulletin of the International Center of Information on Antibiotics) was digitized with the resulting peak data being presented in Table I. The peak data in Table I were entered into the interpreter without any empirical formula information. The sample state entered reflected the fact that the spectrum was taken as a KBr pellet. Table II contains the twenty functionalities and sub-functionalities which the interpreter predicted as most likely to be present in the sample. (The *1*, *2*, and *3* terminology indicated one, two, or three occurrences, respectively, of alpha branching or unsaturation in the alcohol.) Previously this information was essentially all that the user could learn using PAIRS without investing the time necessary to decipher the CONCISE rules for the functionalities in question. The improved version of PAIRS, however, allows the user to ask, for example, "Why was "sulfone" indicated with such a high expectation value?". If the data in Table I are reinterpreted with the decision process for the functionality "sulfone" being traced, the user learns that the high likelihood for a "sulfone" is due to the presence of the 1330 and 1351 cm^{-1} bands of intensity 7 and 6, respectively, the 1121 and 1145 cm^{-1} bands of intensity 7 and 9, respectively, and the presence of more than two bands between 1090 and 1170 cm^{-1} with intensities greater than 7. The actual decision trace is given in Figure 2. Should the user suspect that these bands are due to another functionality, knowledge of how these bands were used in predicting the presence of a "sulfone" may allow the interpreter's prediction of a high likelihood of "sulfone" to be less highly regarded.

Conversely, one may suspect the presence of a particular functionality but discover that the interpreter predicts that functionality with a low expectation value. Knowing the structure of actinospectacin, one would expect that "ketone" should be predicted to be present with a fairly high expectation value. The interpreter, however, returns a value of 0.01 for the likelihood of presence of the "ketone". In this case, the user learns that the low expectation value for "ketone" was based on the absence of any peak with intensity 7 or greater in the carbonyl region between 1571 and 1800 cm^{-1}. In the case of an unknown compound, knowledge of the interpreter's decisions can give the user added insights and ideas, especially when the spectrum is not ideal for a given functionality. The user, in any case, now has the ability to work with the program to see if minor variations in the data would result in different and possibly more reasonable interpretations.

```
┌──────────┐
│ Digitized │
│ Spectrum  │────────┐
└──────────┘        \        ┌──────────────┐        ╭────────────╮
                      \       │    PAIRS     │        │  Chemical   │
                       ┤      │ (Interpreter)│────────│ Functionality│
                      /       └──────────────┘        │ Predictions │
┌──────────┐        /                                 ╰────────────╯
│ CONCISE   │───────┘
│ Rules     │
└──────────┘
```

Figure 1.   Information flow in PAIRS.

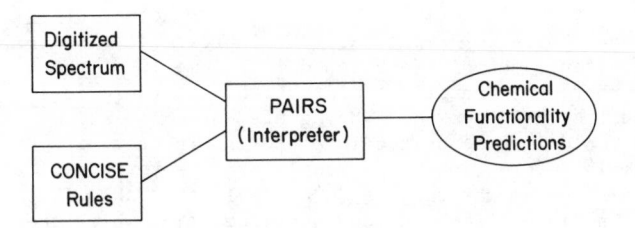

Structure of the antibiotic actinospectacin.

Table I.   Digitized Actinospectacin Spectrum

Peak No.	Position (cm^{-1})	Relative Intensity	Width
1	3527	9	Broad
2	3401	10	Broad
3	3311	10	Broad
4	3254	10	Broad
5	3071	9	Broad
6	2962	9	Average
7	2796	6	Broad
8	2486	2	Average
9	1645	5	Average
10	1629	5	Average
11	1581	4	Average
12	1566	4	Average
13	1460	8	Average
14	1429	6	Average
15	1392	8	Average
16	1351	6	Sharp
17	1330	7	Average
18	1271	2	Average
19	1235	3	Average
20	1215	3	Average
21	1190	6	Average
22	1176	7	Sharp
23	1145	9	Average
24	1121	7	Average
25	1107	8	Average
26	1087	9	Average
27	1078	10	Average
28	1046	9	Average
29	1037	9	Average
30	1024	9	Sharp
31	999	7	Sharp
32	981	3	Average
33	952	4	Sharp
34	936	4	Sharp
35	923	7	Average
36	891	2	Average
37	875	3	Average
38	859	5	Average
39	814	3	Average
40	728	5	Average

Table II. PAIRS Interpretation Results for Actinospectacin

	FUNCTIONALITY	EXPECTATION VALUE
1	ALCOHOL	0.99
2	SULFONE	0.85
3	OLEFIN-(NON-AROM)	0.75
4	OLEFIN-CHR=CH2	0.75
5	ALCOHOL-PHENOL	0.75
6	ALCOHOL-PRIM(*1*)	0.75
7	ALCOHOL-PRIMARY	0.75
8	ALCOHOL-SEC-(*1*)	0.75
9	ALCOHOL-SEC-(*2*)	0.75
10	ALCOHOL-SEC-RING	0.75
11	ALCOHOL-SECONDARY	0.75
12	ALCOHOL-TERT-(*1*)	0.75
13	ALCOHOL-TERT-(*2*)	0.75
14	ALCOHOL-TERT-(*3*)	0.75
15	ALCOHOL-TERT-RING	0.75
16	ALCOHOL-TERTIARY	0.75
17	SULFONAMIDE	0.75
18	SULFONAMIDE-PRIM	0.75
19	SULFONAMIDE-SEC	0.75
20	SULFONAMIDE-TERT	0.75

```
FUNCTIONALITY SULFONE
PASSED INITIAL EMPIRICAL FORMULA TEST
PEAK QUERY
  ANY PEAK(S)  POSITION:  1290 - 1360
  INTENSITY:  7 - 10  WIDTH:  SHARP TO AVERAGE
ANSWER---------YES---------
PEAK QUERY
  ANY PEAK(S)  POSITION:  1110 - 1170
  INTENSITY:  7 - 10  WIDTH:  SHARP TO BROAD
ANSWER---------YES---------
ACTION---------SET SULFONE TO 0.500
  CURRENT VALUE=0.500
PEAK QUERY
  AT LEAST 2 PEAK(S)  POSITION:  1260 - 1360
  INTENSITY:  4 - 10  WIDTH:  SHARP TO AVERAGE
ANSWER---------YES---------
ACTION---------ADD 0.100 TO SULFONE
  CURRENT VALUE=0.600
PEAK QUERY
  AT LEAST 2 PEAK(S)  POSITION:  1260 - 1360
  INTENSITY:  7 - 10 WIDTH:  SHARP TO AVERAGE
ANSWER---------NO---------
PEAK QUERY
  AT LEAST 2 PEAK(S)  POSITION:  1065 - 1170
  INTENSITY:  4 - 10  WIDTH:  SHARP TO AVERAGE
ANSWER---------YES---------
ACTION---------ADD 0.100 TO SULFONE
  CURRENT VALUE=0.700
PEAK QUERY
  AT LEAST 2 PEAK(S)  POSITION:  1065 - 1170
  INTENSITY:  7 - 10  WIDTH:  SHARP TO AVERAGE
ANSWER---------YES---------
ACTION---------ADD 0.150 TO SULFONE
  CURRENT VALUE=0.850
```

Figure 2. Trace of sulfone functionality during interpretation
of actinospectacin.

This point is illustrated by a second example. A vapor-phase spectrum of propionitrile was obtained and its digitization is shown in Table III. For the sake of example, assume the scientist entered the 2246 cm^{-1} peak as average rather than sharp. The interpretation would result in likelihoods of 0.90 for isocyanate and 0.30 for nitrile. Performing the interpretation with the tracing function turned on would quickly show that the rules base the distinction between isocyanate and propionitrile very heavily on the width of the peak in the vicinity of 2260 cm^{-1}. Reinterpreting this spectrum with the correct, sharp width entered for the 2246 cm^{-1} peak results in a nitrile likelihood of 0.50 and isocyanate of 0.40.

Table III. Digitized Propionitrile Spectrum

Peak No.	Position (cm^{-1})	Relative Intensity	Width
1	2246	10	Average
2	2996	8	Average
3	1461	7	Average
4	2950	6	Average
5	1431	5	Average
6	1074	4	Average
7	787	3	Average
8	2892	3	Average
9	1319	2	Average
10	1386	1	Average
11	546	1	Average

Summary

Through the addition of automated spectrum input on instrument-based computers, automated rule generation, and automatic tracing of decision rules, PAIRS has been enhanced to be an even more valuable tool for the spectroscopist. PAIRS is available for distribution from the Quantum Chemistry Program Exchange, Indiana University, Bloomington, IN 47405 (Catalog No. QCPE 497).

Literature Cited

1. Gray, N.A.B. Anal. Chem. 1975, 47, 2426.
2. Woodruff, H.B.; Munk, M.E. J. Org. Chem. 1977, 42, 1761.
3. Woodruff, H.B.; Munk, M.E. Anal. Chim. Acta 1977, 95, 13.
4. Zupan, J. Anal. Chim. Acta 1978, 103, 273.
5. Visser, T.; Van der Maas, J.H. J. Raman Spectros. 1978, 7, 125.
6. Visser, T.; Van der Maas, J.H. J. Raman Spectros. 1978, 7, 278.
7. Leupold, W-R; Domingo, C.; Niggemann, W.; Schrader, B. Fresenius' Z. Anal. Chem. 1980, 303, 337.
8. Woodruff, H.B.; Smith, G.M. Anal. Chem. 1980, 52, 2321.
9. Visser, T.; Van der Maas, J.H. Anal. Chim. Acta 1980, 122, 337.
10. Varmuza, K. "Pattern Recognition in Chemistry"; Springer-Verlag; New York, 1980, No. 2, Lecture Notes in Chemistry Series
11. Woodruff, H.B.; Smith, G.M. Anal. Chim. Acta 1981, 133, 545.
12. Tomellini, S.A.; Saperstein, D.D.; Stevenson, J.M.; Smith G.M.; Woodruff, H.B.; Seelig, P.F. Anal. Chem. 1981, 53, 2367.

13. Farkas, M.; Markos, J.; Szepesvary, P.; Bartha, I.; Szalontai,
    G.; Simon, Z. Anal. Chim. Acta/Computer Techniques and Optimiza-
    tion 1981, 133, 19.
14. Szalontai, G.; Simon, Z.; Csapo, Z.; Farkas, M.; Pfeifer, Gy.
    Anal. Chim. Acta/Computer Techniques and Optimization 1981, 133
    303.
15. Debska, B.; Duliban, J.; Guzowska-Swider, B.; Hippe, Z. Anal.
    Chim. Acta/Computer Techniques and Optimization 1981, 133, 303.
16. Frank, I.E.; Kowalski, B.R. Anal. Chem. 1982, 54, 232R.
17. Zupan, J. Anal. Chim. Acta 1982, 139, 143.
18. Trulson, M.O.; Munk, M.E. Anal. Chem. 1983, 55, 2137.
19. Smith, G.M.; Woodruff, H.B. J. Chem. Inf. Comp. Sci. 1984, 24,
    33.
20. Tomellini, S.A.; Stevenson, J.M.; Woodruff, H.B. Anal. Chem.
    1984, 56, 67
21. Tomellini, S.A.; Hartwick, R.A.; Stevenson, J.A.; Woodruff, H.B.
    Anal. Chim. Acta 1984, 162, 227.
22. Tomellini, S.A.; Hartwick, R.A.; Woodruff, H.B. Appl. Spectrosc.
    1985, 39, 331.
23. Saperstein, D.D.; "A Scheme For Optimized Infrared Interpreta-
    tions", paper # 216, 1985. Pittsburgh Conference & Exposition on
    Analytical Chemistry and Applied Spectroscopy, Feb. 25-March 1,
    1985.
24. DeHaseth, J.A.; Mir, K.A., "A Minicomputer Based Structure Eluci-
    dation Program", paper # 217, 1985. Pittsburgh Conference &
    Exposition on Analytical Chemistry and Applied Spectroscopy,
    Feb. 25-March 1, 1985.

RECEIVED December 17, 1985

Automation of Structure Elucidation from Mass Spectrometry–Mass Spectrometry Data

K. P. Cross[1], P. T. Palmer, C. F. Beckner[2], A. B. Giordani[3], H. G. Gregg[4], P. A. Hoffman[5], and C. G. Enke

Department of Chemistry, Michigan State University, East Lansing, MI 48824

A system has been designed to automate the extraction of structural information from mass spectrometry/mass spectrometry (MS/MS) spectra. Currently operational elements in this system include data bases for MS/MS spectra and molecular structures, spectrum matching programs, and a structure generator. Individual spectra within the complete set of MS/MS spectra are related to the molecular substructures from which they arise. The correlations between individual MS/MS spectra and specific substructures can be determined by identifying the compounds that have matching MS/MS spectra, and then identifying the substructures they have in common. These correlations can supply identified substructures to a molecular structure generator such as GENOA. This empirical scheme assumes no knowledge of the fragmentation process, ion structures, or rearrangements.

The development of mass spectrometry/mass spectrometry (MS/MS) has provided the chemical analyst with a powerful tool for structure elucidation. The primary goal of this project is to develop the full capacity of triple quadrupole mass spectrometry (TQMS) as a tool for routine structure determination. To accomplish this, we have designed and developed computer data bases for spectra and structures (1,2), programs for matching spectra (3), and procedures

[1]Current address: Chemical Abstracts Service, Columbus, OH 43210
[2]Current address: Finnigan MAT, San Jose, CA 95134
[3]Current address: Department of Psychiatry, Mt. Sinai School of Medicine and Bronx Veterans' Administration Medical Center, New York, NY 10029
[4]Current address: Lawrence Livermore National Laboratory, University of California, Livermore, CA 94550
[5]Current address: Lederle Laboratories, American Cyanamid Corporation, Pearl River, NY 10965

0097–6156/86/0306–0321$06.00/0
© 1986 American Chemical Society

for determining spectrum/ substructure correlations. These tools were designed for integration into a complete system for on-line structure determination by MS/MS.

Structure analysis by MS/MS differs from normal MS in that each of the fragment ions from the sample ionization process in the source can be selected, one mass at a time, for further fragmentation and subsequent mass analysis. The ion in the normal mass spectrum selected for analysis is called a parent ion. The fragments of that ion, generally produced by collision-induced dissociation (CID) are called daughters. A mass spectrum of all the daughters of a particular parent ion (called a daughter spectrum) is obtained by holding the first mass analyzer constant at the mass of the selected parent ion and scanning the second mass analyzer. A complete MS/MS spectrum is a three-dimensional array in which there is a daughter spectrum for every mass represented in the normal mass spectrum.

MS/MS data are very explicit; daughter spectra may reveal structural characteristics of isolated portions of the molecule (4), and under certain conditions, all masses in a daughter spectrum are single-event neutral losses from the parent ion. Thus, clear substructure/property relationships can be obtained from MS/MS spectra. These relationships can be used to identify substructures in unknown compounds. Possible compound structures can then be developed from the identified substructures. This approach should facilitate the identification of unknown compounds not previously studied by mass spectrometry.

Data from the TQMS instrument are used in two different ways: 1) to develop a library of spectrum/substructure correlations from studies of known compounds and 2) to use the developed correlations to determine the substructures and thence the overall structures of unknown compounds. The data base required for this process is a library of the spectral characteristics of many substructures, rather than a library of the spectra of all known compounds. In principle, millions of compounds could be identified using a library of only a few thousand spectrum/substructure relationships.

A block diagram of our target system for the automatic elucidation of molecular structure is shown in Figure 1 (5). While the system is not yet complete, the three data bases and a spectrum matching program have been developed and integrated into a comprehensive system to acquire, store, match, and correlate the MS/MS data. Descriptions of their structures and capabilities and examples of their application are included in this paper. Also a molecular structure generator, GENOA (6), has been acquired and implemented, but is not yet integrated into the system. An example of the determination of spectrum/substructure correlations and their application in structure determination through GENOA is also given here.

The flow of data through the system shown in Figure 1 depends on whether the experimental data are from a reference compound for the development of the library or from an unknown compound for analysis. Reference compound spectra are collected in the experimenter's data base and may be archived in the reference data base. They can also be matched against other spectra from other reference compounds by the spectrum matching program. When a match is found indicating that the two compounds have produced an

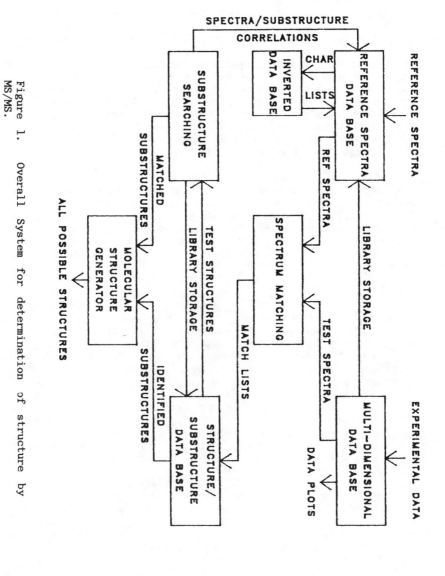

Figure 1. Overall System for determination of structure by MS/MS.

identical ion structure, the molecular structures are compared by
the substructure searching function to determine the substructure(s)
they have in common. These common substructures are candidate
precursors of the common ion. Through rearrangements, it is
possible for more than one substructure to produce a particular ion.
Additional compounds with matching spectra or substructures are
studied until clear spectrum/substructure correlations are produced.
Once the correlations are made, the substructure(s) associated with
a particular spectrum are stored in the structure/substructure data
base, and are logically linked to that spectrum.

The spectra from an unknown compound are matched against the
reference spectra to produce a list of the substructures that are
related to the matched spectra. When this substructural information
has been extracted from the MS/MS spectra, it is entered into the
molecular structure generator called GENOA (6). GENOA, which is
constrained by heuristic chemical rules, uses all available
composition and structure information, including overlapping and
nonunique substructures, to postulate the number and identity of all
possible molecular structures of the unknown compound. If the
resolution of any remaining structural ambiguities is essential to
the experiment, additional information derived from MS/MS or other
sources is fed to GENOA to further reduce the number of output
structures. This structure elucidation scheme combines an
exhaustive and automatic algorithm for the evaluation of the
structural possibilities, the experimenter's chemical intuition, and
the knowledge base of the experimentally determined
spectrum/substructure correlations.

Data Bases

There are three data bases present in our MS/MS information
management system, one for immediate experimental data and two for
archival data. The experimenter's data base has been described
elsewhere (1). One archival data base manages the MS/MS spectra,
while the other manages the structures and substructures. The two
archival data bases are logically linked together so that all
information concerning a particular molecule or substructure is
associated with its spectra.

The MS/MS spectrum data base is capable of storing and
correlating all types of MS/MS spectral data including parent,
daughter, neutral loss, and conventional mass spectra (2). All
spectra are stored in an unabridged format and all spectra for each
compound are logically associated with that compound. Redundant
spectra such as those taken under different operating conditions are
all associated with a single compound registry number thereby
simplifying both the retrieval and maintenance of the data base
information.

The most important feature of the reference spectrum data base
is the provision to generate and store inverted data (data that are
presorted on various secondary elements of the record). The data
present in the spectrum data base may be inverted upon any specified
characteristic, such as m/z value, and then be retrieved using that
characteristic. For instance, a data file inverted about the
daughter m/z value will contain, for each m/z value, a list of
pointers to the reference daughter spectra that have a peak at that

mass.   Hence the pointers to all reference spectra containing a particular m/z value may be very quickly retrieved.   When Boolean algebra operations are performed on inverted data lists, the power of the design increases dramatically.   A prescreen for all reference daughter spectra containing the major features of a test spectrum such as peaks at 43.0 and 57.0 but not 119.0 reduces greatly the number of reference spectra that need to be matched in greater detail.   In addition to a daughter m/z value, spectral data may be inverted about molecular weight, empirical formula, and parent ion m/z value.   Over 30,000 primary spectra and other information are currently stored in the spectrum data base as well as MS/MS spectra corresponding to several specific classes of compounds.

The structure data base was designed to contain both molecular structures and substructures (7).   The MS/MS instrument specifically provides a substructure/property relationship where several daughter spectra may correspond to a single substructure and any daughter spectrum may correspond to more than one substructure.   Even though a simple 1:1 correspondence between daughter spectra and substructures cannot be assumed, there is a basis for a logical link between the MS/MS spectra in the spectral data base and the respective substructures in the structure data base.   This link allows retrieval of structural information from the reference daughter spectra best matching the unknown spectrum.   Structures present in the structure data base may be retrieved via substructure number, Chemical Abstracts Service number, or spectrum data base number, and then drawn.

The structures and substructures are stored unambiguously using a modified version of the Morgan algorithm for encoding molecular structures via connectivity tables.   The version of the algorithm implemented included the modifications described by Wipke and Dyott (8) for the representation of stereochemical isomers.   The notation of the elements was expanded to include all known elements.   Any molecule up to 128 atoms in size (excluding hydrogens) may be included in the data base.   The structure data base contains over 30,000 structures corresponding to the spectra present in the MS/MS spectrum library as well as substructures corresponding to various reference daughter spectra.

Matching MS/MS Spectra

The MS/MS spectra matching program allows the chemist to match any MS/MS spectrum against either MS or MS/MS spectra in the reference spectrum data base (3).   The program uses inverted data organized by m/z value to logically eliminate inappropriate reference spectra. The program first determines the data base frequency (length of the pointer table) of each major peak in the experimental daughter spectrum and then ranks the peaks in ascending order of frequency. Inverted data lists of reference spectra containing peaks are retrieved in this order and logically ANDed together until the number of candidate reference spectra is sufficiently small. Additional reductions in the number of candidate spectra is possible by using molecular weight, parent ion m/z value, and empirical formula may also be invoked to further reduce the number of candidate spectra.   When matching daughter spectra, specifying the parent ion m/z value alone usually produces a sufficiently small number of candidate spectra. Abundance values are not considered and the reference data base is not accessed until intensity-based matching is performed. The short matching times achieved with this design makes it practical to work with unabridged spectra.

Once the number of candidate reference spectra has been reduced to reasonable size (25-100), intensity-based matching is performed to characterize the correspondence between the experimental and remaining candidate spectra. Several different factors indicating the degree to which the spectra match in various respects are determined. The values of these match factors are used to distinguish spectra that arise from identical substructures from those that arise from different substructures.

The various match factors calculated by the matching program are listed in Table I. The overall match factor (PT) is a combination of forward and reverse searching techniques. It takes into account the deviations in intensity of the sample spectrum peaks with respect to the candidate spectrum peaks and vice versa for all peaks in both spectra. The pattern correspondence match factor (PC) is a forward searching match factor which takes into account the intensity deviations of sample spectrum peaks with respect to the candidate spectrum peaks for peaks common to both spectra. This factor detects structural similarities, such as substructures, based on common spectral patterns. NC, NS, and NR give an indication of the number of peaks upon which the match was based and in which direction it was most successful. IS and IR indicate the magnitude of the ion current unmatched in each direction. These match factors are similar to those proposed by Damen, Henneberg, and Wiemann (9).

Because instrument operating conditions can seriously affect the relative intensities of ions in daughter spectra, there was a need to know the range of conditions over which the daughter spectra of identical parent ions could be distinguished from all other daughter spectra. Daughter spectra were collected for several compounds for every combination of a wide range of operating parameters. An acceptable range of standard conditions was defined as that over which the spectrum matching system would provide high match factors for daughter spectra of the same compound.

Of the 32 instrumental parameters on our TQMS, only the collision energy and collision cell pressure were found to significantly affect MS/MS spectra. The acceptable range of collision cell pressure was that found to yield first order fragmentation regardless of the compound type. Since different collision cell pressures are required to obtain first order fragmentation for different compounds, brief kinetic studies are used to determine the fragmentation order, and to ascertain the pressure necessary to provide first order fragmentation. Similarly, we have determined a useable operating range for the collision energy of 15 to 25 eV.

Spectrum/Substructure Relationships

The procedure for obtaining the spectrum/substructure relationships is as follows. For a selected known compound, a daughter spectrum is acquired for every mass value greater than 1% relative intensity that appears in the primary spectrum of that compound. These

Table I.   Match Factor Definitions

PT   An overall match factor that indicates how well the
      intensities of all the peaks in the two spectra match.

      $PT = (\Sigma\ Ys + Yr - 2*\ |Yr - Ys|)\ /\ (\Sigma\ Ys + \Sigma\ Yr)\ *\ 100$

      where $Yi = log_2$ (Intensity/Total Ion Count)

      Ys and Yr correspond to the adjusted abundances at each mass
      in the sample and reference spectra respectively

PC   A pattern correspondence factor that indicates how well the
      intensity of the peaks in common match.

      $PC = (\Sigma\ Ys - |Yr - Ys|)\ /\ (\Sigma\ Ys)\ *\ 100$

NC   The number of peaks common to both the candidate and unknown
      sample spectrum.

NS   The number of peaks remaining unmatched in the unknown
      sample spectrum.

NR   The number of peaks remaining unmatched in the reference
      spectrum.

IS   The percent total ion current of the sample spectrum that
      was unmatched in the comparison due to NS.

IR   The percent total ion current of the reference spectrum that
      was unmatched in the comparison due to NR.

daughter spectra are then matched against a library of daughter spectra from reference compounds.

After the spectral matching process has been completed, the list of compounds with the top matching daughter spectra are identified and retrieved for each daughter spectrum in the reference compound. The molecular structures of the compounds with best matching spectra are drawn and compared for common substructures. The common substructures yield candidate spectrum/substructure correlations. Additional compounds are then tested to confirm or modify each correlation. Once the daughter spectrum is correlated with one or more substructures, this daughter spectrum is stored in the spectrum data base and is linked to the associated substructures stored in the structure data base.

An heuristic program written by Shelley (10) has been adapted for our computer system to display molecular structures and substructures from connectivity tables. Since the molecular structure and substructure representations are stored in a unique, irredundant form, the structure drawings facilitate visual comparison for commonalities.

An example of the spectrum/substructure determination process is illustrated for the reference compound di-n-octylphthalate. Daughter spectra were acquired for every major ion (above 1% relative intensity) that appeared in the conventional mass spectrum (Figure 2) of the reference compound. All the daughter spectra were then matched against the reference daughter spectra of the same parent mass (but from different compounds) in the data base. The results of some of the matches are described below.

The match of the 105+ daughter spectrum of di-n-octylphthalate against the reference library of m/z 105 daughter spectra is presented in Table II. The top four matching spectra all correspond to structure III in Figure 3. Some of the spectra used in this match are shown in Figure 4. Note that the top four matching daughter spectra are very similar; all three contain the same peaks, only the intensity patterns are different (NR, NS, IS, and IR for the three are all zero). There is a large difference in overall match factor values (PT) between daughter spectra representing the correct substructure and that of the next best match.

Table II. Match of 105+ Daughter Spectra vs. Di-n-octylphthalate

PT	PC	NC	NS	NR	IS	IR	Compound
100	100	2	0	0	0	0	Di-n-octylphthate
99	99	2	0	0	0	0	Di-n-pentylphthalate
98	98	2	0	0	0	0	Di-n-butylphthalate
98	98	2	0	0	0	0	Di-n-ethylphthalate
66	93	2	0	2	0	31	4-t-butyl-1,2-benzenediol
60	85	2	0	2	0	20	2-t-butyl-4-methylphenol
38	50	1	1	3	42	29	p-t-butylbenzyl alcohol
36	50	1	1	3	42	52	2-t-butyl-6-methylphenol

The results of the match of the m/z 149 daughter spectrum of di-n-octylphthalate against m/z 149 daughter spectra from other compounds in the reference library is given in Table III. The

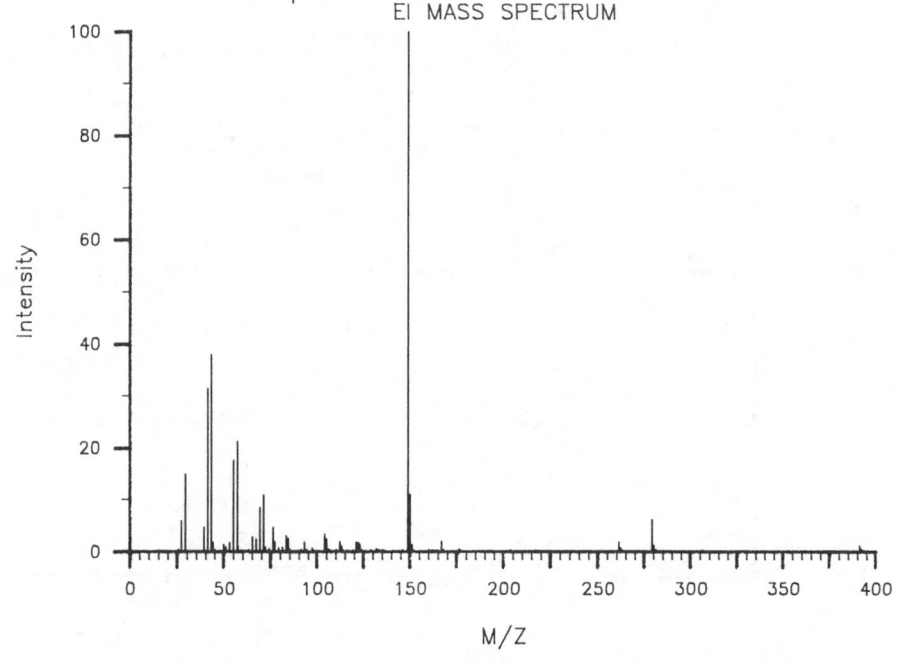

Figure 2. Normal EI mass spectrum of di-n-octylphthalate.

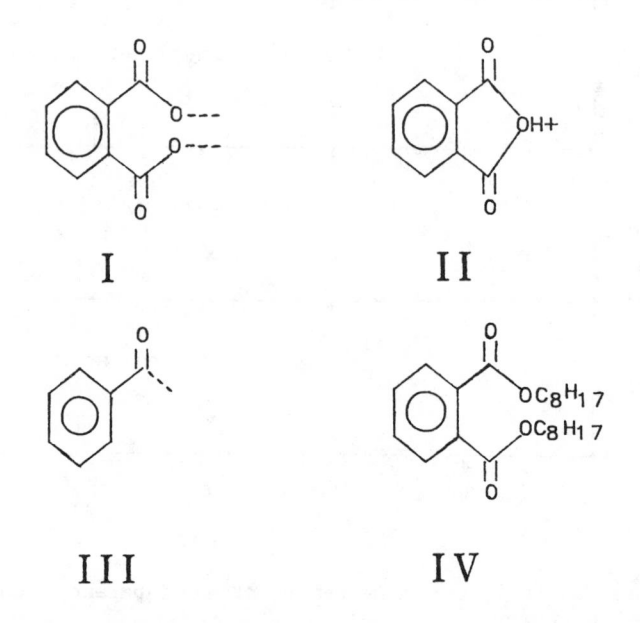

Figure 3. Substructure (I and III), ionic structure (II), and molecular structure (IV) produced by structure drawing program.

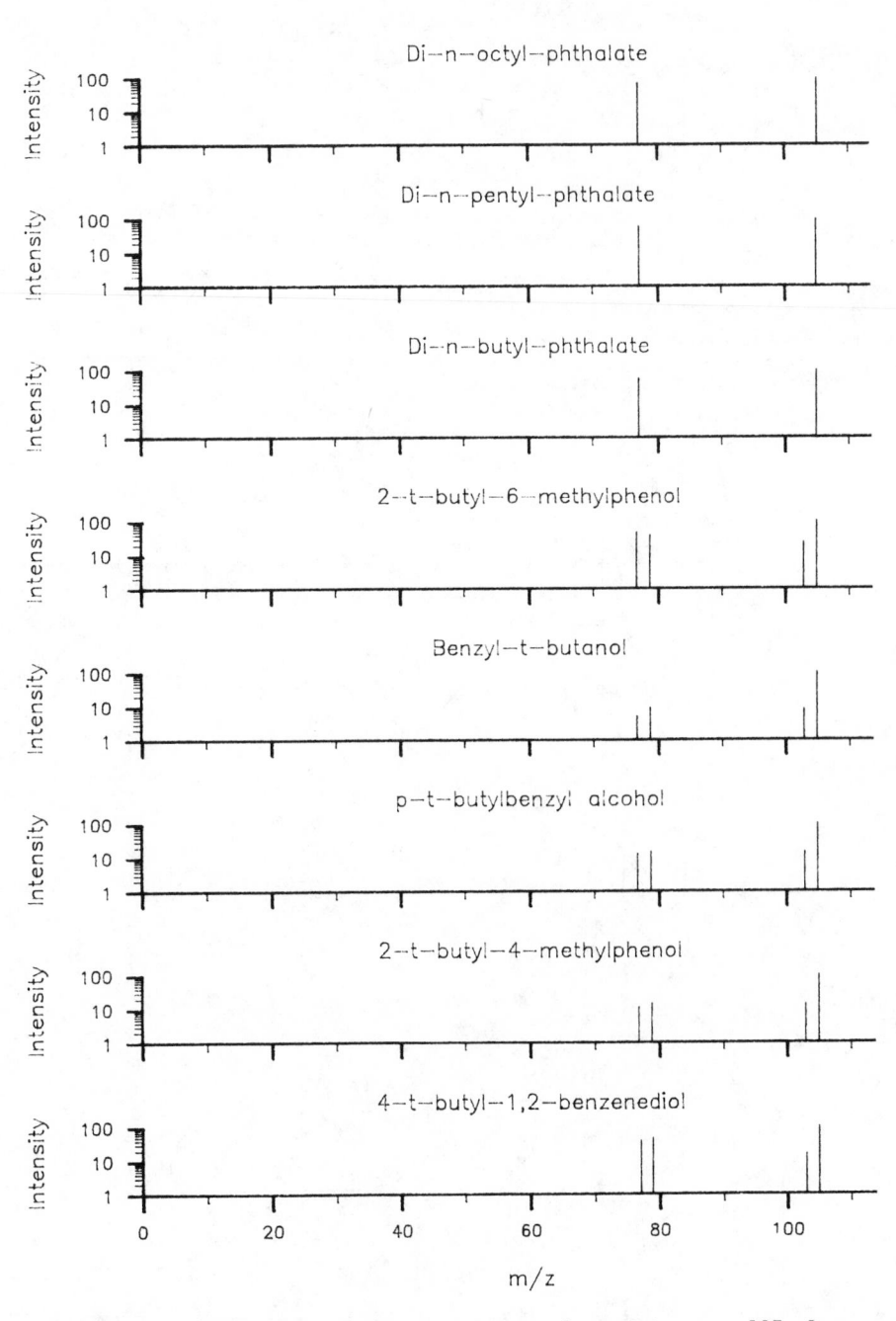

Figure 4.   Selected daughter spectra of parent mass 105 from reference library.

daughter spectra utilized in the matching process are shown in Figure 5. The top four matching spectra all correspond to the same molecular substructure, namely the phthalate substructure (structure I in Figure 3). At this point, it is necessary to make a distinction between a substructure and an ionic structure. The substructure correlated with the top four matching spectra is structure II in Figure 3 whereas the ionic structure of the parent ion m/z 149 is structure II in Figure 3. It is not necessary to know the ionic structure for this empirical approach.

Table III. Match of 149+ Daughter Spectra vs. Di-n-octylphthalate

PT	PC	NC	NS	NR	IS	IR	Compound
100	100	4	0	0	0	0	Di-n-octylphthalate
96	96	4	0	0	0	0	Di-n-butylphthalate
87	86	4	0	0	0	0	Di-n-pentylphthalate
87	86	4	0	0	0	0	Di-n-ethylphthalate
54	57	3	1	7	3	2	2-t-butyl-4-methylphenol
44	56	3	1	10	9	15	p-t-Butylbenzyl alcohol
42	35	1	3	1	19	29	p-t-amylphenol
35	61	3	1	10	3	26	2-t-butyl-6-methylphenol

The compounds yielding the top four daughter spectra are di-n-octylphthalate, di-n-butylphthalate, di-n-pentylphthalate, and diethylphthalate. Once again, only the relative intensities differ between these daughter spectra. It is important that these spectra are properly grouped by the spectrum matching program and that there is a substantial difference between the overall match factors of the matched spectra and those corresponding to unrelated substructures. The difference between the overall match factor of the unknown and the best matching daughter spectra corresponding to a different substructure is 46. Since the overall match factor range is 0-100 and the variance within the similar daughter spectra is 13, a value of 46 represents a good separation. The next best matching daughter spectrum outside of this group of three corresponds to a substructure of 2-t-butyl-4-methylphenol.

From the daughter spectra of di-n-octylphthalate, we were able to determine two spectrum/substructure correlations; the 149+ daughter spectrum to structure I in Figure 3 and the 105+ daughter spectrum to structure III in Figure 3. In order to obtain spectrum substructure relationships for the alkyl portions of the reference molecule di-n-octylphthalate, we would then match other portions of the complete MS/MS spectrum against those of compounds containing alkyl substructures. However, this portion of the reference library has not yet been developed. Thus, to complete the structure elucidation we have used standard methods of spectral interpretation (11). As will be shown, these methods can also lead to useful spectrum/substructure relationships.

Since the substructure represented by the daughter spectrum of the m/z 149 ion was the largest identifiable substructure in the compound, the parent spectrum of m/z 149 was used to obtain data related to the R groups attached to the phthalate substructure (Figure 6). The largest ion (149+) associated with the phthalate

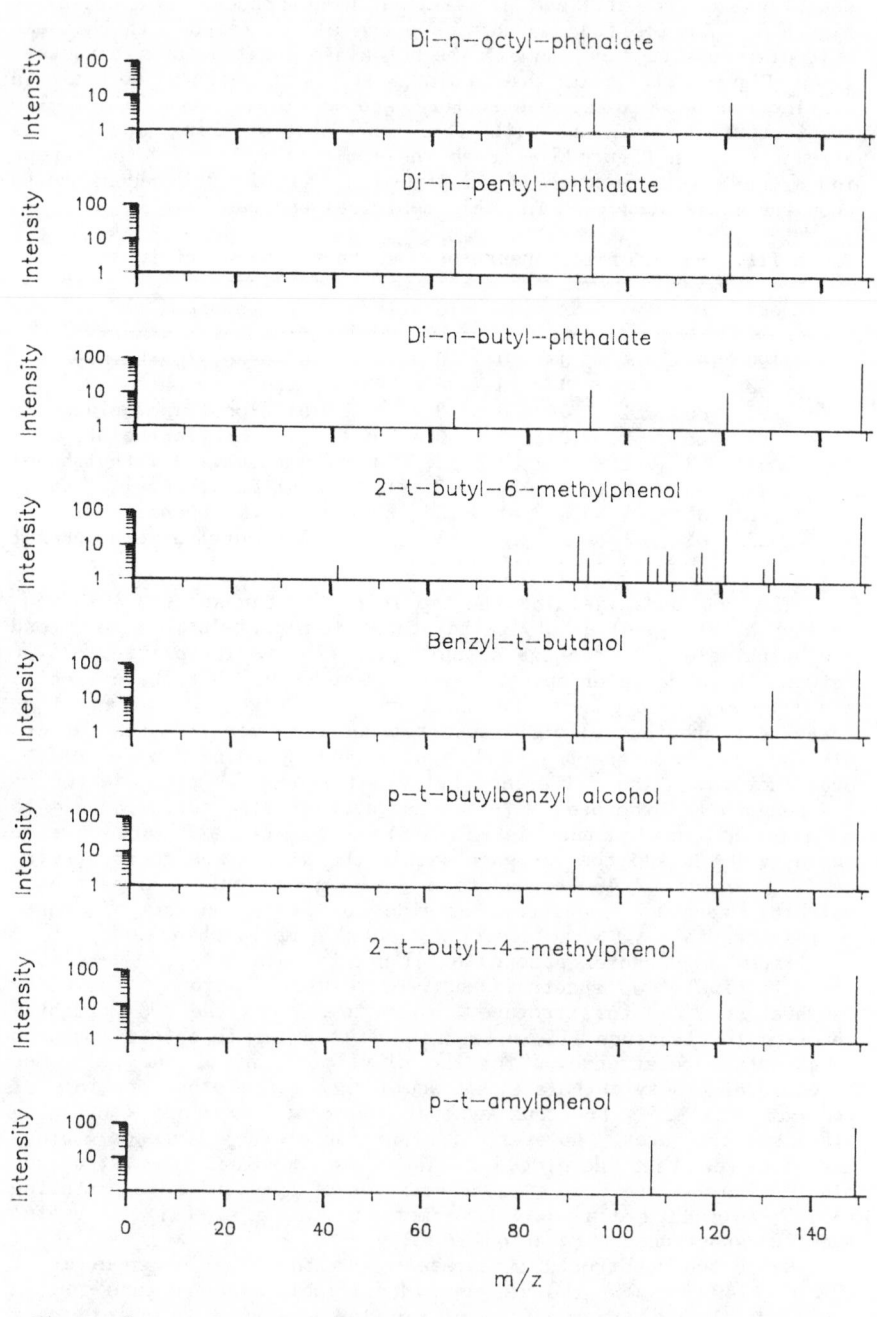

Figure 5.   Selected daughter spectra of parent mass 149 from reference library.

substructure was used since the neutral losses leading to its formation correspond to the groups attached to the phthalate substructure. This parent spectrum need not be acquired from the TQMS directly, since it can be generated from the set of daughter spectra for the unknown. The parent spectrum of m/z 149 has 4 major (non-isotopic) peaks at m/z 167, 261, 279, and 391. This corresponds to neutral losses of 18 (167–149) 112 (261–149), 130 (279–149), and 242 (391–149). The neutral loss of 112 is C_8H_{16}, the loss of 130 is $C_8H_{17}OH$ (which may represent an alkyl group) and the loss of 242 is $C_8H_{17}OC_8H_{17}$ (which is a rearrangement product). The neutral losses in the m/z 149 parent spectrum are thus directly related to the two C_8H_{17} alkyl substructures in the reference compound. The low mass ion series in the primary mass spectrum is also related to the alkyl chains and the unbroken sequences of ions every 16 mass units is indicative of unbranched alkanes (11).

Generation of Molecular Structures

The GENOA program is a constrained molecular structure generator resulting from the Stanford Dendral project (12,13,14) and marketed by Molecular Design Ltd (6). This program generates molecular structures using the overlapping substructural information obtained from the daughter spectrum/substructure relationship and the empirical formula of the compound. Additional spectral and non-spectral information from other sources may also be included. Heuristic rules determine whether a particular generated structure is chemically plausible, and whether or not it is retained. The advantage of the GENOA program is its ability to exhaustively produce all the plausible compounds given the generation constraints. This capability eliminates the possibility that the chemist might overlook any possible compounds. In many cases, the number and types of different structures that are produced suggest the nature of the missing structural data. The experiments needed to acquire such data may then be obtained from the known spectrum/substructure correlations.

An essential piece of information required by GENOA is the empirical formula of the unknown compound. We have developed software that adapts the standard "molecular weight versus possible empirical formulae" table. Utilizing all pertinent MS/MS data, several constraints can be placed upon the empirical formula generator, and it generates all possible empirical formulae consistent with those constraints and the molecular weight. We have been using M+1 daughter spectrum information instead of high resolution mass spectrometry to aid in the determination of the empirical formula of an unknown compound (4,15). The daughter spectrum of the M+1 isotope ion contains peak pairs at adjacent masses representing the distribution of the ^{13}C atom between the ionic and neutral fragments. The relative intensities of these daughter pairs depends on the ratio of carbon atoms lost to carbon atoms retained in fragmention of the M+1 ion to the observed daughter ion. Hence the peak height or area ratios can be utilized to obtain the number of carbon atoms present in the compound. Once the number of carbon atoms is determined by this method, it used as a constraint to the empirical formula generator. The resulting reasonable empirical formulae are given to GENOA.

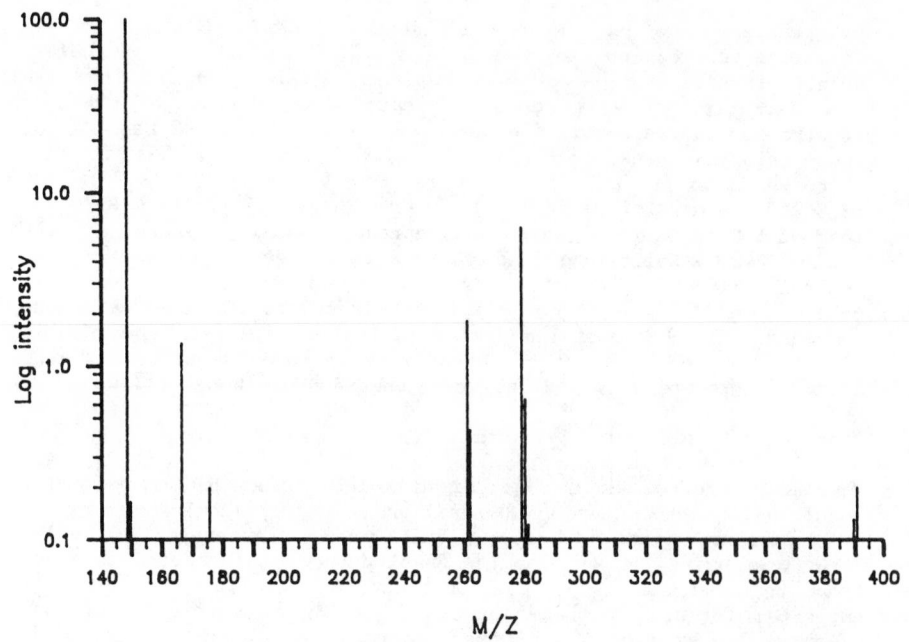

Figure 6.   Parent spectrum of mass 149 from di-n-octylphthalate.

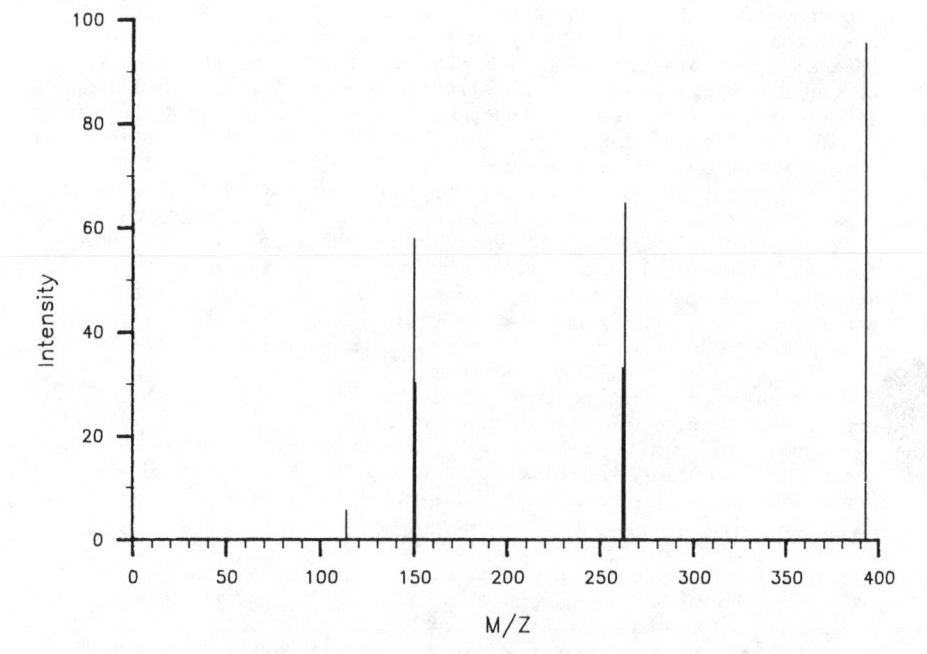

Figure 7.   Daughter spectrum of the ^{13}C-containing protonated
molecular ion of di-n-octylphthalate.

For example, to determine the empirical formula of di-n-octylphthalate, the daughter spectrum of the ^{13}C containing molecular ion (392) was obtained (Figure 7). The relative peak areas of adjacent peak pairs at m/z 149 and 150 is 2:1. This indicates that the M+1 ion is twice as likely to lose a ^{13}C atom as retain it. Thus the ratio of the number of carbon atoms lost to those retained is 2:1. Since the identified phthalate substructure contains 8 carbons, the unknown compound (di-n-octylphthalate) must contain 24 carbon atoms. These data, along with the molecular weight of 390 as determined from the conventional CI mass spectrum of the unknown was fed into the empirical formula generator and the output was one empirical formula: $C_{24}H_{38}O_4$.

Given the phthalate substructure, the two alkyl substructures and the empirical formula, GENOA can now be used to generate all plausible molecular structures. The oxygen in the $C_8H_{17}OH$ group is allowed to overlap with either terminal phthalate oxygen. With this information, GENOA constructs only one molecular structure (structure IV of Figure 3) and it is di-n-octylphthalate. The number of generated structures depends on the completeness of the information provided. If the branching of the alkyl group is not specified, 89 different structures are generated which represent all the isomeric permutations of the alkyl groups. The identities of these generated structures, however, would provide clues as to further needed information. In cases where MS/MS information cannot determine a unique result, additional spectral and non-spectral information may be given to GENOA as structural constraints.

Conclusions

The software tools for structure determination by MS/MS are now at a stage where we can begin to apply them to real elucidation problems. Nearly all of the software tools have been integrated into a comprehensive, interactive system. The system has been successfully used to develop daughter spectra/substructure correlations and extend the MS/MS data bases. The elucidation process is totally empirical and does not assume that structural integrity is maintained in the ionization or fragmentation process. As a result, the ion structures need not be identified. Preliminary results from applying the system to structure determination problems have been very encouraging.

Acknowledgment

This work was supported by NIH grant no. 2R01GM28254.

Literature Cited

1.   Gregg, H.R., Hoffman, P.A., Enke, C.G., Crawford, R.W., Brand, H.R., Wong, C.M., Anal. Chem. 1984, 56, 1121.
2.   Hoffman, P.A., Enke, C.G., presented at 31st Annual Conference on Mass Spectrometry and Allied Topics, Boston, MA (1983); bound p. 556.
3.   Cross, K.P., Enke, C.G., Computers and Chemistry, in press.
4.   Bozorgzadeh, M.H., Morgan, R.P., Beynon, J.H., Analyst 1978, 103, 613.

5. Enke, C.G., presentation at "Applications of AI in Mass Spectrometry", Workshop at 33rd Annual Conference on Mass Spectrometry and Allied Topics, San Diego, CA, May 26–31, 1985.
6. Molecular Design Ltd., 1122B Street, Hayward, CA 94541.
7. Cross, K.P., Beckner, C.F., Enke, C.G., in preparation for submission to Computers and Chemistry.
8. Wipke, T. W., Dyott, T. M., J. Am. Chem. Soc. 1974, 96, 4834.
9. Damen, H., Henneberg, D., Weimann, B., Anal. Chim. Acta 1978, 103, 289.
10. Shelly, C.A., J. Chem. Inf. Comput. Sci.1978, 23, 61.
11. McLafferty, F.W. "Interpretation of Mass Spectra", Univ. Science Books, Mill Valley, CA, 1980.
12. Lindsay, R.K., et al., "Applications of Artificial Intelligence to Organic Chemistry: The Dendral Project", McGraw Hill, New York, NY, 1980.
13. Barr, A., Feigenbaum, E.A., "Handbook of Artificial Intelligence, Vol. II", William Kaufman, Inc., Los Altos, CA, 1981.
14. Carhart, R.E., et. al., J. Org. Chem. 1981, 46, 1708.
15. Todd, P.J., Barbalas, M.P., McLafferty, F.W., Organic Mass Spec. 1982, 17, 79.

RECEIVED January 14, 1986

Artificial Intelligence, Logic Programming, and Statistics in Magnetic Resonance Imaging and Spectroscopic Analysis

Teresa J. Harner, George C. Levy, Edward J. Dudewicz, Frank Delaglio, and Anil Kumar

National Institutes of Health Resource for Multi-Nuclei NMR and Data Processing, Department of Chemistry, Syracuse University, Syracuse, NY 13210

Logic Programming in combination with expert directed statistical analysis makes possible a unique aproach to new expert systems for NMR and other chemical analyses as well as for medical applications of NMR. We have used this approach to begin understanding the behavior of T_1, T_2 and ^1H density in magnetic resonance imaging (MRI). Also, we are utilizing this technique to develop intelligent behavior within our **NMR1** and **NMR2** spectroscopic data reduction systems.

Sets of rules generate a solution space which may be statistical, functional or symbolic (non-numerical). Unlike other expert system environments, the statistical foundations which govern many of the "macroscopic" inferences are included, allowing for modification to the underlying "implicit" statistical bases at any time. The logic programming environment allows modifications to the knowledge-base through automatic and user-generated commands, and lends itself to the development of easily understood natural language interfaces.

Software for NMR applications is now in widespread use and it is therefore important that such packages work well not only in more traditional chemical shift, relaxation or resonance applications but in the more recent context as potential pre-processors of imaging data.

In particular, two systems for NMR spectroscopic analysis, **NMR1**(1) (one-dimensional analysis) and **NMR2**(2) (two-dimensional analysis), are fully operational, while a third system for analysis of magnetic resonance imaging (MRI) parameters, is beginning to emerge. While NMR1 and NMR2 are written in conventional numerically based code (**FORTRAN-77**), the MRI system, **MRI_LOG_ESP** combines the use of **FORTRAN** and the logic programming language, **Prolog**.

Our research on human tissue discrimination methods deriving from MRI parameters is leading to the evolution of prescribed statistical methods for data screening, normalization and discrimination. These are driven by Prolog. In Prolog, sets of logical inferences

0097–6156/86/0306–0337$06.00/0

which carry information about application data types, are capable of distinguishing important features. Procedures pass the data through those analyses which optimize classification.

In this paper we note some of the failures inherent in most current computer-aided (and manual) NMR spectroscopic techniques and reflect on possible solutions via Artificial Intelligence (AI) techniques.

A description of current AI methods which lend themselves to problems of this type is included, as well as a description of applications to NMR spectroscopy and MR imaging. Lastly, there is a brief description of MRI_LOG_ESP in its current preliminary state.

NMR1- Model Computer Software for Spectroscopic Analysis

NMR1 is a graphics-oriented software system containing over 100 options, each allowing the user a large degree of freedom to analyze spectroscopic data in a single dimension. At the core of NMR1 is a set of procedures for data reduction, estimation of initial parameters, and the utilization of a set of convergence methods for baseline conditioning, peak identification and curve fitting.

Curve fitting is currently accomplished using a non-linear minimization (modified Levenberg-Marquardt) algorithm for three-parameter Lorentzians, as well as five additional non-linear peak shapes.

Generally, a user/Chemist may learn a great deal from the displays of the Fourier transformed spectra using the options for analysis available with graphics interaction. Nevertheless there is a great deal of room for improvement. The following list summarizes the most salient current difficulties with traditional computer-aided analysis:

1. Subroutines used to obtain quantitative measures of the parameters associated with overlapping peaks can end in misleading results if an incorrect theoretical line shape has been utilized.

2. Automated initial parameter estimation may not be accurate and then the user will be required to intercede with manual estimation.

3. It may be necessary to manually install or delete peaks; especially when signal-to-noise is low or when the peaks are largely un-resolved. In cases of very small peaks or when overlap between peaks is high, standard algorithms may fail and return unrealistic linewidth values.

4. If initial estimates are too far from correct values, calculations may diverge. Then the process must be restarted with better initial estimates.

5. Because of the underlying mathematical assumptions inherent in statistical modeling of the data, if the assumptions are for any reason incorrect the final fit may be poor or good, **but not significant.** Currently,

software users must have sufficient background in the field to properly interpret the validity of the output from curve fitting and other algorithms.

In addition to these problems, there exist the set of conditions under which the user must manually set all initial parameter estimates. Manual constraints on the parameters may often be the only way to obtain a proper convergence if the true, bounded region of the solution is known by the user, or if one has some specific knowledge of the correct starting values. This may be particularly true if there are additional parameters with complex functional forms such as phase angle. Automated paramater setting which takes into account some of these problems could lead to more consistent results and require less user expertise.

MR Imaging

While analytical spectroscopy has been used for many years in order to obtain information regarding chemical structure, magnetic resonance imaging is a relatively new field. Misleadingly well-resolved images may aid an expert Physician in diagnosing human tissue abnormalities, but as little is understood about the relationships which exist between tissue MRI parameters and tissue health, not to mention secondary factors (genetic, environmental, macro-physiological, etc.), such judgements, accurate or not, are often purely subjective. In similar applications, precedents are well established for the use of expert systems as medical diagnostic tools(3,4,5).

The optimal research strategy involves the systematic search to uncover these relationships at the same time as the development of a computer methodology proceeds. Such software systems will not only give the kind of information about physicochemical structure as have previously designed systems for NMR spectroscopic analysis, but will serve as co-investigators, facilitating through automated procedures, analytical tasks which are normally time-consuming and complex.

Experimental data analysis of tissue parameters and construction of an automated format for MRI research has been proceeding in our laboratory with some success. Statistical Analysis has revealed that, with the proper normality transformations applied to T_1, T_2 and $1H$ density over eight regions of interest in the human brain (left and right sides respectively of Cortical White Matter, Internal Capsule, Caudate Nucleus, and Thalamus), the values within tissue type generally follow a normal distribution(6). This implies that existing discriminant functions may be able to optimally classify data according to tissue type (although initial results also show large overlaps between the normal distributions of several tissue types). Indeed, preliminary results have yielded correct classification percentages between 73 and 86%(7).

As shown in Figure 1, however, statistical analysis alone is only one of the steps towards realizing a fully functional system for MRI tissue discrimination. Experimental data is passed through software (eventually NMR1/NMR2) for pre-processing. Since the statistical analyses must themselves be applied to an ever-increasing number of regions of interest (ROI's) it would be of great use to develop an automated statistical treatment methodology for future research. Also, it is not yet known just what final precision to expect from discriminant analysis, even including secondary descriptor data which should help to obviate overlaps between MRI responses from different ROI's. It seems quite likely that additional, qualitative information, or non-traditional numerical discrimination procedures, might be required at some point to approach an accuracy acceptable for future clinical use of MRI expert systems. Lastly, a general computer system architecture must control these procedures for highest efficacy.

Overview of Proposed AI Techniques

Most researchers are now aware that "Artificial Intelligence" has diverse meanings both inside and outside Computer Science. Such a situation is understandable given the actual variety of methods which could, without being strictly inaccurate, fall under the heading of AI. It is therefore important, if one intends to use AI techniques to develop a computer program architecture, to first define clearly the AI related approaches which one has chosen to apply.

For the present, the following AI tools and techniques comprise the building materials intended for construction of our automated spectroscopic analysis and MRI tissue discrimination systems:

Logic Programming, and specifically, the programming language Prolog, satisfies a specified goal by resolving its premises. For resolution to take place, these premises must in turn become the subgoals for premises which can be satisfied. A goal is specified with a predicate name and a set of arguments whose values must be instantiated for the goal to succeed. The goal-to-premise structure forms sets of clauses which operate upon the principles of first order predicate logic(8).

Decision Trees provide the overall structure for problem resolution in the current system. The outcome of a test at a particular node in the tree is recorded and directs the next decision for branching. If a failure is encountered at all possible branches, the un-resolved problem is passed back up to the node at which there last existed a possible, untested, solution. Prolog lends itself nicely to this structure since its basic architecture includes decision-making via such a "depth-first" search strategy(9)

Model Matching and Similarity Nets(10) are the means by which, at any node in the decision tree, an actual test is made. Within a database, lies a set of model data structures, to which an attempt is made to match the actual input and output format. The initial problem is to find a model which, of all stored models, contains the fewest **differences in structure** between model and actual data set. An example is shown in Figure 2, a system designed to classify human brain tissue type. A data set is entered which contains variable names "Obs","T_1", "T_2" and "^1H density" as headings in the first row. In the first column, it is found, lies a set of integer numbers running from 1 to 4, and within the set itself are "*" 's. Stored within the system are a series of general models which identify data matrices in sets of predicates defining "first row", "column", and other distinguishing characteristics of data sets.

By searching the stored models for such characteristics, the program constructs a model data set which appears to come closest to

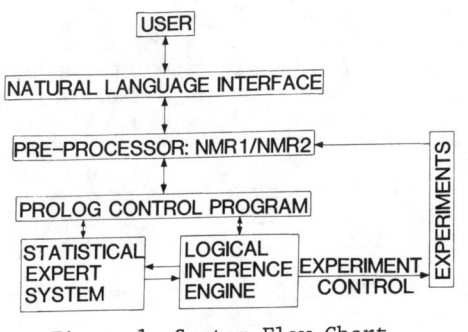

Figure 1. System Flow Chart

ACTUAL DATA SET

Obs	T1	T2	Density
1	663.000	77.0000	96.798
1	775.000	84.0000	107.554
2	659.000	82.0000	99.556
2	*	76.0000	*
3	619.000	79.0000	99.467
3	667.000	79.0000	102.868
4	*	79.0000	*
4	651.000	80.0000	84.752

LOGIC PROGRAMMING CODE

list_of_variable_names[a,b,...,obs,...,T1,..., T2,...,Density,...].

symbol_list[<integers>,<real_numbers>,"*",...].

position_meanings[column(Number,Symbol,Meaning), row(Number1,Symbol1,Meaning1)].

model_data_sets[model1(column(A,B,C),row(D,E,F)),..., modeln(column(I,J,K),row(X,Y,Z))].

MODEL DATA SET - CLOSEST (GENERAL) MATCH

Obs	T1	T2	Density
1	VAL	VAL	VAL
1	VAL	VAL	VAL
2	VAL	VAL	VAL
2	NONVAL	VAL	VAL
3	VAL	VAL	VAL
3	VAL	VAL	VAL
4	NONVAL	VAL	NONVAL
4	VAL	VAL	VAL

Figure 2. Pattern Matching with Prolog

that defined by the structure of the **input** data set. For that particular model, there are now a set of options for action upon the data. The variable name "Obs" is associated with a set of classes, therefore the program knows that 4 classes are represented by the data set. The variable names "T_1", "T_2" and "1_H" density are untransformed names, which information, taken with the fact that "*" has been used in the place of data in certain positions identifies the set as being raw and untransformed. The program will then proceed by "cleaning up" the raw data set, making appropriate transformations and applying a discriminant analysis to the set, under the assumption of four classes.

In actual practice a number of tests must be passed at various nodes before final classification takes place. Also, a prohibitive time would be required to search a large database of models for ones which most closely approximated the actual data set. For this reason the concept of **similarity nets** is introduced. In this case, a more general model is first chosen, one which is clearly not completely absurd. A subset of other models which are variations of this first general model then provides the index for the final choice of model. Such a reduction in the model lists greatly reduces the search space for the closest fit.

While a great many other techniques may be employed to ensure a consistent, and efficient, logic-driven software, the techniques described above are the primary drivers for the effective resolution of goals in the prototype expert system MRI_LOG_ESP. Once a detailed accounting is made of the characteristics which fully describe input and output, providing Prolog code is quite easily accomplished.

Specific Applications:

Imaging

The prototype expert system, MRI_LOG_ESP has been written to aid in classifying tissue type from primary and secondary tissue descriptors and is capable of limited applications. While it would be incorrect to say that the current system is a robust expert system, since it is not yet able to fully make the inferences regarding the input data sets which would lead to automatic tissue classification, the program does successfully enable serial combinations of statistical procedures to be run from a central Prolog controlling program. Commands are parsed so that simple English language structures can be interpreted, and a tracking procedure keeps an automatic log of analyses run and steps taken.

At the experimental interface, we expect to ultimately assemble a data set of approximately 500 patients. Thus far we have only worked on a much smaller data set of 23 individuals. The actual protocol for obtaining this data is described elsewhere(5).

Figure 3 provides a sample session with MRI_LOG_ESP. The controlling code is written in Prolog, while numerically oriented, analytical procedures for classification are written in Fortran-77. The system has been written to run under Data General's AOS/VS operating system (MV series computers) but it is expected to be easily ported to the Digital Equipment VAX VMS environment.

At the front end of the controlling program there are three main branches in the form of predicates:

```
                    WELCOME TO MRILOG
Use the options list below to guide your interaction
while keeping your responses relatively simple, and
you should have few problems working within the system.


************************************************************
*                                                        *
*             GENERAL FEATURES                           *
*                                                        *
*    I.   auto_analysis: based on a user specified       *
*         data file, program determines analyses and     *
*         runs them.                                     *
*   II.   user_driven_analysis: user specifies files     *
*         and run list.                                  *
*  III.   help_file: probably a good place to start.     *
*                                                        *
************************************************************
```

```
What is your general objective,    ----------------------
based on the information          |         NOTE        |
supplied above?                   | Answers can take a  |
                                  | natural language form|
                                  | but user should try |
                                  | to respond within the|
                                  | context of the prompt|
                                   ----------------------
```

```
|: I would like to perform a user driven analysis.
```

```
                      -------------------------------------
                     | REMEMBER: Regular "reset" will     |
                     | delete all the data files accum-    |
Reset? (yes/no/all)  | ulated so far EXCEPT:               |
                     |         <user_analysis(xxx)>        |
                     | and     <auto_analysis(xxx)>.       |
                     | By specifying "all" in the reset    |
                     | command, these will also go.        |
                     | See Users Manual or Help file       |
                     | for further information.            |
                      -------------------------------------
```

```
|: no.
```

Figure 3. An Interactive Session with MRI_LOG_ESP.

Continued on next page

Please either give a clear and
concise description of
your objective, or type in
a list of the procedures
you wish to invoke for data
analysis.

```
---------------------------------------------
|                                             |
| AVAILABLE PROCEDURES FOR ANALYSIS           |
|                                             |
| call name           description             |
|                                             |
| <twod>        - two dimensional             |
|                 graphics.                   |
| <normtest>    - normality testing.          |
| <tran>        - transformation of           |
|                 variables.                  |
| <tran_disc>   - discriminant analysis       |
|                 of transformed vars.        |
| <disc_fun>    - discriminant analysis       |
|                 (untransformed data)        |
|                                             |
---------------------------------------------
```

|: Let's do a normtest.

SEARCHING AVAILABLE ANALYSES FOR YOUR SPECS

The following analyses will be run using data from an
input file. If the list is not correct indicate
that a change is required. Otherwise, type "go".
(or some other affirmative)

RUN LIST:
[normtest]

|: go.

                    INPUT FILE SPECIFICATION: normtest

Will this be new data?
|: no.

You can use any file which contains PROPERLY FORMATTED data.

What is(are) your input file name(s)?
|: I want to examine t1cc and t2cl.

Do you wish to examine file t1cc
|: no.

Do you wish to examine file t2cl
|: no.

                    Figure 3. Continued.

                              Continued on next page

```
Specified input files scanned.
Starting normtest using input file: t1cc
Starting normtest using input file: t2cl

NORMALITY TESTING OF VARIABLES (normtest)

All output has been appended in user_analysis.007

Do you want a printout
|: no.


Please either give a clear and
concise description of
your objective, or type in
a list of the procedures
you wish to invoke for data
analysis.
```

```
-----------------------------------------
|                                                   |
|   AVAILABLE PROCEDURES FOR ANALYSIS   |
|                                                   |
|   call name              description    |
|                                                   |
|   <twod>        - two dimensional      |
|                      graphics.             |
|   <normtest>  - normality testing.     |
|   <tran>         - transformation of    |
|                      variables.           |
|   <tran_disc> - discriminant analysis |
|                      of transformed vars. |
|   <disc_fun>  - discriminant analysis |
|                      (untransformed data) |
|                                                   |
-----------------------------------------
```

```
|: halt.
```

Figure 3. Continued.

- **auto_analysis**

- **user_driven_analysis**

- **help_file**

In the emerging system, **auto_analysis** represents the automated logic-driven search tree which is able to apply the <u>appropriate</u> analysis from logical inference and/or upon receipt from the user of a description of the problem space (data set). The program will then search, sort and classify the data as appropriate to each reasoning technique logically demanded by the problem/data input. In many respects **auto_analysis** represents the "expert system core". For the initial input data set, the user is questioned regarding a final objective. This objective, in combination with the results obtained from each analysis, are what provide deterministic control, as the initial data set is formatted for subsequent analysis and output from previous analysis is itself reformatted (as a result of computer generated interpretation of the output). This reformatting of the output is, once again, determined by the next analysis which the program deams essential to the satisfaction of the primary, user-specified goal.

The capability of determining the order of procedures, based on the user specification of the end goal <u>alone,</u> and to know at what point computation should stop with a result recorded, is the implementation of the theoretical AI techniques described above.

Conversely, **user_driven_analysis** allows the user to specify from one to all of the available analyses and the specific data sets to use for each given analysis. Thus, if predicates **twod** and **normtest** were specified, a "Run_list" would be interpreted consisting of:

[twod,normtest]

It should be noted that all the formatting knowledge required to run auto-analysis is also required to run a user analysis. In this case one is simply overriding the computer's "better judgement" in terms of procedural protocol.

The following predicates comprise the current statistical procedures handled by the system.

1.   **<twod>** - two dimensional graphics

2.   **<threeD>** - three dimensional graphics

3.   **<normtest>** - normality evaluation of variables

4.   **<trans>** - normality transformation of variables

5.   **<disc_fun>** - linear discriminant analysis

These programs, written in Fortran-77 are accessed through systems calls in Prolog. Naturally this list will be greatly augmented in future versions of the program. However this set of statistical analyses is comprehensive in-so-far as a data set may be screened and classified using these routines alone, thereby providing a general but simplified model upon which to base the logical inferences which govern choice of analysis.

At the top level of the program, having provided the "analytical objective" for a given data set, the user is asked to verify that the program has understood the command correctly. If this is verified the program will go on to ask the user for the name of the input file containing data from which it is desired to proceed with testing.

A copy of the results of every test on a set of data is saved and when a single run of a series of tests has been made, the results of each procedure is appended in a single file.

Commands are requested from the user at fairly regular intervals. The program will accept most general sentences which may be construed by the parser to elicit some form of action for which the program has been written.

While several predicates require and accept only affirmative (yes) or negative (no) responses from the user, for the most part, communication with the program is governed by what has been termed a "Context Parser", the main predicate of which has three levels to handle varying levels of linguistic complexity.

The aim of the graphics software, (twod, threeD), is to enable the user to rapidly examine a large number of two- and three-dimensional scatter plots. At present the program is capable of handling up to 120 variables with up to 200 observations for each.

Predicate normtest tests/evaluates normality of the given set of data points (corresponding to any variable), while, disc_fun performs a linear discriminant analysis on groups of data (maximum of 10 groups) with respect to any selected variables (maximum of 20 variables).

There are only two types of output files and output file names. These are:

auto_analysis_out<xxx>

and

user_analysis_out<xxx>

Where "<xxx>" symbolizes a sequence number. As runs are made, each one is placed in a list of output files. As runs are made, whether in auto or user mode, an accumulation of results is inevitable. Also, due to the method by which input and output files are appended, there is some accumulation of "garbage" files. The procedure reset provides a way of deleting unecessary files.

NMR1 and Similar Software (Spectroscopic Analysis)

Precisely how the AI techniques discussed above might improve the current, numerically-based software is still somewhat speculative.

Close inspection of the current failings discussed previously indicates that their source lies in two main areas which are difficult for numerically based algorithms to handle:

1.   The problem of defining the baseline and locating and defining the spectral features (peaks).

2. The correct setting of initial parameters.

For quantitative characterization of molecules including analyses of *in vivo* (metabolic) NMR spectra, the problem of peak overlap and baseline identification (which are particularly problematic for *in Vivo* spectroscopy) is really only one part of the more general experimental problem of discovering spectral differences arising in complex environments. We can greatly increase the efficiency of the overall experimental process *and* solve the peak quantification problem utilizing the data base structure inherent in a logic programming framework.

With sets of rules providing the facts from which a full model can be constructed, the program is informed regarding the origin of the spectrum to be analyzed. A comparison is then made between model and actual spectra. Anomalous features are thus identified.

In (2) above, a second difficulty with spectral analysis is identified which might be alleviated through the use of logic programming methods: The setting of initial parameters. As noted above, this is generally an automated procedure in NMR1 but sometimes does require user intervention. A minimization algorithm which did not derive its information from a pre-set local minimum could be applied at each minimum in the spectrum and then proceed to examine the consistency of the results. Such a method would not be difficult to implement within the framework of a logic-driven, controlling algorithm. By applying this minimization algorithm at multiple minima and by examining consistancy (and not by numerical methods alone!), a determination would be made as to whether the derived minimum was false or global.

Investigation thus far has been made into characteristic *in Vivo* ^{31}P peaks with some thought to localized pattern matching(1). In the coming year we will begin to look at coding characteristic *in Vivo* spectra and developing a Prolog algorithm which analyzes the results of the minimization algorithm. For the most part, it is hoped that MRI_LOG_ESP will provide the "expert system shell" which may be effectively applied to the problems in spectroscopic analysis.

Acknowledgments

The authors acknowledge the collaboration of Dr. Felix Wehrli and co-workers at General Electric Medical Systems and also pilot project funding from NIH (Grants RR-01317, and RR-01831) and the General Electric Company.

Literature Cited

1. Dumoulin, C.L., Levy, G.C.; *Journal of Molecular Spectroscopy*, *113*, 299-310 (1984); Dumoulin, C.L., Levy, G.C.; *Computers and Chemistry*, *5*, 9-18 (1981).

2. Levy, G.C., Delaglio, F., Macur, A., Begemann, J.; *Computer Enhanced Spectroscopy*, in press (1986).

3. Buchanan,B.A., Shortliffe, E.H.; "Rule Based Expert Programs: The MYCIN Experiments of the Stanford Heuristic Programming Project", Addison-Wesley, Reading, MA (1984).

4. Miller,R., Pople, H., Meyers, J.; New England Journal of Medicine, 307, 468-476 (1982).

5. Weiss, S.M., Kulikowski, C.A.; "A Practical Guide to Designing Expert Systems", Rowman and Allanheld, Totowk, NJ (1984).

6. Dudewicz, E.J.; "Statistical Analysis of Magnetic Resonance Imaging Data in The Normal Brain, Part I: Data, Screening, Normality, Discrimination, Variability"; unpublished report, 1985.

7. Levy, G.C., Dudewicz, E.J., Harner, T.J., Wehrli, F.W., Breger, R.; (Submitted), Magnetic Resonance in Medicine, (1985).

8. Kowalski, R.; "Logic for Problem Solving", Computer Science Series, North-Holland Publishing Co., NY (1979).

9. Clocksin, W.F., Mellish, C.S.; "Programming in Prolog",Springer--Verlag,Berlin, (1981).

10. Winston, P. H. ,"Artificial Intelligence",Second Editoin, Addison-Wesley, Reading Massachusetts (1984).

RECEIVED January 24, 1986

27

An Expert System for Organic Structure Determination

Bo Curry

Chemical Systems Department, Hewlett-Packard Laboratories, Palo Alto, CA 94304-1209

We are developing an expert system which interprets
low-resolution mass spectra, infrared spectra, and
other user-supplied information and produces a list of
functional groups present in an unknown organic compound. The input data are interpreted as evidence
supporting the presence or absence of each of the over
900 functional groups and organic substructures represented in the knowledge base. This evidence is then
combined by an "inference engine" to determine the
probability that the group is present. Each type of
input spectra is interpreted by a separate module,
which has private internal data structures; these
modules can use different techniques and even be
written in different computer languages. The modular
architecture was designed to allow new modules interpreting different types of spectra to be easily incorporated into the system. A major goal has been the
reduction of the number of false positive assertions.

An analyst attempting to identify an unknown compound from spectral
data begins by searching libraries of spectra of known compounds
(Figure 1). Programs which rapidly and reliably search spectral
libraries are widely available.(1-2) However, although these
libraries continue to grow, it will remain true that the majority of
compounds encountered in real samples are not represented in the
libraries. These compounds can at present be identified only through
a laborious manual process requiring considerable expertise.
Interpretation of molecular spectra involves four basic steps.
First, major skeletal and functional group components of the molecule are identified, either from assumptions about the compound
origin or from features of the spectra. Second, non-localized
molecular properties such as the molecular weight, elemental composition, and chromatographic behavior are considered. These global
constraints can be used to eliminate unlikely functional groups,
deduce the presence of groups and skeletal units which have no distinctive features in the spectra, and detect multiple occurrences of

0097-6156/86/0306-0350$06.00/0
© 1986 American Chemical Society

functional groups. Complete candidate structures are then generated
by assembling the functional groups subject to the global constraints. More data may be collected to narrow down the number of
candidates. Finally, the candidate structures are tested for compatibility with all original data. Final confirmation is obtained by
synthesis of the candidate compound and comparison with the unknown.

We are developing an expert system to automate the first step
of this process, the interpretation of molecular spectra and identification of substructures present in the molecule. The automatic
interpretation of spectra would by itself provide a useful tool for
an organic chemist who may not be an expert spectroscopist. Also,
reported algorithms for the assembly of candidate structures from
known substructures, such as the GENOA program,(3-6) rely on the
input of accurate and specific substructures in order to function
correctly and efficiently. Identification of substructures is thus a
logical starting point.

Information about substructures present in an unknown can be
obtained from a wide variety of sources, and one of our major objectives has been to allow all available data to be used by the program.
Programs have been described in the literature which interpret C-13
and 1-H NMR spectra,(7-13) low and high-resolution mass spectra,
(14-15) infrared spectra,(16-23) MS-MS spectra,(24) and 2D-NMR
spectra.(25) The methods employed may be generally classified as
rule-based methods or pattern-matching methods. Rule-based methods
apply interpretation rules to discrete features of the spectra.(26)
These rules are usually empirical correlations having physical significance , expressed in a form similar to that used by human interpreters. Rule-based systems maintain a relatively detailed internal
representation of their knowledge, and can explain their conclusions
in a language intelligible to the user. Pattern-matching methods
attempt to classify the spectrum based on some global measure of
"spectral distance" from spectra of known compounds.(27) Any physical
knowledge used by the algorithm is embodied in its distance measure,
which may be a complicated function of many features of the spectra.
The classification decision is made from a statistical analysis of
the distance from representative members of the classes being distinguished. Explanations of the system's conclusions are are usually
limited to reporting the computed spectral distances. Whichever
method is employed, the output is in the form of a list of suggested
substructures, chosen from a predefined set, with confidence factors
variously computed.

The choice between rule-based and pattern-matching approaches
depends not only on the predilection of the experimenters, but also
on the nature of the data being interpreted. The reported NMR interpreters all use rule-based methods. The pattern-matching algorithm
used in the STIRS program (14) appears to be the most successful at
interpreting low-resolution mass spectra of general organic compounds. Both rule-based and pattern-matching techniques have been
applied to the interpretation of infrared spectra. The rule-based
methods seem to be the most successful.(16-23) We have therefore
designed our program to allow each type of spectrum to be interpreted
by the most efficient method; different methods can even be simultaneously applied to the same spectrum.

When the unknown is present in sub-microgram amounts, as is
often the case when it has been isolated chromatographically, the

primary structural techniques are mass spectrometry, infrared spectroscopy, and various methods of determining elemental composition.
We have therefore concentrated our initial efforts on interpreting
these types of data, while recognizing the need to be able to use
data from other sources, such as NMR, when they are available. A
skilled chemist can often correctly identify an unknown of moderate
size (molecular weight < 200) using only the IR spectrum, the low-
resolution mass spectrum, and some knowledge of the sample origin.
Even when a precise identification is not possible, a generic classification of the compound type is useful and often sufficient. A
program which interprets IR and mass spectra is therefore a useful
analytical tool in its own right, and provides the basis for development of more comprehensive capabilities in the future.

In our present system, infrared spectra are interpreted using a
rule-based approach, while mass spectra are interpreted by the STIRS
algorithm. The abilility to use different techniques for different
types of data implies a modular architecture, in which the "expert"
responsible for the interpretation of each spectrum maintains its own
rules and data structures (Figure 2). It is important, however, that
the interpretation of the various spectra be mutually consistent.
Information obtained from the mass spectrum, for example, should
affect the way the infrared spectrum is assigned. Conversely, the
interpretation of mass spectral lines must be consistent with the
presence of functional groups known to be present from other sources.
This requires a means of communication among the parts of the program
responsible for the interpretation of different types of data. Consistency also requires a means of combining evidence from different
sources. When data from different sources contradict each other, the
individual modules should be able to reinterpret their data so as to
resolve the contradiction.

As in any classification problem, there is a tradeoff between
the rate of recall, or proportion of correct substructures detected,
and the reliability, or avoidance of false positive assertions. It
is rather the exception than the rule for an observation to have a
single, unequivocal explanation. When reasonable alternative interpretations are possible, a decision must be made about what to
report. At one extreme, all possibilities could be asserted, ensur-
ing 100% recall (i.e. no substructure which is actually present will
fail to be detected) at the cost of a high rate of false positives.
At the other extreme, ambiguous data could be ignored, which guaran-
tees no false positives, although many substructures which are
present will be missed. We have taken a middle road between these
extremes by developing a measure of the "best" or most probable
interpretation, taking into account all of the data available. When
the best choice is not clearcut, the disjunction of the competing
alternatives is explicitly asserted. The goal has been to minimize
the rate of false positives, while reporting the most specific
possible interpretation of the data.

An important feature of expert systems is the accessibility to
the user of the knowledge base and the reasoning process. Both the
terminology used by the program and its interpretation of data have
chemical significance. Each conclusion reached by the program can be
traced by the user to the original data. When alternative explanations for an observation are possible, the choice is visible to the

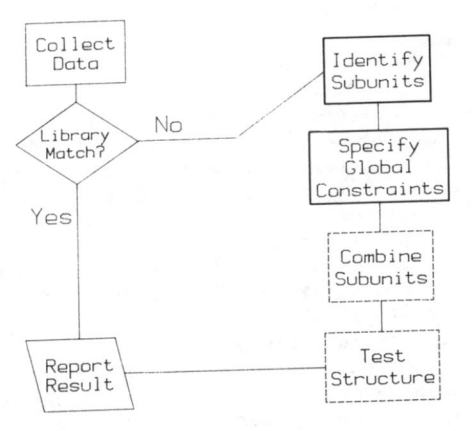

Figure 1. Flow chart for identification of an organic compound.

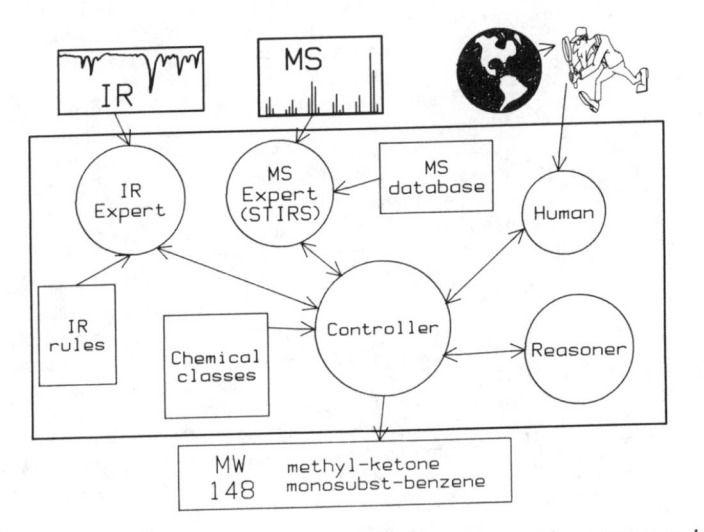

Figure 2. Schematic drawing of the interpreter. The program is represented by the area inside the solid rectangle. Program modules are drawn as circles, and their associated databases as rectangles. All of the modules have read access to the Chemical Classes database.

user. If the program has made an error, the user can correct it, thereby modifying the original conclusions.

Program Description

The architecture of our current system is shown schematically in Figure 2. The design is modular, with a Controller module, a Reasoner module, a database of over 900 organic substructures, and a separate "Expert" module assigned to each kind of input data. The Controller module controls the progress of the calculation by considering each of the substructures which has not yet been eliminated, beginning with the most general. It requests each of the Expert modules to supply it with evidence supporting or denying the presence of the substructure currently being considered. This evidence is collected and passed to the Reasoner. When no more evidence can be collected, the analysis is finished.

The Reasoner combines evidence from all sources and makes deductions from this evidence. The combination of evidence results in a single "confidence level" for each substructure. These confidence levels designate the degree to which the evidence supports the presence of the substructure in the unknown compound. They range from -100% (substructure definitely absent), through 0% (no information), to +100% (substructure definitely present). The confidence levels are ultimately derived from statistical analysis of representative spectral libraries. Details of the generation and propagation of confidence levels will be described in a separate report.(28)

Each Expert module is permitted to use any convenient method to carry out its mission of interpreting its assigned data. The Experts use private rules and data structures, and communicate with the Controller module both by suggesting the presence of substructures, and by evaluating the likelihood of substructures under consideration. Each Expert can read the current confidence level associated with each substructure, and thus has access to information generated by other Experts or deduced by the Reasoner.

Communication among these modules is accomplished in two ways. First, the chemical database, besides storing the chemical knowledge of the program, serves as a "blackboard" on which the progress of the computation is recorded.(29) Only the Controller and Reasoner modules are allowed to write on the blackboard, but all modules can read it. In this way the conclusions of each Expert module are available to all the others to guide their interpretation. Second, the Controller module controls the overall path of the analysis by sending messages to the individual Experts. The only requirement of a new Expert module being added to the system is that it be able to respond appropriately to these messages.

The current prototype system includes three Expert modules, the IR Expert, the STIRS Expert, and the Human. All modules are written in Lisp. The IR Expert is a rule-based infrared interpreter which we have developed. The STIRS Expert is an interface to the STIRS program, a pattern-matching mass spectrum interpreter developed by McLafferty and coworkers at Cornell University, which is written in Fortran.(14) The interface translates the output of STIRS into a form palatable to our program, and handles the message-passing protocol required by the Controller. The Human module controls communication with the user. It allows user-supplied elemental or substructure information to influence the course of the analysis. The power of the modular approach is shown by our ability to integrate the results of three interpretation methods which differ profoundly in their internal details.

The Chemical Database. The chemical knowledge of the system is embodied in a database of over 900 organic substructures, arranged in a hierarchy (Figure 3). With each of these substructures is associated a connection table, stability information, and a probability of occurrence denoting how common the group is. This information may be used by the Expert modules when deciding among possible interpretations.

As the analysis progresses, evidence is accumulated supporting the presence or absence of defined substructures. The evidence is combined by the Reasoner module to form a belief function, which describes the degree to which each substructure is currently believed. This information is stored in the chemical database, where it is available to the Expert modules and to the Controller as it decides the course of the analysis. As the belief function evolves, the current state is displayed graphically to the user, who may halt the analysis, query the current state, and redirect the course of the analysis by supplying evidence for or against a substructure.

IR Expert Module. The IR Expert's rule base consists of over 1000 correlations between observed infrared bands and vibrational modes of specific substructures. Associated with each rule is a wavenumber range, an intensity range, and two confidence levels. Four intensity levels are allowed. The intensity levels are defined on an approximate semilog scale, relative to the most intense peak in the spectrum: WEAK = 2 - 5%, MEDIUM = 5 - 15%, STRONG = 15 - 40%, VSTRONG = 40 - 100%. The program does not attempt to assign bands weaker than 2% of the strongest band. Each IR rule is equivalent to the pair of propositions:

a) IF a band of intensity I appears in the region x1 - x2 cm-1, THEN it is due to the vibrational mode M of substructure S, AND

b) IF no band of intensity I appears in the region x1 - x2 cm-1, THEN the substructure S is not present in the unknown.

About 800 of these rules were chosen by testing all the IR correlations we could find in the literature,(30-32) mostly for condensed phases, against the EPA gas-phase library of 2300 compounds. (33-34) About 30% of the literature correlations were not generally satisfied by the library spectra, and were discarded. Another 200 rules were discovered by searching for patterns in compound classes in the library which could reasonably be attributed to expected vibrational modes of those classes. Statistics were generated for the probability that each of the IR rules would be satisfied for compounds which contained, or did not contain, the substructure specified by the rule. These statistics were used to compute two confidence levels for each rule, corresponding to the confidence in the two propositions a) and b) implied by the rule.

Messages. As noted above, the expert modules communicate their results to the user and to the Controller by responding to messages sent by the Controller. There are six messages to which each Expert module is required to respond:

The ALIVE? message asks the Expert if it is available for consultation in this analysis. The receiving Expert resets its internal state, and responds TRUE if it has data, FALSE if it doesn't.

The SUGGESTIONS message asks the Expert to report any substructures it believes, on its own, to be present or absent. The report takes the form of a list of items of evidence, each supporting the presence or absence of a particular chemical group.

The SPECIALIZE message asserts the hypothetical presence of a chemical group, and asks the Expert which subgroups may be present. For example, the message "SPECIALIZE carbonyl" would cause the receiving Expert to return evidence for or against the presence of ketone, aldehyde, ester, amide, and other specific types of carbonyl, under the assumption (for the moment) that the compound does in fact contain a carbonyl group.

The TEST message asks the Expert to return any evidence it may have against the presence of the group being tested.

The REEVALUATE message is sent when a piece of evidence supplied by an Expert has been contradicted. It asks the Expert to modify or retract the evidence, if possible. Many infrared correlations have known exceptions in specific cases. For example, a nitro group on a benzene ring raises the expected frequency ranges of the hydrogen wags. If the presence of a nitro group is known or suspected, the aromatic wag assignments must be reevaluated.

The EXPOUND message asks the Expert to print out, for the user's benefit, the reasons supporting a piece of evidence. Each piece of evidence originated initially in some feature of the data. The degree of detail supplied in response to this message depends on the individual Expert. The IR Expert, for example, can report the infrared bands which were assigned to a particular vibrational mode of a substructure, as well as possible alternative assignments. The STIRS Expert reports the incidence of the substructure among the best hits in different STIRS data classes.

Example: 4-phenyl-2-butanone

The results of the interpretation of the gas phase IR and low-resolution mass spectra of 4-phenyl-2-butanone are given in Figure 4. This compound, with a molecular weight of 148, is typical of the size and complexity of compounds which our program handles well. The IR spectrum was taken from the EPA gas-phase IR library, and the mass spectrum from the Registry of Mass Spectral Data.(35)

The program was run three times: first with only the STIRS results, second with only the results of the IR interpretation, and finally with both spectra together. All functional groups reported by the program with confidence levels > 10% are listed. In addition, STIRS correctly determined the molecular weight.

The most specific defined functional groups actually present in the unknown are benzyl, monosubstituted-benzene, X-CH2CH2-X (where the "X" represents any group other than -H or -CH2-), and methyl-ketone. That is, the program would have achieved a perfect score had it reported these substructures and no others. In fact, the program was unable to determine the correct environments of the ketone and -CH2- groups, although it reported only one incorrect substructure.

These results are consistent with our goal of reducing the rate of false positives, at the cost of failing to report the most specific possible substructures which are actually present. If the low-confidence report of the presence of benzyl and X=C-CH3 groups is accepted (Figure 4), the reported results suffice to uniquely determine the complete structure.

The effects of the low-level combination of evidence are illustrated by two features of the output. First, the confidence level for the ketone group increases from 19% for the IR-only interpretation to 30% for the combined interpretation, despite the fact that STIRS had nothing to say about the presence of a ketone or even of a carbonyl. This is explained by the increased confidence in monosubstituted-benzene derived from the combined spectra, which causes a fingerprint line tentatively assigned to an ester C-O stretch to be reassigned to a phenyl vibration. Reducing the likelihood of an ester group increases the likelihood that the C=O stretch is due to a ketone group. Secondly, the contradiction between STIRS' assertion of methyl-benzene and the IR denial not only reduces the belief in methyl-benzene, but also allows the assertion of benzyl and unsaturated-CH3 (X=C-CH3). These substructures were not suggested by either spectrum taken alone.

A slightly abridged explanation offered by the program for its belief in methyl-benzene is shown in Figure 5. There is both positive and negative evidence. The positive evidence comes primarily from STIRS, and the negative evidence results from the failure to observe a medium intensity C-H stretching band expected for methyl-benzene. A small amount of positive support for methyl-benzene is also supplied by the IR Expert, showing that conflicts can occur between different features of a single spectrum. The degree to which each piece of evidence is in conflict with other evidence is noted. The explanation facility traces the final belief back to primitive pieces of evidence supplied by the Expert modules. The Experts are then responsible for explaining how the evidence depends on the observed spectrum. STIRS is unable to do more than report which of its data classes supported the substructure and with what probability. The IR Expert module, on the other hand, can give a richly detailed description of the assignment of the spectrum.

Results

We have evaluated our prototype system at several levels. Each Expert module has been tested individually. Detailed results of tests of the STIRS program have been published by McLafferty et al.(36) The IR Expert module was tested extensively against the EPA library.

The effects of competition among the IR rules were explored by using the complete system, with the STIRS module disabled, to interpret the spectra of 1807 compounds from the library. For the test, we selected 500 of the 900 chemical substructures which both are chemically interesting and display at least one distinctive infrared band. Some of the selected substructures were subsets of others: for example, alcohol, phenol, and primary alcohol were all in the test set. As expected, some functional groups displaying very distinctive infrared bands were detected much more reliably than others. Figure 6 shows the reliability, false positive and recall rates for a few selected functional groups.

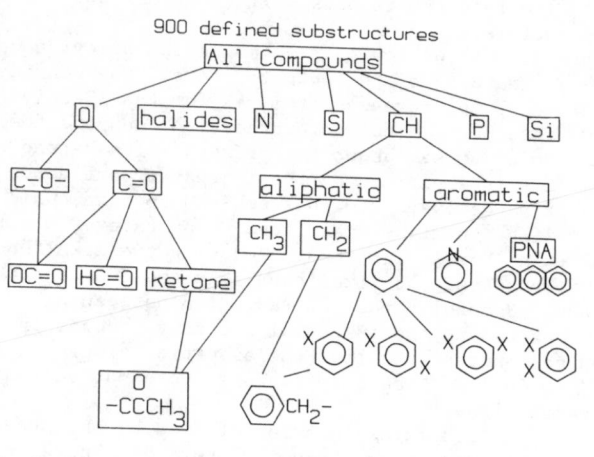

Figure 3. A subset of the chemical substructures database, showing the hierarchical ordering.

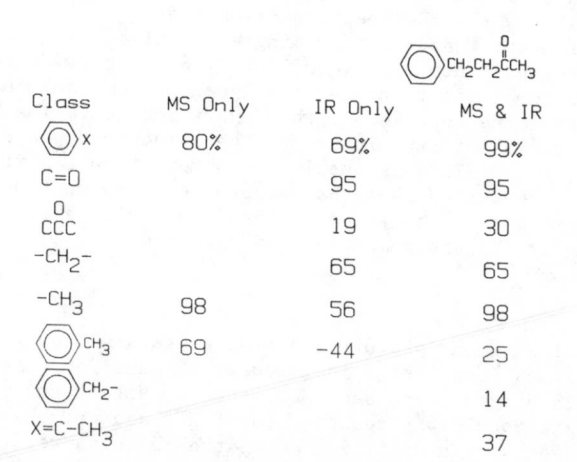

Figure 4. Substructures reported for 4-phenyl-2-butanone at > 10% confidence, for three runs of the interpreter using different data sets.

Why *methyl-benzene*?

36% POSITIVE:

    41% from STIRS (conflict 27%)
    8% from IR band at 2933 cm-1
       assuming unsaturated-C-CH3 (37%)
       (conflict 27%)

11% NEGATIVE:

    23% because of failure to satisfy
      C-H-sym-methyl-benzene-1
      IR band 2860-2883 m
      (conflict 45%)

Figure 5. Sample of the explanations provided by the program for its conclusions. More detail about the source of the reported conflict, the assignments of IR bands, or the data classes responsible for the STIRS evidence can also be provided.

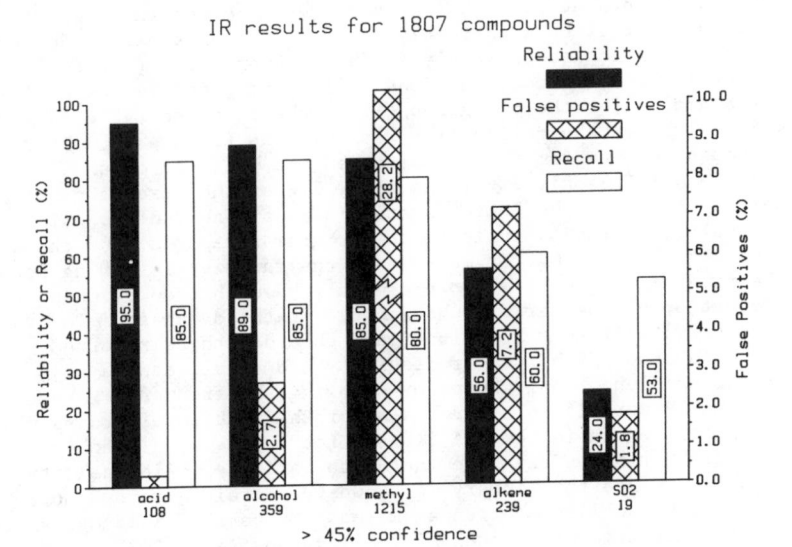

Figure 6. Statistics for 5 selected substructures of the 500 tested on the EPA IR database. Values of the Reliability, False Positives, and Recall (see text) are compared at the 45% confidence level. The number of compounds in the database containing each substructure is given beneath the substructure name. Note the expanded scale used to plot the False Positive measure.

The "recall" is the probability that a substructure present in the unknown will be reported, while the "reliability" is the probability that a reported substructure is actually present.(36) These functions are defined as:

Recall(S) = Number_correctly_reported(S) / Total_number_present(S)

Reliability(S) = Number_falsely_reported(S) / Total_number_reported(S)

for all compounds in the database containing substructure S. Both measures are functions of the confidence level (CL) threshold above which we count a substructure as "reported". All substructures are reported at CL > -100%, while none are reported at CL > +100%. We have arbitrarily chosen CL > 45% as a threshold in Figure 6.

An alternative measure of reliability often used is the "false positive" rate, defined as:

FP(S) = Number_falsely_reported(S) / Total_number_present(NOT S) ,

which is related to the recall and reliability measures by:

$$FP(S) = \frac{Total\_number\_present(S) * Recall * Reliability}{Total\_number\_present(NOT\ S) * (1 - Reliability)}$$

This is the probability that a compound which does not contain substructure S will be incorrectly reported to contain it. For substructures which occur rarely in the database, the (1 - FP) rate will be considerably greater than the reliability, and may be misleading. For example, for the SO2 group (1% of the database), the FP rate was < 8%, although the reliability was only 25% (Figure 6). That is, although the program falsely asserted the presence of an SO2 group (with > 45% CL) only 8% of the time, 3/4 of the assertions of SO2 were incorrect. The latter statistic is probably of more interest to an analyst trying to evaluate the program's reports. On the other hand, the FP is a better measure of the raw discriminating power of the program, since it would presumably be unchanged by changing the proportion of the target substructure in the database. The two measures serve different functions, and should both be reported.

The tradeoff between reliability and recall can be adjusted for individual functional groups by changing the frequency ranges allowed for the IR correlations. For some of the functional groups which are well represented in the EPA library (e.g. esters, alcohols) we have manually optimized the rule ranges to maximize (3*Reliability + Recall). Since the library is known to contain errors, and is skewed towards the smallest (often anomalous) members of homologous series, we have not tried to do this for all groups (e.g. SO2). Further testing on larger libraries will allow further refinements of the IR rules.

Many of the errors observed result from the consistent confusion of two particular functional groups. For example, although the presence of a methyl group was erroneously reported (at >45% confidence) for 30% of the 400 compounds which lack methyl groups, a methyl group was reported for only 1 of the 33 compounds lacking both CH3 and CH2 groups. Conversely, the presence of a methylene group was never incorrectly asserted for compounds which lack methyl groups. Examination of the reasons for the confusion confirm that the C-H stretching and HCH deformation vibrations, whose frequency and intensity ranges are similar for methyl and methylene, are often misassigned. Such consistent confusion between similar substructures can be dealt with by assigning the bands to a generic -CH2X group, and deciding between methyl and methylene only after the nearby environment has been determined.

Average results for 500 IR-active substructures are shown in Figure 7 at four different confidence levels. The average compound in the database contains 8.1 of the 500 substructures. At a confidence level of > 45%, only 1.4 (of 492) incorrect substructures are reported, while 4.6 of 8.1 substructures actually present are reported. In other words, a "typical" analysis will report 6.0 substructures at > 45% confidence, of which 4.6 are correct. 3.5 substructures actually present in the compound will fail to be reported. In an actual analysis, infrared data is combined with other types of data, so that many of the substructures undetected by infrared would be found by other techniques.

We have analyzed over 100 unknown compounds using both the mass spectrum and the IR spectrum in combination. The combination of the two techniques gives substantially better results than does either technique alone. As expected, many functional groups are preferentially detected by one technique or the other. For example, ketone groups are rarely detected in the mass spectrum, but are usually correctly interpreted from the infrared. Chlorine and bromine, on the other hand, are easily detected in the mass spectrum but often missed by the infrared interpreter. Also, because of the interaction between the two interpretation methods, substructures are frequently detected by the combined techniques which are not found by either technique alone. This can occur as a result of resolving a contradiction between the two Experts, as in the example above, or because one Expert is able to further specialize a result suggested by the other. For example, in the interpretation of bis-2-chloro-ethyl-ether, the IR Expert alone fails to detect the presence of chlorine. When chlorine is suggested by the STIRS Expert, however, the IR Expert correctly reports the -CH2Cl group. A few substructures, such as non-terminal olefins, are not reliably detected in either mass or infrared spectra. For such groups, other techniques (NMR, UV absorption, Raman) are necessary.

In many cases, the results of the IR and mass spectrum interpretation are sufficient to allow a complete molecular structure to be deduced. In preliminary tests on 12 unknown compounds of molecular weight 100-200, the author, using the results reported by the program but without access to the original spectra, was able to correctly identify 9 of the unknowns.

These results are encouraging, and suggest that our system in substantially its present form could serve as a useful tool for an analytical chemist, as well as eventually providing a framework for completely automated identification of organic compounds.

Conclusions

We have developed an expert system which can interpret various kinds of data and report functional groups present in an unknown organic

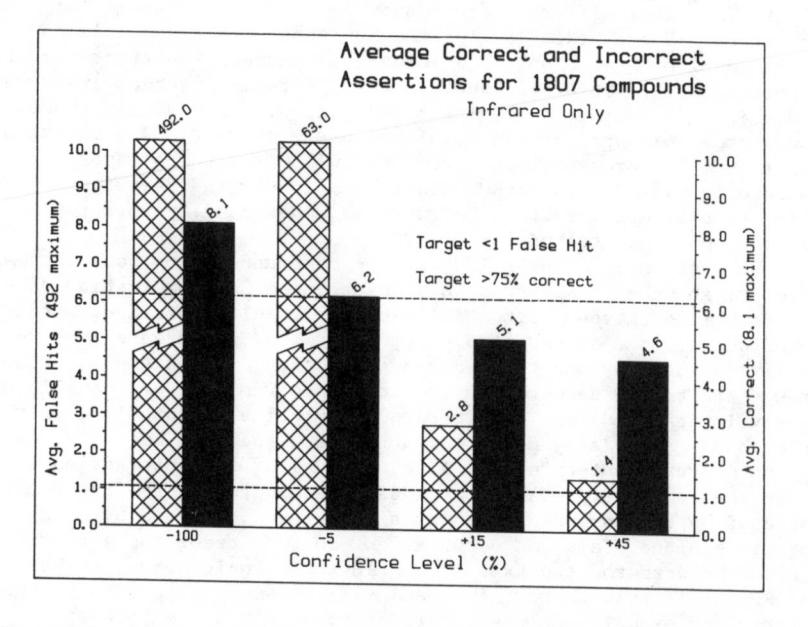

Figure 7. Average number of substructures reported correctly
(solid color) and incorrectly (hatched) at four different
confidence levels, for IR data only. A total of 500
substructures were considered, of which an average of 8.1 were
present in each compound.

compound. The program employs a modular construction, which allows
each type of data to be interpreted in the most efficient way. The
conclusions derived by different modules are able to influence each
other at a low level.

The program knows the chemical relationships between functional
groups, and can use this knowledge in its reasoning process.

The reasoning process is accessible to the user, so that each
conclusion can be traced back to the original data responsible for
it. Choices made by the program can be isolated and overridden by
a knowledgeable user.

Contradictions arising among evidence from different sources
are resolved in a natural way, using knowledge about the effects of
perturbations and common interferences on the spectra.

A rule-based infrared spectra interpreter has been developed as
a major module of the program. This module has been tested as a
stand-alone system, and in conjunction with STIRS. The low rate of
false positive assertions is encouraging, and work continues to
reduce this rate still further by incremental refinement of the
knowledge base.

In its present form, our system can provide significant assistance to a chemist trying to identify an unknown organic compound.
Research is in progress to extend the capabilities of the program
both by expanding the number of different data sources it can handle
(NMR, UV/visible absorption spectra) and by incorporating a "molecule
builder" which assembles complete candidate structures, where possible, from the suggested substructures.

Acknowledgments

I would like to thank Reed Letsinger and others in the Expert Systems
Department at HP Labs for helpful discussions and technical assistance.

Literature Cited

1. Hippe, Z.; Hippe, R. Appl. Spectrosc. Reviews 1980, 16, 135-186.
2. Bally, R. W.; van Krumpen, D.; Cleij, P.; van't Klooster, H. A.
    Anal. Chim. Acta 1984, 157, 227-243.
3. Masinter, L. M.; Sridharan, N. S.; Lederberg, J.; Smith, D. H.
    J. Am. Chem. Soc. 1974, 96, 7702-7723.
4. Carhart, R. E.; Smith, D. H.; Gray, N. A. B.; Nourse, J. G.;
    Djerassi, C. J. Org. Chem. 1981, 46, 1708-1718.
5. Nelson, D. B.; Munk, M. E.; Gash, K. B.; Herald, D. L.
    J. Org. Chem. 1969, 34, 3800.
6. Shelley, C. A.; Hays, T. R.; Munk, M. E. Anal. Chim. Acta
    Computer Techniques and Optimization 1978, 103, 121-132.
7. Fujiwara, I.; Okuyama, T.; Yamasaki, T.; Abe, H.; Sasaki, S.
    ibid 1981, 133, 527-533.
8. Szalontai, G.; Simon, Z.; Csapo, Z.; Farkas, M.; Pfeifer, Gy.
    ibid 1981, 133, 527-533.
9. Debska, B.; Duliban, J.; Guzowska-Swider, B.; Hippe, Z. ibid
    1981, 133, 303-318.
10. Dubois, J.-E.; Carabedian, M.; Dagane, I. Anal. Chim. Acta
    1984, 158, 217-233.
11. Gribov, L. A.; Elyashberg, M. E.; Koldashov, V. N.; Plentnjov,
    I. V. ibid 1983, 148, 159-170.

12. Small, G. W.; Jurs, P. C. Anal. Chem. 1984, 56, 1314-1323.
13. Gray, N. A. B. Artificial Intelligence 1984, 22, 1-21.
14. Haraki, K. S.; Venkataraghavan, R.; McLafferty, F. W. Anal. Chem. 1981, 53, 386-392.
15. Buchs, A.; Schroll, G.; Duffield, A. M.; Djerassi, C.; Delfino, A. B.; Buchanan, B. G.; Sutherland, G. L.; Feigenbaum, E. A.; Lederberg, J. J. Am. Chem. Soc. 1970, 92, 6831.
16. Ishida Y.; Sasaki, S. Computer Enhanced Spectrosc. 1983, 1, 173-184.
17. Varmuza, K. Anal. Chim. Acta 1980, 122, 227-240.
18. Zupan, J. ibid 1978, 103, 273-288.
19. Visser, T.; van der Maas, J. H. ibid 1980, 122, 363-372.
20. Smith, G.; Woodruff, H. B. J. Chem. Inf. Comp. Sci. 1984, 24, 33.
21. Gray, N. A. B. Anal. Chem. 1975, 47, 2426.
22. Delaney, M. F.; Denzer, P. C.; Barnes, R. M.; Uden, P. C. Anal. Lett. 1979, 12 963-978.
23. Bink, W. G.; van 't Klooster, H. A. Anal. Chim. Acta 1983, 150, 53-59.
24. Cross, K. P.; Giordani, A. B.; Gregg, H. R.; Hoffman, P. A.; Beckner, C. F.; Enke, C. G. "An Automated Structure Determination System for MS/MS Data", 190th ACS National Meeting, Chicago, IL (1985).
25. Christie, B. D.; Munk, M. E. "Computer-assisted Structure Elucidation Using 2-Dimensional NMR Data", 190th ACS National Meeting, Chicago, IL (1985).
26. Buchanan, B. G.; Shortliffe, E. H. "Rule-based Expert Systems"; Addison-Wesley: Menlo Park, CA, 1984.
27. Jurs, P. C.; Isenhour, T. L. "Chemical Applications of Pattern Recognition"; Wiley: New York, NY, 1975.
28. Curry, B., manuscript in preparation.
29. Charniak, E.; McDermott, D. "Introduction to Artificial Intelligence"; Addison-Wesley: Menlo Park, CA, 1985.
30. Bellamy, L. J. "The Infrared Spectra of Complex Molecules"; Chapman and Hall: London, 1975.
31. Nyquist, R. A. "The Interpretation of Vapor-Phase Infrared Spectra", vol. 1; Sadtler Research Labs: Philadelphia, PA, 1984.
32. Socrates, G. "Infrared Characteristic Group Frequencies"; John Wiley and Sons, Ltd.: New York, NY, 1980.
33. Griffiths; et al., GC-IR Subcommittee of the Coblenz Society Evaluation Committee, Appl. Spectrosc. 1979, 33, 543.
34. de Haseth, J., Chemistry Dept., Univ. of Georgia, Athens, GA, personal communication.
35. "Registry of Mass Spectral Data"; Electronic Data Division, Wiley: 605 Third Ave., New York, NY 10158.
36. Dayringer, H. E.; McLafferty, F. W. Org. Mass Spectrosc. 1976, 11, 543-551.

RECEIVED December 17, 1985

Concerted Organic Analysis of Materials and Expert-System Development

S. A. Liebman[1], P. J. Duff[1], M. A. Schroeder[1], R. A. Fifer[1], and A. M. Harper[2]

[1]U.S. Army Ballistic Research Laboratory, Aberdeen Proving Ground, MD 21005-5066
[2]Chemistry Department, University of Texas at El Paso, El Paso, TX 79968-0513

A prototype multilevel expert system network has been developed for application to materials characterization. Selected analytical instruments generate databases which are treated and interpreted within an analytical strategy toward a desired goal. Using a commercial expert system shell, TIMM, a linked network of expert systems, EXMAT, has been developed. The expertise of a chemometrician is embedded within the network at the data analysis and interpretation stages as a linked expert system, EXMATH. For general chemical analysis, expert systems capable of symbolic and numeric processing appear necessary to provide integrated decision structures using data generated from appropriate instruments and sensors. Final implementation of EXMAT will demonstrate the potential significance of artificial intelligence (AI) in analytical chemistry with varied intelligent laboratory and process instrumentation.

Requirements for high-performance materials have focused on the ability to relate structure/composition to end-use behavior. Analytical instrumentation designed over the past decade has made impressive advances in defining the composition of complex polymeric systems, including detailed description of polymer chemical microstructure. Concerted organic analysis has been followed since the early '70s (1-3), including multivariate profile analysis of gas chromatographic (GC) patterns (4), and computer simulations of GC/spectral patterns to aid interpretation (5). Work reported in 1968 (6,7) included automated data acquisition and computer-aided interpretation from multiple analytical spectrometers (mass, nuclear magnetic resonance (NMR), infrared (IR), and ultraviolet (UV)). The four spectrometers were tied to individual computers which fed data into a central

0097-6156/86/0306-0365$06.00/0

computer programmed for structure elucidation based on a
combination of all four types of data. Most recently,
applications of computer models that describe relationships
between chemical, physical, and mechanical responses were
described by Kaelble (8). Many correlations between chemical
structure and polymer composite performance have been established
over the past decade within the industrial R&D community.
Kaelble's work emphasizes the significance of modern
characterization methods for this purpose.

Concurrently, pattern recognition programs were developed as
interpretive aids along with comprehensive experimental design,
factor analysis, and other statistical approaches within the
chemometrics field (9-15). Only within the past few years has the
precision and high reproducibility of appropriate key
instrumentation made possible realistic applications for materials
analysis. Microprocessor-based chromatographic, pyrolysis/
concentrator, thermal, and spectral instrumentation are combined
with chemometric tools to provide chemically significant
information as "intelligent" instruments become available (16).
These advances are allied to highly automated hardware common in
clinical labs and computer-controlled process equipment (17-19).

Automated calibration and data-handling methods have been
integral parts of commercial analytical systems for many years, as
well as embedded software to automate complex pneumatic/electronic
sequences in concentrator and chemical reactor instrumentation
using on-line GC analysis (20). Recently, a commercial high
pressure liquid chromatograph (HPLC) system (21) demonstrated
adaptive intelligence to optimize separations for complex sample
mixtures. The optimization program, OPTIM II, initially queries
the chromatographer and then performs a sequence of automated
steps. Likewise, library search algorithms (22-28), pattern
recognition (29-36), and optimization (37-42) methods have
developed in numerous laboratories.

The well-known DENDRAL and META-DENDRAL programs (43) are
noted as the major AI success in chemical applications over the
past decade. However, advances in analytical technology and
computer capabilities have led to new approaches (44-56).
Information fusion from selected instrumental tools often is a
more productive route than exhaustive data analysis from a single
source. Furthermore, combination of chromatographic separation
with spectral, thermal, and microchemical analyses can be
realistically achieved in many laboratories. Generalizing and
documenting this trend using an AI approach seemed appropriate at
this time.

Results and Discussion

General. We have studied the characterization of multicomponent
materials by combining modern analytical instrumentation with a
commercially available AI expert system development tool.
Information generated from selected analytical databases may be
accessed using TIMM, ("The Intelligent Machine Model,") available
from General Research Corp., McLean, VA. This Fortran expert
system shell has enabled development of EXMAT, a heuristically-
linked network of expert systems for materials analysis.

An important aspect of our AI application is the attention paid to including well-established Fortran programs and database search methods into the decision structure of an expert system network. Only certain AI software tools (such as TIMM) effectively handle this critical aspect for the analytical instrumentation field at this time (57-60). The ability to combine symbolic and numeric processing appears to be a major factor in development of multilevel expert systems for practical instrumentation use. Therefore, the expert systems in the EXMAT linked network access factor values and the decisions from EXMATH, an expert system with chemometric/Fortran routines which are appropriate to the nature of the instrumental data and the information needed by the analyst. Pattern recognition and correlation methods are basic capabilities in this field.

TIMM - The Intelligent Machine Model. The expert system shell, TIMM, is a frame-like system which employs an analogical partial match inferencing procedure, similar to a forward-chaining process when the explicit linking method is followed. Partial match inferencing, as proposed by Hayes-Roth and Joshi (61) means matching on a subset of clauses in individual rules. Analogical inferencing uses similarity, as well as exactitude, to match rule clauses. Thus, TIMM effectively uses incomplete and approximate knowledge in a supervised learning format. The created expert system is divided into two sections: (1) a decision structure with ordered input factors and values, and (2) the knowledge base containing rules that are displayed to the user in an "if, then" format. A set of test conditions is compared to those contained in the knowledge base and a weighted similarity metric is applied. A variation of the nearest neighbor search algorithm is used for pattern-matching.

Heuristically-linked individual expert systems (ES) are prepared using implicit and/or explicit linking methods to permit processing of "microdecisions" that are part of more complex "macrodecisions". The prototype EXMAT was developed using an implicit linking procedure wherein the decision choices of one ES become the first ordered factor/values of another ES. Prior to linking, each system is independently built, trained, exercised, checked for consistency and completeness, and then generalized. Terse or verbose explanations may be included, as well as decision confidence levels that are trained into the system by the domain experts. TIMM is domain independent, permitting expert systems to be readily developed in fields wherein expertise exists. EXMAT was developed within this protocol with the important advantage that TIMM ES can be embedded in routines which are basic to the analysts' problem-solving capability and accessed using the advanced REASON subroutine developed by General Research Corporation.

EXMAT - A Linked Network of Expert Systems for Materials Analysis. Seven individual expert systems comprise EXMAT: (1) problem definition and analytical strategy; (2) instrumental configuration and conditions; (3) data generation; (4) chemometric/search algorithms; (5) results; (6) interpretation; (7) analytical goals. Dynamic headspace (DHS)/GC and pyrolysis GC (PGC)/concentrators

interfaced to Fourier transform infrared (FTIR) and mass spectral
(MS) detectors, combined with HPLC, thermal, and elemental
analyses have been chosen in this approach for composite materials
characterization. Generation of databases in the prototype EXMAT
system will focus on the specific domain of propellants and
polymer composites. However, the general concept of integrating
information from relevant databases emulates the actions of a
pragmatic problem-solver in many domains. Clearly, the specific
analytical strategy, instrumental configurations, databases, and
interpretive aids must be developed accordingly (8). EXMAT
illustrates the inherent potential of combining intelligent
instrumentation with AI symbolic processing in a problem-solving
format.

Figure 1a outlines the decision and control structure of
TIMM; Figure 1b, the expert systems network; and Figure 1c, the
overall decision structure of EXMAT. Expert System (ES) #1 is
given (in part) in Figure 2 showing the decision choices and
factors/values needed to establish the problem definition and
analytical strategy. Analytical systems included in the strategy
specifically emphasize those tools available in the Ballistic
Research Laboratory and which have a proven capability of
generating precise, reproducible data on a wide variety of
materials. Therefore, the analyst may select the combination of
instrumentation (chromatographic, spectrometric, thermal, or
elemental) dependent on the scope of the problem, nature of the
information needed, details of the samples involved, and the
available analytical tools and methods. There are approximately
85 rules in the knowledge base of ES #1 at this time, four of
which are shown in Figure 3.

Figure 4 outlines a portion of ES #2 for choice of the
specific instrumental configuration and conditions which are
indicated by the decisions and factors provided in ES#1. This is
a critical step, since the databases generated in ES #3 must be
directly correlated to the specific instrumental configuration and
conditions in ES #2 for the concerted analysis of samples,
references, etc.; e.g., pattern comparisons between analyses with
specialty GC detectors (FID-flame ionization, TCD-thermal
conductivity, NPD-nitrogen/phosphorus, PID-photoionization). This
stage focuses on the attributes of modern analytical
instrumentation: flexible, modular, microprocessor/computer-
controlled hardware that can be readily interfaced for efficient
data-acquisition and handling. ES #2 also emphasizes varied
sample processing, such as pyrolysis and dynamic headspace, in
order to analyze materials which cannot be introduced directly
into the chromatographic or spectral systems. Also, instrumental
methods designed for trace organic analysis or for sample-limited
studies are important capabilities. The instrumental
configurations are grouped into six major systems - Sys 1-GC, Sys
2-FTIR, Sys 3-MS, Sys 4-HPLC, Sys 5-Thermal, and Sys 6-Elemental.

ES #3 dictates the selected databases and sample-tracking
mechanism that are based on the decisions of ES #2. For example,
data obtained using a direct FTIR method as suggested in the
decisions of ES #1 and #2 would be put into the FTIR database
under D conditions. However, a sample examined with the GC-FTIR
configuration would be entered into the GC-FTIR database with

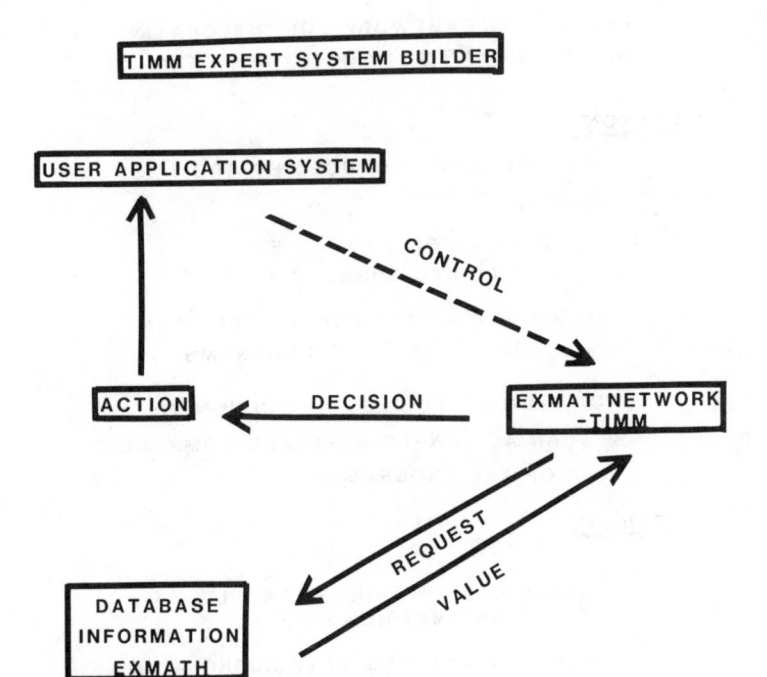

Figure 1a. Expert system shell - TIMM.

o EXMAT

   ESTABLISH FRAMEWORK FOR INTEGRATING ES
   DECISIONS AND ACTIONS TO-BE-TAKEN

o REASON

   A SUBPROGRAM PRODUCED BY GRC
   ENABLES

   (A) APPLICATION TO CALL ES
       AS A LINKED SUBROUTINE

   (B) AN ES TO CALL AND UTILIZE DATA
       FROM OTHER FILES/PROGRAMS

   (C) TIMM ES TO PASS A DECISION TO
       AN ACTION-TO-BE-TAKEN COMPONENT
       OF THE PROGRAM

o EXMATH

   CALLS USER DEFINED DATA FILE
   FOR CHEMOMETRICS

   DECISION INVOKES MATH SUBROUTINE AND
   ACCEPTS EXMAT DECISIONS

   CONVERTS MATH RESULTS TO FACTOR VALUES
   FOR INPUT TO TIMM EXPERT SYSTEMS

   Figure 1b.   An expert system network.

A LINKED NETWORK OF EXPERT SYSTEMS FOR MATERIAL ANALYSIS

ES #1 ANALYTICAL STRATEGY FOR DEFINED PROBLEM

ES #2 INSTRUMENTAL CONFIGURATION/CONDITIONS

ES #3 DATABASE GENERATION

ES #4 DATA TREATMENT

ES #5 DATA RESULTS

ES #6 DATA INTERPRETATION

ES #7 ANALYTICAL GOAL

Figure 1c. Development of EXMAT.

```
DECISION:
--------

         ANALY STRATEGY
            Choices:
            GC/SYS1
            FTIR/SYS2
            MS/SYS3
            LC/SYS4
            TA/SYS5
            EL/SYS6

FACTORS:
--------

         SCOPE
           Type of Values: Unordered Descriptive Phrases
           Values:
             SCREEN
             TIME/FUND LIMIT
             QUAL/QUANT
             QUANT
             PURITY
             VOLATILES
             TRACE DETECT
             KINETICS
             MECHANISM
             CORRELATION
             R&D

         SAMPLE AMT
           Type of Values: Linearly-Ordered Descriptive Phrases
           Values:
             UNLIMITED
             GM
             MG
             MICROGM
             TRACE

         SAMPLE FORM
           Type of Values: Unordered Descriptive Phrases
           Values:
             POWDER
             BULK
             SEMISOLID
             LIQUID
             FILM/LAMIN
             FIBER
             MULTIMEDIA
```

Figure 2.   Partial decision structure of ES #1 analytical strategy.

```
Rule 17
-------
    If:
        SCOPE               IS SCREEN
        SAMPLE AMT          IS GM
        SAMPLE FORM         IS MULTIMEDIA
        SAMPLING PROCESS    IS RANDOM
        SAMPLE HISTORY      IS UNKWN
        INSTR. AVAIL        IS NO LC
    Then:
        ANALY STRATEGY      IS GC/SYS1(50)
                               FTIR/SYS2(50)


Rule 18
-------
    If:
        SCOPE               IS TRACE DETECT
        SAMPLE AMT          IS MG
        SAMPLE FORM         IS POWDER
        SAMPLING PROCESS    IS STATIC
        SAMPLE HISTORY      IS DEGRADATION
        INSTR. AVAIL        IS NO LC  .
    Then:
        ANALY STRATEGY      IS GC/SYS1(30)
                               MS/SYS3(70)


Rule 19
-------
    If:
        SCOPE               IS QUANT
        SAMPLE AMT          IS TRACE
        SAMPLE FORM         IS FILM/LAMIN
        SAMPLING PROCESS    IS STATIC
        SAMPLE HISTORY      IS UNKWN
        INSTR. AVAIL        IS NO METHOD
    Then:
        ANALY STRATEGY      IS MS/SYS3(100)


Rule 20
-------
    If:
        SCOPE               IS TRACE DETECT
        SAMPLE AMT          IS TRACE
        SAMPLE FORM         IS FILM/LAMIN
        SAMPLING PROCESS    IS RANDOM
        SAMPLE HISTORY      IS DEGRADATION
        INSTR. AVAIL        IS NO ELEM
    Then:
        ANALY STRATEGY      IS GC/SYS1(20)
                               MS/SYS3(80)
```

Figure 3.   Typical rules in ES #1.

DECISION:

```
        EXPTL CONFIG
           Choices:
              GCSYS1/A
              GCSYS1/ABC
              FTIRSYS2/D
              FTIRSYS2/ABCD
              MSSYS3/E
              MSSYS3/ABCE
              LCSYS4/FIK
              LCSYS4/GJK
              LCSYS4/FIL
              LCSYS4/GJL
              TASYS5/M
              TASYS5/N
              TASYS5/O
              TASYS5/P
              ELSYS6/Q
              ELSYS6/R
```

FACTORS:

```
        ANALY STRATEGY
           Type of Values: Unordered Descriptive Phrases
           Values:
              GC/SYS1
              FTIR/SYS2
              MS/SYS3
              LC/SYS4
              TA/SYS5
              EL/SYS6

        GC CONFIG
           Type of Values: Unordered Descriptive Phrases
           Values:
              DIRECT GC/FID/TCD
              DIRECT GC/FID/NPD
              DHS/FID/TCD
              DHS/FID/NPD
              PGC/FID/TCD
              PGC/FID/NPD
              DHS/PGC/FID/TCD
              DHS/PGC/FID/NPD

        FTIR CONFIG
           Type of Values: Unordered Descriptive Phrases
           Values:
              DIRECT
              MICROSAMPLING
              DRIFT
              ATR
              VARIABLE T
              DHS/FTIR
              GC-FTIR
              DHS/GC-FTIR
              PGC-FTIR
              DHS/PGC-FTIR

        MS CONFIG
           Type of Values: Unordered Descriptive Phrases
           Values:
              RIC
              SIM
              PYROL/MS
              DHS/MS
              GC-MS/FID
```

Figure 4.   Partial decision structure of ES #2 configuration.

conditions designated AD, ABD, or ABCD. In the fully documented ES form, the conditions A/B/C/etc. will be described in appropriate detail for the user and accessed by using the "verbose version" from the menu. Format for instrumental database generation and management was aided by the work reported earlier by R. Crawford, C. Wong, and coworkers at Lawrence Livermore Laboratory (62,63). Additionally, data report transfer from GC data stations to the host VAX-VMS system was aided by recent work reported from Argonne National Laboratory (64,65).

Data treatment in ES #4 incorporates chemometric methods available for chromatographic or spectral analysis: preprocessing of data, normalization, smoothing, deconvolution, optimization, fingerprinting, pattern recognition, factor analysis (eigenvector and canonical methods), and other appropriate routines. The latter have been purchased or incorporated from the literature; e.g., PAIRS, an infrared interpretive program by H. Woodruff and coworkers (66), and the MS library/search programs provided by Hewlett-Packard for their MS systems. These searches provide a "hit list" from the respective libraries and some additional options for spectral interpretation.

Our linked pattern recognition expert system, EXMATH, operates on given databases via the preprocessing, data manipulation, classification, factor analysis, or plotting packages as driven within EXMAT. For example, the classification package includes linear discriminant analysis, regression analysis, principal component catagory analysis, nonlinear mapping and nearest neighbors analysis. The factor analysis package provides loading extraction, factor scores, factor rotations, and canonical correlation analysis.

The results of data treatment are documented and evaluated in ES #5 and the interpretation in ES #6 is guided by the analyst's constraints and requirements. For instance, simple visual pattern comparisions may be acceptable for sample identification, or a combined database (GC-FTIR/GC-MS), (PGC/FTIR), (GC/TA), etc., analysis may be required. Judgmental decisions must be trained into the system as to depth of analysis, its acceptability and reliability (e.g., the hit quality index (HQI) of the MS search combined with that from the FTIR search may confirm within a 95% confidence level the GC peak or sample identity).

Finally, ES #7 incorporates the interpretive results of these treatments to direct the analyst toward the designated analytical goal(s) via implicit/explicit linking mechanisms. The final goal (structure, composition, mechanism, kinetics, correlation, experimental design analysis, or library extension) is approached by incorporating the earlier decision/choices of ES #1-6 for evaluation in the decision structure of ES #7. Some procedures may be straightforward; e.g., a screening analysis with a single instrument/configuration generates a sample pattern that visually matches a known reference to the satisfaction of the analyst. Other studies involving several instrumental systems (in our scenario...chromatographic, spectral, thermal, or elemental), may require feedback from several interpretive results. Since TIMM is easily modified, the final form of EXMAT will likely be improved over that described for this prototype; i.e., including explicit and implicit linking mechanisms.

EXMATH - An Expert System for Pattern Recognition. A prototype
expert system for pattern recognition and data analysis, EXMATH,
has been developed to embed a chemometrician's expertise into an
accessible form for researchers. The selected library of
subroutines developed over the past ten years comprise a portion
of the EXMATH program to permit an integrated expert systems
approach (Figures 5 and 6).

For each analytical system, expert systems drivers were
written which control data input to and operation of the
algorithm. A second, more intelligent set of drivers: (1) receive
input in the form of a decision from the external expert system
network; (2) collect the necessary subroutines for a heuristic
algorithm to solve data questions; (3) inspect the validity of the
input data; (4) drive the algorithm; and (5) transfer the results
via GRC's REASON algorithm back to the external expert network for
future decision-making. For example, if a least squares
regression on the data file is called by the external expert
network, the EXMATH system inspects the input data, drives the
regression under jacknifing protocols, and collects variable,
residual, and fit correlation results for analysis by the other
expert system modules. The procedure is implemented and executed
without any mathematical expertise from the user.

Summary

Development of a linked network of expert systems, EXMAT, has been
described for application to materials characterization. Selected
instrumentation which are common to modern laboratories generate
databases that are treated and interpreted within an analytical
strategy directed toward a desired goal. Extension to other
problem-solving situations may use the same format, but with
specialized tools and domain-specific libraries. Importantly, a
chemometrician's expertise has been embedded into EXMAT through
access to information derived from a linked expert system,
EXMATH. Figures 7 and 8 outline this multilevel expert systems
approach developed for application of selected analytical
instruments to the field of materials science.

Additionally, use of a commercial AI shell for expert system
development has been demonstrated without the need to learn
computer programming languages (C, Pascal, LISP or any of its
variations), nor to have an intermediary knowledge engineer.
Although this development effort of 4-5 man months was on a
minicomputer, adaptation of EXMAT to the microcomputer version of
TIMM is anticipated. The completed implementation of EXMAT will
support the belief that AI combined with intelligent
instrumentation can have a major impact on future analytical
problem-solving.

In general, it appears that expert systems which combine
symbolic/numeric processing capabilities are necessary to
effectively automate decision-making in applications involving
analytical and process instrumentation/sensors. Furthermore,
these integrated decision structures will likely be embedded (67-
69) within the analytical or process units to provide fully
automated pattern recognition/correlation systems for future
intelligent instrumentation.

EXAMPLE-ALGORITHM BUILDING-EXSPDS

PURPOSE- EMULATION OF SPSS PROCEDURE FOR
DISCRIMINANT ANALYSIS

- USED IN ANALYSIS OF VARIANCE MODE

- PRODUCES DATA MAPPING OF SPACE
OF SAMPLE REPLICATE VARIATION ABOUT
SAMPLE MEANS

OPERATION- INSPECTS INPUT DATA FOR

(1) PRIOR PREPROCESSING
(2) NECESSITY OF RANK REDUCTION
PRIOR TO ANALYSIS

- SCALES DATA IF NEEDED

-PERFORMS FACTOR ANALYSIS REDUCTION IF NEEDED

-COMPUTES SAMPLE MEANS AND ARRANGES DATA AS
A TRAINING SET OF MEAN VECTORS AND TEST SET
OF REPLICATE VECTORS

- PROJECTS BY FACTOR ANALYSIS OF MEAN VECTORS

- REPRODUCES VARIABLE WEIGHTS FOR PROJECTION
AND FURTHER ANALYSIS

Figure 5.   EXMATH – heuristic design.

EXTGRT   - INSPECTS INPUT DATA MATRICES
                      FOR PREPROCESSING TASKS

                   - LOCATES TARGET AND MERGES/SORTS FILE
                      FOR INPUT TO DATA ANALYSIS

                   - PERFORMS FACTOR ANALYSIS IF NEEDED

                   - LEAST SQUARES ROTATION TO
                      TARGET OR HYPOTHESIS

                   -RECONSTRUCTION OF MEASUREMENT
                      INFORMATION   MATRIX TO REFLECT
                      CORRELATIONS

WHY?

(1) DECONVOLUTION OF COMPONENTS IN MIXTURES

(2) HYPOTHESIS TESTS ON INTERPRETATION OF RESPONSES

(3) SPECTRAL MATCHING TO REFERENCE RESPONSES

(4) LEAST SQUARES REGRESSION MODELING WITH
     "NOISE FILTERING"

(5) DETERMINATION OF FUNDAMENTAL PHYSICAL FACTORS
     UNDERLYING SAMPLE MEASUREMENT RESPONSES

Figure 6.   Target rotation – subroutine in EXMATH.

EXMAT

A LINKED NETWORK OF EXPERT SYSTEMS

WITH PATTERN RECOGNITION AND SEARCH PROGRAMS FOR MATERIALS CHARACTERIZATION

COMPONENTS   **ATTRIBUTES**

1. DATABASE MANAGEMENT

    A. STORAGE OF PARAMETERS AND DATA ON SAMPLES FOR SELECTED INSTRUMENTAL TECHNIQUES

    B. RETRIEVAL OF SELECTED SAMPLES FORMING A DATA SET FORMATTED FOR MULTIVARIATE ANALYSIS

    C. CREATE, ADD, DELETE, HELP AND SHOW FUNCTIONS

2. EXPERT SYSTEMS AND EMBEDDING SUBPROGRAMS-TIMM

    A. FORTRAN SOURCE CODE

    B. EMBEDDING OF TIMM SYSTEM WITHIN USER PROGRAMS

    C. CAPABLE OF HANDLING METRIC AND NON-METRIC INFORMATION

    D. HEURISTIC DESIGN

Figure 7. EXMAT – a linked network of expert systems.

Continued on next page

COMPONENT   ATTRIBUTES

3. PATTERN RECOGNITION   A. HEURISTIC DESIGN
   EXPERT SYSTEM -
   EXMATH   B. SUPERVISED AND UNSUPERVISED
                              PATTERN RECOGNITION, FACTOR
                              ANALYSIS, PLOTTING

                           C. EXPERTISE INCLUDES DATA
                              PREPROCESSING AND
                              EVALUATION OF RESULTS

                           D. USER INTERVENTION FOR
                              DATABASE MODIFICATIONS

                           E. IMPLEMENTABLE AS JACKNIFING
                              PROCEDURE

4. SPECTRAL SEARCH AND
   MATCH ALGORITHIMS   A. PAIRS INFRARED SPECTRA
   PARTIAL INTERPRETATION
   AIDS   B. PBM-MASS SPECTRA

Figure 7. Continued.

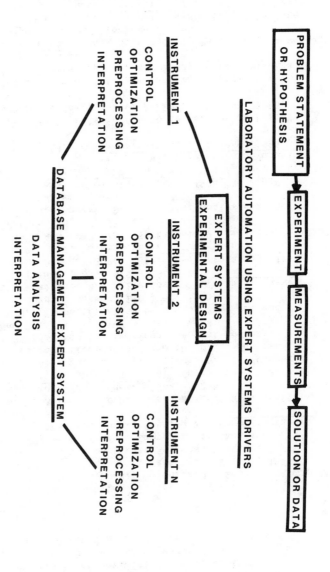

Figure 8. Multilevel expert systems outline. Automated experimental design and decision-making.

Acknowledgments

Contributions to data transfer and computer interfacing by
J. Romanski, K. Fickie, and R.M. Cahoon at the Ballistic Research
Laboratory are gratefully acknowledged. We appreciate the
cooperation provided by the General Research Corporation and
discussions with M.J. Aiken during this development effort.

Literature Cited

1. Liebman, S.A. Amer. Lab., 1971, 18.
2. Liebman, S.A.; Ahlstrom, D.H.; Quinn, E.J.; Geigley, A.G.;
    Meluskey, J.T. J. Polym. Sci., Part A-1, 1971, 9, 1843.
3. Liebman, S.A. ACS 6th Northeast Regional Meeting,
    Burlington, VT, 1974, Sympos. Computers in Chemistry.
4. Liebman, S.A.; Ahlstrom, D.H.; Hoke, A.T. Chromatographia,
    1978, 11, 427.
5. Liebman, S.A.; Ahlstrom, D.H.; Starnes, Jr., W.R. Schilling,
    F.C. J. Macromol. Sci., Chem., 1982, A17(6), 935.
6. Sasaki, S.; Abe, H.; Ouki, T. Anal. Chem., 1968, 40, 2221.
7. Yamasaki, T.; Abe, H.; Kudo, Y.; Sasaki, S. In "Computer-
    Assisted Structure Elucidation"; Smith, D.H., Ed.; ACS
    SYMPOSIUM SERIES, No. 54, American Chemical Society,
    Washington, DC, 1977; p. 108.
8. Kaelble, D.H. "Computer-Aided Design of Polymers and
    Composites"; Marcel Dekker, Inc., NY, 1985.
9. Jurs, P.C.; Kowalski, B.R.; Isenhour, T.L. Anal. Chem.,
    1969, 41, 21; Ibid., 690, 695.
10. Pichler, M.A.; Perone, S.P. Anal. Chem., 1974, 46, 1790.
11. Kowalski, B.R. Ed., "Chemometrics: Theory and
    Applications"; ACS SYMPOSIUM SERIES No. 52, American Chemical
    Society, Washington, DC, 1977.
12. Malinowski, E.R.; Howery, D.G. "Factor Analysis in
    Chemistry", J. Wiley & Sons, NY, 1980.
13. Delaney, M.F. Anal. Chem. Fund. Rev., 1984, 56, 261R.
14. Harper, A.M.; Meuzelaar, H.L.; Metcalf, G.S.; Pope, D.L. In
    "Analytical Pyrolysis"; Proc. 5th International Symposium,
    Voorhees, K.J., Ed.; Butterworths Publ., London, 1984,
    Chapter 6.
15. Harper, A.M. In "Pyrolysis and GC in Polymer Analysis";
    Liebman and Levy, Eds., Marcel Dekker, Inc., NY, 1985;
    Chapter 8.
16. Harper, A.M.; Liebman, S.A. Chemometrics Research
    Conference, Gaithersburg, MD, May 1985, to be published in
    NBS Research Journal.
17. Beckman Instruments, Inc., Spinco Div., Palo Alto, CA 94304;
    Spin Pro Expert System, Brochure SB-664.
18. Kraus, T.W.; Myron, T.J., Control Engineering, 1984, 106.
19. Proc. 9th Annual Advanced Control Conf., Purdue University,
    W. Lafayette, IN, 1983.
20. Chemical Data Systems, Div. of Autoclave Engineering, 7000
    Limestone Rd., Oxford, PA 19363. Sample Concentrators, Models
    320, 330; CDS Geochemical Research System, Model 820; Model
    8000 Series Micro-Pilot Plant Systems.

21.   Spectra-Physics, Autolab Div., San Jose, CA Liquid
      Chromatograph, Model SP8100 XR; Technical Bulletin D/S-01,
      12/84.
22.   Delaney, M.F.; Uden, P.C.   J. Chromatogr. Sci., 1979, 17,
      428.
23.   Delaney, M.F.; Warren, Jr., F.V.; Hallowell, Jr., J.R.   Anal.
      Chem., 1983, 55, 1925.
24.   Delaney, M.F.; Hallowell, Jr., J.R.; Warren, Jr., J.R.   J.
      Chem. Inf. Comput. Sci., 1985, 25, 27.
25.   Stauffer, D.B.; McLafferty, F.W.; Ellis, R.D.; Peterson,
      D.W.   Anal. Chem., 1985, 57, 1056; Ibid., 899 and refs.
26.   Kalchhauser, H.; Robien, W.   J. Chem. Inf. Comput. Sci.,
      1985, 25, 103.
27.   Fein-Marquart Assoc., Inc., 7215 York Rd., Baltimore, MD
      21212; Mass Spectral Info. System (MSIS); MASCOT: software
      pkg. MS data on a PC.
28.   Sadtler Research Laboratores, Inc., Spring Garden St.,
      Phila., PA 19122.
29.   Lysyj, I.; Newton, P.R.   Anal. Chem., 1972, 44, 2385.
30.   Kanal, L.   IEEE Trans. Info.Theory, 1974, Vol. IT-20, 697;
      and Proc. IEEE, 1972, 60, 1200.
31.   Kowalski, B.R.   Chemtech, 1974, 300.
32.   Byers, W.A.; Perone, S.P.   Anal. Chem., 1980, 52, 2173.
33.   Moncur, J.G.; Bradshaw, W.G.   J. High Resol.Chromatogr. & CC,
      1983, 6, 595.
34.   Frankel, D.S.   Anal. Chem., 1984, 56, 1011.
35.   Fredericks, P.M.; Osborn, P.R.; Swinkels, D.A.J.   Fuel, 1984,
      63, 139; and BHP Tech. Bulletin No. 27, 1983.
36.   Infometrix, Inc., 2200 Sixth Ave., Seattle, WA 98121.
37.   Deming, S.N.; Morgan, S.L.   Anal. Chem., 1973, 45, 278A.
38.   Deming, S.N.   Amer. Lab., 1981, 13, 42.
39.   Deming, S.N.; Morgan, S.L.   "INSTRUMENTUNE-UP: A Computer
      Program for Optimizing Performance of Common Lab.
      Instruments"; Elsevier Scientific Software, Amsterdam, The
      Netherlands, 1984.
40.   Galjch, J.L.; Kirkland, J.J.; Squire, K.M.; Minor, J.M.   J.
      Chromatogr., 1980, 199, 57.
41.   Sabate, L.G.; Diaz, A.M.; Tomas, X.M.; Gassiot, M.M.   J.
      Chromatogr. Sci., 1983, 21, 439.
42.   Statistical Designs, 9941 Rowlett, Suite 6, Houston, TX;
      "Software for Experimental Design and Optimization".
43.   Barr, A.; Feigenbaum, E.A.; Eds.   Handbook of AI, William
      Kaufman, Los Altos, CA, Vol. II, 1981.
44.   Cooper, J.R.; Johlman, C.; Laude, D.A.; Brown, R.S.; Wilkins,
      C.L.   Proc. Pitts. Conf. Anal. Chem. & Spectr., New Orleans,
      LA, 1985; and Anal. Chem., 1984, 56, 1163; Ibid., 57, 1044.
45.   Greene, W.W.; Isenhour, T.L.   Proc. Pitts. Conf. Anal. Chem.
      & Spectr., New Orleans, LA, 1985; and Anal. Chem., 1983, 55,
      1117.
46.   Borman, S.A.   Anal. Chem., 1982, 54, 1379.
47.   Hayes-Roth, F.; Waterman, D.A.; Lenat, D.B.; Eds.   "Building
      Expert Systems", Addison-Wesley Publ. Co., Reading, MA, 1983.
48.   Buchanan, B.G.; Shortliffe, E.H.   "Rule-Based Expert
      Systems"; Addison-Wesley Publ. Co., Reading, MA, 1984.
49.   Lenat, D.B.   Sci. Amer., 1984, 204.

50. Kinnucan, P. High Tech., 1984, 30; Ibid., 1985, 16.
51. Third Annual Conf. on Applied AI, Boston, MA, 1985, DPMA/Tech. Training Corp., and Embedded Computer Software Conf., Boston, MA, 1984.
52. Dessy, R.E. Anal. Chem., 1984, 56, 1200; Ibid., 1313.
53. Klass, P.J. Aviation Wk. & Space Tech., April 22, 1985, 46.
54. Harmon, P.; King, D. "Expert Systems", J. Wiley & Sons, Inc. NY, 1985.
55. Bramer, M. & D. "The Fifth Generation", Addison-Wesley Publ. Co., Reading, MA, 1984.
56. Pearl, J. "Heuristics-Intelligent Search Strategies for Computer Problem-Solving", Addison-Wesley Publ. Co., Reading, MA, 1984.
57. Selected AI Literature: ICS Applied AI Reporter, Univ. Miami, Intell. Computer Systems Res. Institute, Coral Gables, FL 33124; The AI Magazine, Amer. Assoc. AI (AAAI), Menlo Park, CA 94025; and Expert Systems, Internat. J. of Knowledge Engineering, Croall, Ishizuka, Waterman, Eds., Learned Information, Inc., Marlton, NJ
58. Michie, D.; Muggleton, S.; Riese, C.; Zubrick, S. First Conf. AI Applns., Denver, CO, 1984 RuleMaster-A Second-Generation Knowledge Engineering Facility Radian Tech. Rpt., MI-R-623, Radian Corp., P.O. Box 9948, Austin, TX 78766.
59. SRL DEXPERT, Systems Research Labs., Inc., Dayton, OH 45440-3639; Integrates LISP algorithms into Fortran or Ada systems.
60. Proc. Workshop on Coupling Symbolic and Numerical Computing in Expert Systems, sponsored by AAAI, Aug 1985, Boeing Computer Services, Bellevue, WA.
61. Hayes-Roth, F. In "Pattern-Directed Inference Systems", Waterman and Hayes-Roth, Eds., Academic Press, NY, 1978.
62. Crawford, R.W.; Brand, H.R.; Wong, C.M.; Gregg, H.R.; Hoffman, P.A.; Enke, C.G. Anal. Chem., 1984, 56, 1121.
63. Wong, C.M. Energy & Techn. Rev., Lawrence Livermore National Laboratory, University CA, Livermore, CA, 1984, 8.
64. Demirgian, J.C. J. Chromat. Sci., 1984, 22, 153.
65. Demirgian, J.C.; Eikens, D.I. Proc. Pitts. Conf. Anal. Chem.& Spectr., 1985.
66. Tomellini, S.A.; Hartwick, R.A.; Woodruff, H.B. Appl. Spectr., 1985, 39, 331. Quantum Chemistry Program Exchange, Univ. Indiana, Bloomington, IN, Program #426.
67. Wilson, J.W.; Levine, J.B. Business Wk., June 10, 1985, 82.
68. Robinson, P. BYTE, June, 1985, 169.
69. Yianilos, P.N. Electronics, 1983, 56, 113.

RECEIVED December 17, 1985

INDEXES

Author Index

Abbott, Seth, 278
Bach, René, 278
Beckner, C. F., 321
Bellows, James C., 52
Bertz, Steven H., 169
Burnstein, Ilene, 244
Cabrol, Daniel, 125
Cachet, Claude, 125
Corbett, Michael, 244
Cornelius, Richard, 125
Cross, K. P., 321
Curry, Bo, 350
Delaglio, Frank, 337
Dolata, Daniel P., 188
Dudewicz, Edward J., 337
Duff, P. J., 365
Edelson, David, 119
Ehrlich, Steven, 244
Enke, C. G., 321
Evens, Martha, 244
Ferrin, Thomas E., 147
Fifer, R. A., 365
Garfinkel, David, 75
Garfinkel, Lillian, 75
Gasteiger, J., 258
Giordani, A. B., 321
Gough, Alice, 244
Gregg, H. G., 321
Griffith, Owen Mitch, 297
Hahn, Mathew A., 136
Hansch, Corwin, 147
Harner, Teresa J., 337
Harper, A. M., 365
Hawkinson, Lowell B., 69
Heffron, Matt, 297
Hemphill, Charles T., 231
Herndon, William C., 169
Hoffman, P. A., 321

Hohne, Bruce A., 87
Houghton, Richard D., 87
Huang, Conrad, 147
Hutchings, M. G., 258
Johnson, Peter, 244
Karnicky, Joe, 278
Keith, L. H., 31
Klein, Teri E., 147
Knickerbocker, Carl G., 69
Kulikowski, Casimir A., 75
Kumar, Anil, 337
Langridge, Robert, 147
LaRoe, William D., 231
Levinson, Robert A., 209
Levy, George C., 337
Liebman, S. A., 365
Löw, P., 258
Martz, Philip R., 297
Moore, Robert L., 69
Moseley, C. Warren, 231
Palmer, P. T., 321
Pavelle, Richard, 100
Renkes, Gordon D., 176
Riese, Charles E., 18
Saller, H., 258
Schroeder, M. A., 365
Smith, Allan L., 111
Smith, Dennis H., 1
Smith, Graham M., 312
Soo, Von-Wun, 75
Stuart, J. D., 18,31
Tomellini, Sterling A., 312
Trindle, Carl, 159
Wang, Tunghwa, 244
Wilcox, Craig S., 209
Wipke, W. Todd, 136,188
Woodruff, Hugh B., 312

Subject Index

A

Abstraction, 189
Actinospectacin
  digitized spectrum, 315,317t
  PAIRS interpretation, 315,318t
  structure, 315-316

Actinospectacin--Continued
  trace of sulfone functionality during
    PAIRS interpretation, 315,318f
Actions, definition, 94,95t,96
Agricultural formulations
  requirements, 87
  structure of the expert
    system, 89,91-97

Production by Meg Marshall
Indexing by Deborah H. Steiner
Jacket design by Pamela Lewis

Elements typeset by Hot Type Ltd., Washington, DC
Printed and bound by Maple Press Co., York, PA

RECENT ACS BOOKS